Music and Technoculture

Music and Technoculture

EDITED BY RENÉ T. A. LYSLOFF AND
LESLIE C. GAY, JR.

Afterword by Andrew Ross

WESLEYAN UNIVERSITY PRESS
Middletown, Connecticut

Published by Wesleyan University Press, Middletown, CT 06459

© 2003 by Wesleyan University Press

All rights reserved

Printed in United States of America

5 4 3 2 1

Library of Congress Cataloging-in-Publication Data

Music and technoculture / edited by René T. A. Lysloff and Leslie C. Gay, Jr. ; afterword by Andrew Ross.
   p.   cm. — (Music/culture)
ISBN 0-8195-6513-X (alk. paper) — ISBN 0-8195-6514-8 (pbk. : alk. paper)
   1. Music and technology. 2. Music—20th century—History and criticism. 3. Music—21st century—History and criticism. 4. Music—Social aspects. 5. Ethnomusicology. I. Lysloff, René T. A. II. Gay, Leslie C. III. Series.
   ML197.M78 2003
   780'.06—dc21   2003004628

Contents

Acknowledgments ix

CHAPTER ONE
Introduction: Ethnomusicology in the Twenty-first Century 1
René T. A. Lysloff and Leslie C. Gay, Jr.

CHAPTER TWO
Musical Life in Softcity: An Internet Ethnography 23
René T. A. Lysloff

CHAPTER THREE
A Riddle Wrapped in a Mystery:
Transnational Music Sampling and Enigma's "Return to Innocence" 64
Timothy D. Taylor

CHAPTER FOUR
"Ethnic Sounds":
The Economy and Discourse of World Music Sampling 93
Paul Théberge

CHAPTER FIVE
Technology and the Production of Islamic Space:
The Call to Prayer in Singapore 109
Tong Soon Lee

CHAPTER SIX
Plugged in at Home:
Vietnamese American Technoculture in Orange County 125
Deborah Wong

CHAPTER SEVEN
Technology and Identity in Colombian Popular Music:
Tecno-macondismo in Carlos Vives's Approach to *Vallenato* 153
Janet L. Sturman

CHAPTER EIGHT
The Nature/Technology Binary Opposition Dismantled
in the Music of Madonna and Björk 182
Charity Marsh and Melissa West

CHAPTER NINE
Before the Deluge: The Technoculture of Song-Sheet Publishing
Viewed from Late-Nineteenth-Century Galveston 204
Leslie G. Gay, Jr.

CHAPTER TEN
Stretched from Manhattan's Back Alley to MOMA:
A Social History of Magnetic Tape and Recording 233
Matthew Malsky

CHAPTER ELEVEN
Tails Out: Social Phenomenology and
the Ethnographic Representation of Technology in Music Making 264
Thomas G. Porcello

CHAPTER TWELVE
"There's not a problem I can't fix, 'cause I can do it in the mix":
On the Performative Technology of 12-Inch Vinyl 290
Kai Fikentscher

CHAPTER THIRTEEN
Sounds Like the Mall of America:
Programmed Music and the Architectonics of Commercial Space 316
Jonathan Sterne

CHAPTER FOURTEEN
Consuming Audio: An Introduction to Tweak Theory 346
Marc Perlman

CHAPTER FIFTEEN
Fairly Used: Negativland's *U2* and
the Precarious Practice of Acoustic Appropriation 358
David Sanjek

Afterword: Back to Basics with the Roland 303 379
Andrew Ross

List of Contributors 383

Index 387

Acknowledgments

Several people and institutions deserve our thanks for their support in publishing this book. We are grateful to our friends and colleagues who share our fascination with music and technoculture for their intellectual engagement in conversation, colloquia, and symposia, and ultimately for their contributions to this collection. We are especially grateful, too, for their patience throughout this process. We want to note the importance of the Society of Ethnomusicology in providing venues for the several conferences, symposia, and publications that first gave voice to many of the ideas presented here.

We also want to acknowledge the support of our home institutions, the University of California, Riverside, and the University of Tennessee. Both institutions have been generous in their support of our work; especially noteworthy is the direct support for this publication from the University of Tennessee's Exhibit, Performance and Publication Expenses Fund and the University of California Academic Senate Research Fund.

We also thank the University of Illinois Press for making several essays available for publication here: Tong Soon Lee's "Technology and the Production of Islamic Space: The Call to Prayer in Singapore," Thomas Porcello's "Tails Out: Social phenomenology and the Ethnographic Representation of Technology in Music Making," and Jonathan Sterne's "Sounds Like the Mall of America: Programmed Music and the Architectonics of Commercial Space," all of which appeared previously in the journal *Ethnomusicology*. Additionally, we thank this press for allowing the publication of Leslie Gay's essay "Before the Deluge: The Technoculture of

Song Sheet Publishing Viewed from Late Nineteenth-century Galveston," first published in *American Music*.

Finally, we want to thank the staff at Wesleyan University Press, especially Suzanna Tamminen and Leonora Gibson, for their guidance and work in seeing this book through its publication.

Music and Technoculture

CHAPTER ONE

Introduction
Ethnomusicology in the Twenty-first Century

René T. A. Lysloff and Leslie C. Gay, Jr.

In 1995, for the 40th Annual Meeting of the Society for Ethnomusicology (SEM), held in the beautiful Hotel Biltmore in Los Angeles, we organized a Pre-Conference Symposium on music and technoculture. Although several ethnomusicologists have been concerned with such issues in the past,[1] this pre-conference setting marked the first time scholars had come together to speak specifically about the impact of technology—particularly audio, communications, and informational technology—on cultural practices involving music. We numbered about ninety participants with fifteen presentations. The symposium was publicized using an image that poked gentle fun at the SEM logo (a stylized image of a nude and obviously male pre-Columbian flute player that is printed on all SEM publications, including the flagship journal *Ethnomusicology*). Our participants, not all necessarily members of SEM, sported name badges with the logo parody, clearly distinguishing them from the attendees of another Pre-Conference Symposium held at the same time—on the music and scholarship of Béla Bartók. There couldn't have been two more contrasting events in the same hotel.

Our aim in organizing the symposium was to explore some new understandings of musical culture in which the concept of "culture" itself is radically reconfigured in terms of globalizing, increasingly sophisticated media and information technologies. We wanted to establish an *ethnomusicology of technoculture*, by which we mean an ethnographic study of musical culture with emphasis placed on technological impact and change. Our purpose in arguing for an ethnomusicology of technoculture was to break from past conventions of examining only folkish or high art "traditions" of music, on the one hand, and to go beyond discussing the music "industry" in the study of popular culture, on the other. The playful transformation of the Society for Ethnomusicology logo, transformed into a pre-Columbian electric guitar player sporting sunglasses, who was affectionately called "Ethno-Techno-person" (shown above), is a spoof aimed at our field, suggesting how much the world has changed over the past fifty years or so since ethnomusicology has become institutionalized in the American academy. The ethnographic Other is now fully plugged in, and the ethnomusicologist is no longer the only person in the field with high-tech equipment (video camera, tape recorder, etc.). Even the process of transforming the SEM logo into Ethno-Techno-person, printing the image on flyers and conference badges, placing it on a website, and so forth, involved learning sophisticated computer programs, a testament to the new kinds of skills and knowledge that ethnomusicologists must know to keep the discipline relevant in a changing world. Thus, while its meaning and use may differ from place to place, and while it might be either oppressive or liberating, researchers and their ethnographic Others have both long fully embraced media technology—lock, stock, and circuit board.

On the Concept of "Technoculture"

According to several essays by Andrew Ross, the term *technoculture* refers to communities and forms of cultural practice that have emerged in response to changing media and information technologies, forms characterized by technological adaptation, avoidance, subversion, or resistance.[2] Ross (1991: 3) argues that "it is important to understand technology not as a mechanical imposition on our lives but as a fully cultural process, soaked through with social meaning that only makes sense in the context of familiar kinds of behavior." An ethnomusicology of technoculture, extending Ross's argument, is concerned with how technology implicates cultural practices involving music. It includes not only technologically based musical countercultures and subcultures but behaviors and forms of knowledge ranging from mainstream and traditional institutions, on the one hand, to contemporary music scholarship, on the other. As Paul Théberge (1993: 151) points out, "electronic technologies and the industries that supply them are not simply the technical and economic context within which 'music' is made, but rather, they are among the very *preconditions* for contemporary musical culture, thought of in its broadest sense, in the latter half of the twentieth century" (italics ours). Thus, by examining technocultures of music, we can overcome the conventional distinction, even conflict, between technology and culture, implicit *especially* in studies of "traditional" musics in the field of ethnomusicology.³

Nevertheless, common sense tells us that technology, rather than being part of our lived experience, only mediates it. Until recently, this view has been apparent in much of the ethnomusicological literature and the various recording methods used to represent the soundscape of the musical Other. Ethnomusicologists are generally encouraged to use technology in research, bringing to the field cameras, cassette tape players, microphones, handy-cams, and so on. Upon return from the field, we listen to our material on our stereos, view it on our VCRs and color TVs, write about it on our personal computers, and use it to teach in our classrooms. We might say, then, that technology privileges the researcher, distancing him or her from the object of research—whether music/sound or human behavior—and allowing him or her to control it. Indeed, the sound document becomes a true object: isolated from the noisy chaos of real life in the field, it becomes analyzeable, frameable, manipulable, and, ultimately, exploitable.

The technologically privileged position of the ethnomusicologist is largely assumed in the literature. After all, the history of ethnomusicology is intimately linked to the history of audio recording, a technology tacitly regarded as culturally neutral by most scholars and uncritically utilized in music research (see Shelemay 1991: 279). Writing on "field technology" in the graduate textbook *Ethnomusicology: An Introduction,* Helen Myers (1992: 84) complains about the declining standards of recording in the field: "In the 1990s we face the danger that professional ethnomusicologists, by opting for convenience, are preserving the sights and sounds of music of our time on domestic equipment designed originally as dictation machines or for amateur enthusiasts to make home movies." That is, ethnomusicology is entirely dependent on audio technology and, in Myers's view, the quality of the audio equipment researchers bring to the field has a direct impact on the resultant quality of the ethnographies they produce after they return. We need, she asserts, higher standards in field recordings, since "pedestrian recordings" (presumably made with mediocre equipment and/or poor training in using recording technology in the field) result in "equally pedestrian standards of writing" (ibid.). This privileging of recording technology and the curatorial positioning of ethnomusicology was not lost on Richard Middleton (1990: 146), who viewed it as "yet another result of the colonial quest of the Western bourgeoisie, bent on preserving other people's musics before they disappear, documenting 'survivals' or 'traditional' practices, and enjoying the pleasures of exoticism into the bargain." After noting that ethnomusicologists focus almost exclusively on the music of "oriental" high cultures and "folk" or "primitive" societies (an argument no longer completely fair), he continues with the following critical observation: "The primary motives for ethnomusicological exoticism, then, obviously lie in value judgements about 'authenticity' in musical culture" (ibid.). Indeed, concerns about cultural authenticity almost always loom over the work of ethnomusicologists, even when the term "authentic" is carefully avoided in much of the literature.

It is perhaps the technologically privileged perspective that provides ethnomusicology with the authority to authenticate a music—after all, the ethnomusicologist was *there,* in the field, with a portable audio tape deck, a VTR, or a still camera, with recorded documentation to prove it. Using expensive recording equipment and often heavily funded by grants, we go to the field to bring something back: photographs, audio and video recordings, experience, knowledge, etc. As James Clifford (1988: 230) points out, "'cultures' are ethnographic collections." One might argue that ethnographic fieldwork, despite our best intentions, is a largely one-sided acquisitive enterprise dependent on—and invested in—differences of power, along with geographical and social distance, between researcher and native (see further Clifford 1988: 230–36). In the conventional discourse of music ethnography, distance and difference are encoded in words like "culture," "tradition," "customs," and, indeed, "fieldwork." Although there are exceptions to this rule, we might assume that "the field" remains inhabited by people comfortably remote from our academic lives, socially and geographically.

While study of music in or as culture—the definition of ethnomusicology—affirms the importance of cultural context and background, a rapidly growing market of non-Western music recordings has fostered the emergence of an aesthetic-ethnographic divide. Aging rock stars and self-styled curators of tribal musics now compete with ethnomusicologists and folklorists to preserve the world's musics through archives of recordings, some emphasizing intrinsic beauty and others ethnographic relevance. Clifford refers to this as the "art-culture system," in which "the concrete, inventive existence of tribal cultures and artists is suppressed in the process of either constituting authentic, 'traditional' worlds or appreciating their products in the timeless category of 'art'" (Clifford 1988: 200). The "art-culture system" has noticeably impinged on the field of ethnomusicology only in the last few decades, with the emergence of a world music industry, yet even early on—certainly with the advent of recording technology—ethnomusicologists had to weigh aesthetic content against ethnographic context. In most cases, evaluation of musical authenticity has generally been informed not just by a concern for ethnographic reality but also by a desire for aesthetic interest.[4] This whole business becomes politicized when it is applied to recorded musical Others: that is, the argument that an "authentic" performance is a good thing often becomes conflated with the assertion that a good performance must *sound* "authentic." The technologically privileged ethnomusicologist thus finds him- or herself caught up in a web of conflicting notions of cultural authenticity, intellectual authority, and aesthetic value. Consider the following passage from the same essay by Helen Myers (1992: 53): "The ethnomusicologist's concern for context should extend even to recording techniques. In fieldwork, it is essential to remember one is recording not only a sound source but also its context, the sound field. The ethnomusicologist's dream of placing all the performers in a professional recording studio . . . robs the performance of its natural ambience: audience, traffic, animals, conversation, discussion, cooking, eating, drinking—life."

In this passage, an aesthetics of authenticity blurs with a quasi-anthropological discourse on contextualized realism. The researcher must decide between the musicological "purity" of studio production and the ethnological "realism" of the field recording in a "natural" setting. Our point here is not to debate the merits of one kind of recording over another but to problematize the ethnomusicologist's position vis-á-vis audio technologies—and to suggest the implicit and often contested notions of authenticity and authority related to such technologies when music recordings are read as cultural "texts."

On the other hand, when "natives" use electronic devices or enjoy mediated performances, technology is now considered intrusive and often rendered invisible by the researcher. Electric amplifiers, spotlights, tape recorders, and similar technologies do not fit well in the romanticized narrative of exotic music traditions in distant lands. An example of this is a documentary film made several years ago of Javanese shadow theater. The film crew insisted on the use of an oil lantern in the performance, instead the usual electric lamp, for purposes of "authenticity." Another example took place in the 1970s, when an ethnomusicologist recorded Javanese court gamelan music in Yogyakarta and Surakarta for Nonesuch Records. He did not place microphones in such a way as to feature the vocalists, as Javanese engineers usually do in studio recordings, radio broadcasts, and live performances (see Sutton 1996). Instead, he placed them so that the vocal parts remained backgrounded as simply another layer in the complex texture of traditional Javanese music—according to current American views of "authentic" gamelan sound—thus rendering invisible any suggestion of electronic amplification. This nostalgic view of "authenticity" and the hostility toward technological intrusion are not far from the view held by fans at the Newport Folk Festival in 1965 who booed Bob Dylan when he went "electric." As Simon Frith (1986: 265–66) notes in reference to popular music, "the implication is that technology is somehow false or falsifying," that it is "unnatural" (creating an artificial presence in performance), "alienating" (coming between performers and their audiences), and somehow "opposed to art" (emptying musical performance of creativity and expressiveness). This may explain why ethnomusicologists, at least until very recently, have been reluctant to concern themselves with mass-mediated and experimental or contemporary musics. Even as we use media technology in our everyday and professional lives, we generally fail to recognize that these devices—the radio, the stereo system, the TV—have all become heavily loaded with ideological and cultural baggage. Thus, it is especially easy to forget that technology can take on different meanings across cultural, subcultural, and class boundaries and may be used in entirely new contexts.

Technology and Culture

> American culture is technoculture, from the boardroom to the bedroom. This is not to say that there is just one American culture; there are many, yet each is a technoculture. Truckers and cyberpunks, rap musicians and concert pianists, even hippies and the Amish all employ technologies in such a way that cultural activity is not intelligibly separate from the utilization of these technologies. The Amish have their wagons and farm equipment, the hippies their Volkswagen buses. The rap DJ has his or her turntable, which is employed differently from the turntable of a commercial radio DJ; the cyberpunk has a computer complete with modem, and this utilization differs from the accountant at his or her computer console. (Menser and Aronowitz 1996: 10)

As Michael Menser and Stanley Aronowitz point out, there are many technocultures in America. However, we want to emphasize that technology is implicated in the formation and perpetuation of cultures not only in America but throughout the world. Furthermore, while Menser and Aronowitz are correct in pointing out that people use technologies in ways such that the "cultural activity is not intelligibly separate from the utilization of these technologies," we believe that technologies are created and/or used in particular cultural contexts for specific purposes.

To help clarify this place of technology within society and, again, following Menser and Aronowitz (1996: 15), we suggest three methodological distinctions for understanding the place of the technological in society: the *ontological,* the *pragmatic,* and the *phenomenological.* The *ontological* determines what technology *is,* the focus being on the technological object itself (i.e., the turntable, the amplifier, the CD, etc.). The *pragmatic* is concerned with how technologies are used and the practices and forms of knowledge that arise from their use (such as deejaying, computer hacking, etc.). Finally, the *phenomenological* focuses on how the technological impacts human experience in ways not directly tied to the function of a particular technology (for example, how the automobile has taken on other meanings beyond its simple utility, like upward social mobility, masculinity, youthfulness, fashion, etc.). With these methodologies, we look beyond nuts and bolts or DRAM chips, beyond the hard edges of physical objects, to view technology as a culturally saturated component of human activity. Indeed, our definition of technology encompasses not only the technological artifact but the ways in which technologies are used and conceived (Théberge 1997; Bijker et al. 1987). To study technoculture, then, we must examine technologies not just as things—autonomous or neutral "devices"—but as material culture that people use and experience in ways meaningful to their particular needs and circumstances.

With the above methodologies in mind, we posit three kinds of culturally determined agency closely related to technology: *interaction, knowledge,* and *experience. Interaction* means engaging with the technological in a direct way: driving a car, playing a piano, spinning a record, etc. *Knowledge* involves understanding the significance of a device or phenomenon, learning what it can or is supposed to *do*. This can range from fully understanding its constitution (its scientific principles or construction) to simply intuiting its intended use (or potential uses). For example, when a small child bangs on a piano, he or she has come to realize a causal relationship between striking a key and making a sound but may not yet realize the significance of that relationship. The child's mother, on the other hand, knows that the piano can make meaningful sounds, that it produces music (even if she does not play—i.e., physically interact with—the piano herself). *Experience* involves understanding a technology in terms of both the past and the present (as well as, perhaps, the future). This means thinking about the technological in relation to other aspects of life. Much of our knowledge of the piano comes from learning about the ways it has been used through history, knowing that it is part of the Western musical tradition, and knowing its class and, perhaps, gender implications (see Parakilas 2000; Loesser 1954). In other words, people may purchase a piano not only because they want to make music but because they understand that owning a piano may say something about their social status, educational background, and so forth. Technologies like the piano are thus imbued with meaning through their utilization and, because of this, their meanings may change over time and place. At the same time, because these same technologies may offer other possibilities of meaning, they often also generate new desires, needs, and uses.

Using these methodologies (ontological, pragmatic, and phenomenological) and recognizing the kinds of human agency related to particular devices (interaction, knowledge, and experience), we might understand how using technology is implicated in the ongoing work of producing culture. These concepts can help us make sense of the new uses and meanings that various communities find for existing technologies—and how such technologies may change in meaning as they cross national, linguistic, and cultural boundaries. Thus, we begin with the obvious but seldom explored premise that *technologies become imbedded in cultural systems and social institutions, which, in turn, are reconfigured by those same technologies.* Technologies are acculturated in societies through human agency—through their utilization by people—and, in turn, generate new needs and social spaces for those people. Without the meaning conferred through use, technologies become dead objects, empty artifacts: our books turn into useless scraps of paper, our homes into piles of wood, and our computers into bits of wire and plastic within hulks of metal.

That is, a device like the piano takes on meaning not only because it is part of a "cultural activity" but also as a result of its own history, both as an idea and as a concrete object. That history is long and complex, a kind of genealogy weaving through other genealogies of related musical technologies. In other words, the "pianoforte" (a keyboard instrument that can be played either loudly or softly without having to change settings) was prefigured in instruments like the organ and harpsichord (Loesser 1954). Historically speaking, we might argue that older technologies often suggest possibilities for subsequent ones.[5] More importantly, however, certain technologies are created to serve specific needs and desires, but they also generate new needs and desires that, in turn, give rise to new technologies.

It is the *use* of technology that draws the interest of students of culture. However, this is not to say that technology has meaning only through its use. Clearly, new devices do not simply come into existence culturally neutral, handed over to humanity like the gift of fire from Prometheus. They are developed for explicit reasons by people living in specific historical moments and social contexts. Indeed, technologies are often designed with particular social effects in mind. One might even say that many, if not all, technologies are inherently political. As Langdon Winner (1999: 34) points out, "If we examine social patterns that comprise the environments of technical systems, we find certain devices and systems almost invariably linked to specific ways of organizing power and authority. The important question is: Does this state of affairs derive from an unavoidable social response to intractable properties in the things themselves, or is it instead a pattern imposed independently by a governing body, ruling class, or some other social or cultural institution to further its own purposes?"

Looking at extreme examples like the nuclear bomb and sophisticated military equipment, the answer to Winner's question is obvious. Access to the World Wide Web, we all know, is bound up in the crude economics of who can afford to own a home computer and subscribe to the Internet. While TV advertising might say you can be "anyone" on the Internet, in all probability you are male, middle class, and white. With other forms of technology, particularly early and basic technologies like the printing press, combustible engine, magnetic recording tape, etc., the matter becomes far more complex, bound up in attendant questions of how such technologies have been deployed and by (or for) whom. Indeed, technologies can also be used to support (see Gay, this collection) or subvert long-established sociopolitical and economic institutions.

We believe, then, that technology emerges from and gives rise to particular social configurations. The creation, distribution, and use of technology have social consequences. However, recall our earlier argument that it is not new technologies as such that challenge old social configurations and create new ones but the people that engage with and use these technologies. For example, MP3 (MPEG 1–Audio Layer 3), a compression software for near-CD quality digital audio files, makes audio recordings easily available through the Internet. This file format has quickly acquired notoriety for undermining the long-standing marketing networks of the popular music industry (Garofalo 1999: 349). With the increasing popularity of MP3, which can be used in combination with other software such as Napster and Gnutella, tools that link online communities of music consumers with vast databases of music files distributed over the Internet, individuals now have easy access to thousands of music titles and gigabytes of digital information. The conflicts, and the passions, surrounding the emergence of these technologies are indeed large, with Napster claiming over 20 million users and some 500,000 available song titles (see Evangelista 2000; Hamersley 2000; Snider 2000). In response, the music industry (the Recording Industry Association of America), joined by several performers including the members of the rock band Metallica, filed lawsuits attempting to outlaw its use through existing copyright laws.

MP3 and Internet software such as Napster have been crucial in the establishment of new possibilities for the exchange of music.[6] What is particularly interesting about these technologies is that they give greater control of music selection to the end user, allowing people to freely share music online and, in this way, undermine the power that the popular music industry has over its consumers (see Lewis 2000; Segal 2000). Such technologies further distort the landscape by allowing unsigned artists to distribute their own recordings on a scale heretofore unthinkable, establishing new musical careers and reinvigorating past pursuits. They circumvent the established recording industry altogether by making "the world your hard disk [and] everyone a publisher," says David Post (quoted in Snider 2000). Or, as Andrew Sullivan views the phenomena, only partly tongue-in-cheek, these technologies form the center of a social non-revolution, an emerging "e-topian" or "dot-communist" world where "mine-thine" distinctions are erased for those with a computer, a modem, and a phone line (Sullivan 2000).

Similarly, René Lysloff, in this collection, explores the "virtual community" surrounding digital music modules, or "mods." This community of musicians focuses on the production and exchange of digital music files and related software (called "trackers," a designation that also identifies mod users). He argues that what is radical about the "mod scene" (as community members refer to it) is not the network of relationships it engenders but the creative ways in which members construct and maintain these relationships using new technologies. Like Napster's emerging social collectivities, mod trackers have created a community over the Internet through their common interest and purpose—the sharing of digital music—unrestrained by geographical proximity in a way that challenges conventional forms of social relationships and commercial music distribution. Unlike most Napster users, however, mod trackers do more than just share prerecorded music; they are active as composers, manipulating sound samples with virtual mixing boards to create their own mod files. In effect, they have established a significantly large and active community made up of both composers and their fans, all interacting with one another online. Lysloff shows that this community is formed around a complex social hierarchy, elaborate prestige system, and energetic exchange network—a (sub)culture of simulation grounded in the "virtual materiality" of graphical interface computing—and its political economy is based upon the notion that the free exchange of data and communications, not money, garners the most social value.

Localizing Global Technologies

As we have already mentioned, technologies also become saturated with social meaning as they acquire a history of use. The automobile, for example, becomes a symbol of middle-class mobility in the United States as a result of its particular history. In other words, technological meaning is historically grounded and, as a result, becomes located within a larger social imaginary. New technologies, on the other hand, are volatile. Their social meaning is not established: a new technological device might be known to only a few people or be economically unfeasible, its social affects debated, its intended use subverted. Thus, we argue that the technological device, whether it is a quill pen or a personal computer, gains meaning through human agency. Because of human agency, technology can be politically oppressive yet also liberating; it might build community while simultaneously causing social alienation; and it can benefit the sick and infirm but also damage the environment or even destroy life. It is at the intersection of human agency and technological artifact where meaning is most often contested, where we find resistance, subversion, co-option, coercion, and domination. This is why technological devices often become politically charged with conflicting ideologies of identity, community, knowledge, and power.

Such conflicts over identity, community, and power regularly emerge within what Mark Slobin has called the global industrial interculture of commodified music (1993: 61). Slobin's observations draw upon Arjun Appadurai's (1990) mapping of a global cultural economy across several "landscapes," including "ethnoscapes," "mediascapes," and "technoscapes," concepts pertinent to this study. Several essays in this collection address these themes, illustrating much about this global industrial interculture, its use of media and information technologies, and its overlapping, often colliding domains that reflect both local traditions and international conventions. Timothy Taylor uses the story of Enigma's appropriation of an aboriginal Taiwanese song for their recording "Return to Innocence" to illustrate and critique notions of globalization. Taylor argues that colonialism and imperialism first shaped the globe into a lopsided configuration in which Western forms and signs were disseminated throughout much of the world. However, with today's globalization, driven by multinational capitalism and its information and media technologies, forms and signs from elsewhere flow to Europe and the United States with increasing speed. Such junctures between peripheries and metropoles show how within this postmodern world sharp distinctions between local and global are often elusive—Taylor borrows the term "glocalization" from Paul Virilio to emphasize this point. This configuration remains lopsided, though: Enigma's recording illustrates the value to the music industry of indigenous music and musicians who by and large remain unidentified and unpaid for their contributions. Paul Théberge makes a related argument in his essay, describing an unrestricted marketplace of non-Western music fragments recorded and packaged as basic compositional material for "world music" recordings. The use of technologies of reproduction—digital synthesizers and samplers—not only makes possible the appropriation of the expressive forms of non-Western peoples but through the discourse surrounding the technology, justifies their commodification, exoticization, and exploitation.

Although, as these essays illustrate, technologies often transcend political, linguistic, and even cultural boundaries, much of their associated meaning is locally grounded—again, through the specifics of history and circumstance.

Take the light bulb, for instance. It is a simple but absolutely crucial form of technology throughout the world. One might say that its meaning is simple: it provides light. However, in Indonesia (specifically Java), artificial light has a different meaning than it does in the United States—at least economically. Light bulbs are priced according to wattage (the stronger the light bulb, the more expensive it is). Since electricity is considered a luxury in Java, ideas regarding the amount of wattage necessary to provide adequate lighting are very different from those in the United States. Bright lights are considered wasteful and reserved only for the very wealthy. Even local governments conserve electricity, using few streetlights (most streets are dimly illuminated by nearby houses), often with dim neon tubes, to light up an area. At night, small towns and villages seem gloomy and murky by Western standards: houses are lit by 25-watt light bulbs and street vendors still use candles or gas lanterns. Larger stores are closed by 9:00 or 10:00 P.M., although small kiosks often remain open. Even local festivals seem rather poorly illuminated. One might speculate that the West has used electrical light to extend daytime into nighttime, and that this has had an impact on our work habits, our shopping patterns, and our activities related to amusement and pleasure. Lighting in the United States was indeed developed not only to simply illuminate the dark but also to simulate daylight.[7] In Java, on the other hand, the light bulb simply provides enough illumination to get around in the dark. Thus, following Clifford Geertz (1983), we argue that all technological knowledge is local knowledge, though we recognize that technologies may reshape notions of the local, extending or transcending old boundaries through changes in transportation and communications. This is why, for example, radio broadcasting and sound amplification take on different meanings in different parts of the world.

Tong Soon Lee's essay in this collection presents an especially telling example of local adaptation of media technologies: how the use of the radio to broadcast the Islamic call to prayer has reconfigured the Muslim "acoustic community" in Singapore. As a correlate to the transformation of Singapore into a more multi-ethnic, urban environment, Islamic communities moved into locales now shared with non-Muslims. The subsequent loss of the sacred acoustic environment of the call to prayer as amplified from the old mosques was regained through electronic mediation, transforming public displays of Muslim identity and religious practice into private ones. Within this technoculture, Muslims have "traditionalized" the media technology of radio broadcasting even as they redefine its social meaning.

Technologies, even the most oppressive and alienating, are thus constantly being reinterpreted in ways that make sense of local circumstances and that intersect with local interests, often subverting their original intent.

Thus, rather than viewing technologies solely as imposed from the West and accepted passively elsewhere, like Penley and Ross, we see technologies within social and political processes "by which people . . . make their own independent sense of the stories that are told within and about an advanced technological society" (1992: xv). Such a view allows "technocommodities" to be transformed into social, economic, and political resources, emerging from the unorganized daily activities of a populace (Penley and Ross 1992: xvi). Indeed, it is here where meaning may be contested and expressive innovation emerges in the use of technology.

Deborah Wong's essay, advancing this argument, illustrates how Vietnamese Americans reshape familiar technologies of the mass media, from television and video to CDs and CD-ROM, to actively construct their own history and memory of Vietnam and engage in political resistance. As Wong argues, media technologies do not necessarily equate with multinational hegemony but rather can open pathways of resistance through localized adaptations and new meanings. Likewise, Janet Sturman follows the Colombian musical genre *vallenato* in her essay as it moves from rural, regional practices to its emergence within a cosmopolitan arena of Latin American popular music to illustrate how Colombian and Latin American technocultural identities contrast with those of North America and Europe. Here, her distinctly Latin American concept, *tecno-macondismo*, which draws from Brunner's adaptation of Gabriel García Márquez's utopian world of *Macondo*, embraces a nostalgia for nature together with "whatever technology transcends the isolation of a regional position." Instrumental to this concept are the links among several media technologies that Colombians make, illustrated through the work of Colombian singer and *telenovela* star Carlos Vives. In another essay, Melissa West and Charity Marsh advance a feminist view of technology by questioning the relationships between nature, technology, and gender. Through an analysis of the Icelandic performer Björk and the "mega pop star" Madonna, nature and technology—cultural categories often defined as opposites within popular music genres—become integrated.

The Ripple Effect and Regimes of Technologies

Each new technology seems to recede into the background of human experience with increasing speed. Perhaps it is because earlier technologies prefigure subsequent ones: that is, the uses of earlier devices become both concretely and metaphorically subsumed by more recent devices. For example, the technology of writing may have changed the way people communicate with one another, how they understand the past, and how they tell stories to one another, but it also prefigured the printing press and other, even seemingly disparate, technologies. Indeed, many kinds of abstract knowledge (such as mathematics, geometry, physics, and so forth), necessary for so many technological changes over the past several hundred years, are based on textual literacy—i.e., writing technology (see Finnegan 1988). We might see this as a kind of ripple effect, where one kind of technological innovation gives rise to other forms of innovation.

Defined in terms tellingly close to today's world, the technocultural community discussed in Leslie Gay's essay illustrates that technological adaptations of the past often "ripple" forward and continue to shape our lives. Looking back to Galveston, Texas, in the late-nineteenth-century United States, Gay examines song sheet publishing and performance as part of a technoculture built on the technologies of literacy, printing, manufacturing, and marketing. Galveston is shown here as a point within an emerging transregional public culture, a reterritorialized locale bound by specific adaptations and meanings in which access to technologies create and maintain divisions among ethnic groups and socioeconomic classes.

Thèberge (1997) points out two aspects important to technological innovation and the ripple effect of technological change: "continuous" and "transectorial" innovation. Thèberge argues, in the case of the first, that the increasing pace of technological innovation derives from "dependencies between small, creative firms and large, scale-intensive corporations" (1997: 68). The second concept, transectorial innovation, argues that "innovations generated within a specific industrial sector find subsequent application in other, often unrelated sectors" (1997: 28). Thus, like the technological adapations generated by continuous innovation, transectorial innovation results in an interdependence among often disparate industry firms, multinational corporations, and manufacturing sectors (1997: 59). Indeed, these sort of alliances are instrumental to the technocultural world detailed by Gay.

Today, many societies are heavily dependent on such transectorial and interdependent regimes of technologies. On the one hand, certain forms of technology are used in a variety of devices. Integrated circuitry, for instance, is fundamental to a constellation of technologies, from computers to microwave ovens, stereo systems, and even advanced weaponry and surveillance devices. On the other hand, societies bring together arrays of disparate technologies for particular purposes. For example, tourism involves more than simply getting on a plane to go somewhere else—the tourist draws from a host of communication, transportation, and media technologies. Here, we argue that the meanings we derive from technology are situational, fluid, and polysemantic. For this reason, many technologies are often seen as both good and bad: for example, audio technology gave rise to the recording industry, turning music into a highly profitable commodity and alienating its audience from the social processes of musical performance. Yet that same technology also provides us with access to a far greater amount and variety of music. It allows the individual to build a library of recordings, copy or compile recordings, and even create new musical forms out of past recordings.

Several other essays show interdependent and transectorial regimes of technologies, notably Wong and Struman, both discussed earlier, and Matthew Malsky, who addresses how the introduction and early history of magnetic tape recordings—built, in part, out of a link between telephony and military technologies—shaped a "cultural logic of recorded sound." For both the social elite—in such art music composers as Vladimir Ussachevsky—and mass audiences—with tape machines marketed by the likes of Sears, Roebuck & Co.—magnetic tape and its use began to dislodge (musical) sound from its source. This dislocation required new strategies for understanding sound and affirming what is "real." Drawing meaning from tape-recorded sound and its reproduction demanded an understanding of a set of linked signs, situated within the technology's sonic characteristics, its fidelity vis-à-vis other audio technologies and unmediated performance, and an affinity with the listener's changing expectations. Thomas Porcello, in his essay, extends this discussion to the phenomenon of magnetic tape print-through. Here, these audio ghosts, dependent upon a deficiency of tape, become metaphors to critique formalist theories of music as they relate to performance, experience, and text. Moreover, the time-disordering operations found in a recording studio, which, like print-through, disrupt the flow of musical time, become a means of exploring the temporal conditions of research and ethnographic representation.

We do not want to argue for technological determinism or imply that technology shapes society in a predestined way or that its trajectory is inevitable. Nor, again, do we believe that technology is neutral—that it "develops" independently of its cultural context. New technologies are created, and existing technologies change, according to what Mackenzie and Wajcman (1999: 19) call "path dependence." They state that "the history of technology is a path-dependent history, one in which past events exercise continuing influences. Which of two or more technologies eventually succeed is not determined by their intrinsic characteristics alone, but also by their histories of adoption." That is, it is not the best technology that wins out, but the one that is best adopted by it users. They go on to point out how, for example, the *qwerty* keyboard is the standard for personal computers, with letters of the alphabet placed in the now familiar keyboard configuration to overcome the mechanical limitations of earlier typewriters (see also Bukatman 1994). The *qwerty* keyboard illustrates how past contingencies in the development of particular technologies, often locally conceived for short term results, may persist in new technologies even when they no longer bring about optimal results. Unanticipated results may also emerge from unforeseen pathways, to take Mackenzie and Wajcman's metaphor further. Kai Fikentscher provides such an example through his examination of the deejay's significance in the redefinition—the discovery of a new sociotechnological pathway—for audio reproduction technologies, i.e., the re-invention of the turntable and audio sampler as music performance instruments. Connecting to Malsky's essay on magnetic tape, Fikentscher points to changes in the notion of the music text but foregrounds the shift of audio recordings from fixed forms to elastic materials open to deejay elaboration in performance, especially within dance clubs.

Technological Control, Resistance, and Subversion

> The things we call "technologies" are ways of building order in our world. Many technical devices and systems important in everyday life contain possibilities for many different ways of ordering human activity. Consciously or not, deliberately or inadvertently, societies choose structures for technologies that influence how people are going to work, communicate, travel, consume, and so forth over a very long time.... The issues that divide or unite people in society are settled not only in the institutions and practices of politics proper, but also, and less obviously, in tangible arrangements of steel and concrete, wires and transistors, nuts and bolts.
> (Winner 1999: 32–33)

Langdon Winner's interlinked structures of technologies, particularly those that have emerged in the late twentieth century, are especially omnipresent for audio and other music technologies within consumer spaces, where technological permutations often drive consumption even as consumption pushes permutation. Thus it is not surprising to see Paul Théberge argue, for instance, that popular music musicians today are shaped first as "consumers of technology," in which musical practices align with consumer practices (1997: 6). Moreover, in our high temple of consumerism, the American shopping mall, as Jonathan Sterne shows us in his essay, the links between consumption and audio technologies are built into the mall's infrastructure, managed like the air conditioning and lighting. Within this "architectonic" structure, the acoustic spaces of programmed music (such as *MUZAK*) connect too with consumer practices within the mall, defining territory and motivating shoppers.

Théberge also argues that "the ability of the consumer to define, at least partially, the meaning and use of technology is an essential assumption" (1997: 160). This is the case as outlined by Marc Perlman in his essay on audiophiles. Perlman illustrates a technoculture of these consumers and how social meaning is constructed and personal emotion imbued into audio technology for this predominately male, rarefied class. Particularly important to Perlman is how the process of "tweaking" audio components—that is, modifying them—also personalizes them, shifting impersonal commodities into highly personal possessions even as, or because, such tweaks are nullified by scientifically sanctioned uses and meanings.

Winner's concern for societies and their structures of technologies is important, as Sterne and Perlman's essays show, but we believe it is only half of the story. Technology is not just about "building order in our world." In most cases, as in that of MP3 and Napster software, the development of new technologies is also implicated in political, social, or economic *control*. Simon Frith, for example, argues that recordings, from their inception near the beginning of the twentieth century, shifted new authority to performers by reducing the importance of composers for the popular music industry and gave a commercial voice to musically illiterate blues and "hillbilly" musicians (1986: 269–70). From these new voices (and technology), African American and southern European American communities also increased their political and economic power.[8] Recording companies, however, motivated more by profit than altruism, also struggle to control these communities through their marketing policies, creating music categories—"race" records within the era of Jim Crow laws in the United States, for instance—to delimit and restrict communities. Thus, recording companies provide us with limited choices of music not because they don't want us to have as many options as possible but because they simply want to control those options, constraining our choices to what they can actually provide (with the lowest production expenditure, and mostly conforming to prevailing social norms). David Sanjek, in this collection, foregrounds such conflicts over technology, its meaning, and its control. He gives an account of the satiric group Negativland, their recording *U2*, an audio collage that combines samples from the Irish rock band by the same name with outtakes from Casey Kasem's "Top 40" radio program, and the legal melee this recording fostered. Through this account, Sanjek assesses notions of the romantic anti-hero, virtuosity and narrative for sampled music, and the contradictions of copyright, political protest, and public speech. Accordingly, he reminds us of the social implications of technology, not just as, borrowing from Andrew Ross, hardware extensions to the body, but as part of intentional linguistic processes that include debate, even defiance, and the possibility of altering technological (and political) authority (1991: 3).

To close, we agree with Winner's argument that socio-technological structuring is implicitly political. He points out that "if our moral and political language for evaluating technology includes only categories having to do with tools and uses, if it does not include attention to the meaning of the designs and arrangements of our artifacts, then we will be blinded to much that is intellectually and practically crucial" (1999: 32). Thus, if we have not developed a discourse about technological meaning, how can we reach informed decisions about the aesthetic, social, and political consequences of technological change? It is our aim, then, that this collection advances such a discourse.

Technologies are brought into existence both to control and to expand human potential, creating ongoing tensions between cynical exploitation and utopian cooperation. The essays in this collection show that it is in the *use* of technology that meaning is found, and that it is at this intersection of human agency and the technological artifact that meaning is also contested—this is where we find the potential for resistance, subversion, co-option, coercion, and domination. Thus, technologies related to music are far from neutral, and they are never fully controlled by any single constituency. They are sites of continuous social and political struggle—in the field of ethnomusicology, such struggles are acted out in terms of cultural ownership, musical authenticity, and intellectual authority. While it might be true that many technologies were developed in the interests of industry and corporate profit, and for the purposes of domination and exploitation, these same technologies continue to become more widely disseminated, and the powers that produce them increasingly decentralized. In other words, although many technologies may be a facet of larger hegemonizing and homogenizing forces, their accessibility and availability provide people with more means to cope with and even resist or subvert those same forces. As all forms of music become increasingly implicated in the globalization of advanced technologies, an ethnomusicology of the twenty-first century will have to adapt to changing ideas of musical authenticity, cultural representation, and intellectual authority. It will be the work of the ethnomusicologist and other music scholars to analyze and explain the cultural negotiations involved with the global intersections—and local understandings—of the world's musics, popular desires, and technological possibilities.

Notes

1. Charles Keil (1994) was perhaps the first ethnomusicologist to examine music and the cultural implications of media technologies from a crosscultural perspective. His article "Music Mediated and Live in Japan" first appeared in the journal *Ethnomusicology* in 1984. That same year, Roger Wallis and Krister Malm (1984) broke new ground for ethnomusicology through their crosscultural comparison of the recording industries in several economically small countries; implicit to their study are the technological foundations of these industries. More recently, a book by Peter Manuel (1993) examines popular music and audio cassette technology in India. And Steven Feld (1994), building from the work of R. Murray Schafer (1994 [1977]), theorizes on the dislocation of music from its producers as engendered by audio and recording technologies, and the implications for cultural (mis)representations. Moreover, both of us have earlier addressed the relationships among music, culture, and technology (see Lysloff 1997 and Gay 1998).

Other scholars, particularly those from popular music studies, have taken up the theme of music and technology, although not necessarily from an ethnographic perspective. For example, Chapple and Garofalo (1977) address technology vis-à-vis struggles for market dominance among media corporations; Hosokawa (1984) investigates the socially transformative nature of personal audio technologies like the Walkman; Goodwin (1990 [1988]) argues the consequence of digital technologies for music-making and concepts of creativity; Walser (1993) connects electric guitar and amplification technologies to heavy metal sound; and Rose (1994) explicitly links rap and hip-hop to the technological terrain of the urban United States.

2. Parts of this section are based on Lysloff 1997.

3. Early ethnomusicologists (and comparative musicologists) sometimes were concerned with the technology of musical instruments within so-called "primitive" societies, treating these technologies as indicators of cultural development within social evolutionary schemes. See, for example, Hornbostel (1933) and Wachsmann (1961).

4. See, for example, Bruno Nettl's remarks (1983: 316): "The concept of the 'authentic' for a long time dominated collecting activities, became mixed with 'old' and 'exotic' and synonymous with 'good.'"

5. We do not want to suggest an implicit assumption of linear development for the piano, or any other technology. Linearity in the development of technologies, a belief often tied to other problematic concepts such as the notion of singular invention, mostly follows from retrospective constructions where "rational paths" toward specific goals subsume an artifact's untidy narrative of cultural adaptation (see Bijker 1995: 6–7).

6. Here, one new aspect of music with broad implications, as Shuhei Hosokawa (1990) argues, is its reconceptualization as another from of digital information technology.

7. This conception of light extends to much of Europe as well. For the 1889 Paris Exposition, architect Jules Bourdais proposed a tower with electric arc-lights powerful enough to illuminate all of Paris, turning, it was believed, "night into day." Bourdais's proposal for the *Tour Soleil* lost out to a competing proposal by Gustave Eiffel not because it was technically unfeasible but because of safety concerns over the dazzling effect of arc-light technology (Schivelbusch 1988: 3, 128–34). See also Wuebe E. Bijker's discussion of the development of flourescent light technology and the "high-intensity daylight fluorescent lamp" (1995: 199–267).

8. Charles Keil similarly argues that so-called "folk" musics (Keil prefers the

phrase "people's music"), such as blues and polka, materialized in the United States as part of its media soundscapes, that is, supported by media technologies, not separate from them (Keil 1985; see also the chapter "Technology and Authority" in Frith 1996).

References

Appadurai, Arjun. 1990. "Disjuncture and Difference in the Global Cultural Economy." *Public Culture* 2 (2): 1–24.
Baudrillard, Jean. 1983. *Simulations.* New York: Semiotext(e).
Bijker, Wiebe E. 1995. *Of Bicycles, Bakelites, and Bulbs: Toward a Theory of Sociotechnical Change.* Cambridge: MIT Press.
Bijker, Wiebe E., Thomas P. Hughes, and Trevor J. Pinch. 1987. "General Introduction." In *The Social Construction of Technological Systems: New Directions in the Sociology and History of Technology.* W. E. Bijker, T. P. Hughes, and T. J. Pinch, eds. Cambridge: MIT Press.
Bukatman, Scott. 1994. "Gibson's Typewriter." In *Flame Wars: The Discourse of Cyberculture.* M. Dery, ed. Durham: Duke University Press.
Chapple, Steve, and Reebee Garofalo. 1977. *Rock 'n' Roll Is Here to Pay: The History and Politics of the Music Industry.* Chicago: Nelson-Hall.
Clifford, James. 1988. *The Predicament of Culture.* Cambridge: Harvard University Press.
Druckery, Timothy, and Gretchen Bender. 1994. *Culture on the Brink: Ideologies of Technology.* Seattle: Bay Press.
Evangelista, Benny. 2000. "Net Future Of Music On Trial; Industry, Napster To Plead Their Cases." *San Francisco Chronicle,* July 26, 2000, D1.
Feld, Steven. 1994. "From Schizophonia to Schismogenesis: On the Discourses and Commodification Practices of 'World Music' and 'World Beat.'" In *Music Grooves: Essays and Dialogues.* C. Keil and S. Feld, eds. Chicago: University of Chicago Press.
Finnegan, Ruth. 1988. *Literacy and Orality: Studies in the Technology of Communication.* New York: Basil Blackwell.
Frith, Simon. 1986. "Art Versus Technology: The Strange Case of Popular Music." *Media, Culture & Society* 8 (3): 263–79.
———. 1996 *Performing Rites: On the Value of Popular Music.* Cambridge: Harvard University Press.
Garofalo, Reebee. 1999. "From Music Publishing to MP3: Music and Industry in the Twentieth Century." *American Music* 17 (3): 318–53.
Gay, Jr., Leslie C. 1998. "Acting Up, Talking Tech: New York Rock Musicians and Their Rhetoric of Technology." *Ethnomusicology* 42 (1): 81–98.
Geertz, Clifford. 1983. *Local Knowledge: Further Essays in Interpretive Anthropology.* New York: Basic Books.
Goodwin, Andrew. 1990. "Sample and Hold: Pop Music in the Digital Age of Reproduction." In *On Record: Rock, Pop, and the Written Word.* S. Frith and A. Goodwin, eds. New York: Pantheon Books. First published in 1988.
Hamersley, Ben. 2000. "Music for a Song." *The Times,* June 12, 2000.
Hornbostel, Erich M. von. 1933. "The Ethnology of African Sound Instruments." *Africa* 6 (2): 129–57, 277–311.
Hosokawa, Shuhei. 1984. "The Walkman Effect." *Popular Music* 4: 165–80.
———. 1990. *The Aesthetics of Recorded Sound.* English Summary. Tokyo: Keisó Shobó.
Jackson, Kenneth T. 1985. *The Crabgrass Frontier: The Suburbanization of the United States.* New York: Oxford University Press.

Keil, Charles. 1994. "Music Mediated and Live in Japan." In *Music Grooves: Essays and Dialogue*. C. Keil and S. Feld, eds. Chicago: University of Chicago Press. First published in 1984 in *Ethnomusicology* 27 (1): 91–96.

———. 1985. "People's Music Comparatively: Style and Stereotype, Class and Hegemony." *Dialectical Anthropology* 10: 119–30.

Lewis, Peter H. 2000. "State of the Art; Napster Rocks the Web." *New York Times*, June 29, 2000, G1.

Loesser, Arthur. 1954. *Men, Women, and Pianos: A Social History*. New York: Simon and Schuster.

Lysloff, René T. A. 1997. "Mozart in Mirrorshades: Ethnomusicology, Technology, and the Politics of Representation." *Ethnomusicology* 41 (2): 206–19.

Mackenzie, Donald, and Judy Wajcman, eds. 1999. *The Social Shaping of Technology*. 2nd ed. Buckingham: Open University Press. First published in 1985.

Manuel, Peter. 1993. *Cassette Culture: Popular Music and Technology in North India*. Chicago: University of Chicago Press.

Menser, Michael, and Stanley Aronowitz. 1996. "On Cultural Studies." In *Technoscience and Cyber Culture*. S. Aroniwutzm, B. Martinsons, and M. Menser, eds. New York: Routledge.

Middleton, Richard. 1990. *Studying Popular Music*. Milton Keynes: Open University Press.

Myers, Helen, ed. 1992. *Ethnomusicology: An Introduction*. New York: W. W. Norton.

Parakilas, James, ed. 2000. *Piano Roles: Three Hundred Years of Life with the Piano*. New Haven: Yale University Press.

Penley, Constance, and Andrew Ross, eds. 1992. *Technoculture*. Minneapolis: University of Minnesota Press.

Rose, Tricia. 1994. *Black Noise: Rap Music and Black Culture in Contemporary America*. Hanover, N.H.: Wesleyan University Press / University Press of New England.

Ross, Andrew. 1991. *Strange Weather: Culture, Science, and Technology in the Age of Limits*. New York: Verso.

———. 1992. "New Age Technoculture." In *Cultural Studies*. L. Grossberg, C. Nelson, and P. Treichler, eds. New York: Routledge.

Schafer, R. Murray. 1994. *The Soundscape: Our Sonic Environment and the Tuning of the World*. Rochester, Vt.: Destiny Books. Previously published as *The Tuning of the World*, Philadelphia: University of Pennsylvania (1980), and New York: Knopf (1977).

Schivelbusch, Wolfgang. 1988. *Disenchanted Night: The Industrialisation of Light in the Nineteenth Century*. transl. A. Davies. Oxford: Berg.

Segal, David. 2000. "Napster Looking for a Groove; Online Rock-and-Roll Is Here to Stay, but Its Pioneer Needs a Plan." *Washington Post*, July 26, 2000, E01.

Shelemay, Kay Kaufman. 1991. "Recording Technology, the Record Industry, and Ethnomusicological Scholarship." In *Comparative Musicology and Anthropology of Music: Essays on the History of Ethnomusicology*. B. Nettl and P. V. Bohlman, eds. Chicago: University of Chicago Press.

Slobin, Mark. 1993. *Subcultural Sounds: Micromusics of the West*. Hanover, N.H.: Wesleyan University Press / University Press of New England.

Snider, Mike. 2000. "Napster's Siren Song Entices Loyalty But Even Some Fans See Site as Scene of Crime." *USA Today*, July 26, 2000, 3D.

Sullivan, Andrew. 2000. "The Way We Live Now: 6-11-00: Counter Culture; Dotcommunist Manifesto." *New York Times*, June 11, 2000, 30 (section 6).

Sutton, R. Anderson. 1996. "Interpreting Electronic Sound: Technology in the Contemporary Javanese Soundscape." *Ethnomusicology* 40 (2): 249–68.

Théberge, Paul. 1997. *Any Sound You Can Imagine: Making Music/Consuming Technology.* Hanover, N.H.: Wesleyan University Press / University Press of New England.

———. 1993. "Random Access: Music, Technology, Postmodernism." In *The Last Post.* S. Miller, ed. Manchester: Manchester University.

Wachsmann, Klaus P. 1961. "The Primitive Musical Instruments." In *Musical Instruments through the Ages.* A. Baines, ed. New York: Walker and Company.

Wallis, Roger, and Krister Malm. 1984. *Big Sounds from Small Peoples: The Music Industry in Small Countries.* New York: Pendragon Press.

Walser, Robert. 1993. *Running with the Devil: Power, Gender, and Madness in Heavy Metal Music.* Hanover, N.H.: Wesleyan University Press / University Press of New England.

Winner, Langdon. 1999. "Do Artifacts Have Politics?" In *The Social Shaping of Technology.* D. MacKenzie, and J. Wajcman, eds. Buckingham: Open University Press.

CHAPTER TWO

Musical Life in Softcity
An Internet Ethnography

René T. A. Lysloff

Introduction: Ghosts in the Machine

For me, the Internet is one gigantic ghost town, a great dreamlike metropolis where I always seem to arrive the moment everybody else has left.[1] It is a deserted metropolis where I find traces of life everywhere, but no people, no living bodies. The urban center of this vast, software-driven city is made up of hundreds of thousands of elaborate websites that stand as deserted monuments to organizations, businesses, and public institutions. We can all shop and bank there, but we do these things separately. We can commune with one another, but we can never actually gather together. Its suburbs are countless personal homepages—an immense neighborhood of haunted houses containing the tastes, hobbies, and lifestyles of countless missing people we've never seen, never met. It continues to grow in size as more and more people from the offline realms come to shop, find pleasure or fortune, or search for information.

Synchronous message systems seem like electronic Ouija boards in eerie rituals of cyber-seance. Ghostly messages appear on the monitor, typed by unseen hands, as if from the other side of the grave. Yet it is not the dead but the living that supernaturally communicate with us. The people we write to are "there" but not *really there*—we write not to *them* but to virtual addresses in this ghost town that we all visit individually, alone in front of our computers. Many of the people that we encounter we will never see or touch, never hear their voices; we may not be sure *where* they really are when they write to us—are they at home, in an office, at a cyber-café? We may not even be sure what part of the world they live in, what state or country. On our computers, they seem to be phantoms built on words alone.

This is Softcity, what William J. Mitchell also calls the "City of Bits," a metaphorical place of infinite space that exists without location or materiality. The World Wide Web is an equally appropriate name, since the Internet is also a vast network of information, relationships, desires, possibilities. But, like a web, it ensnares us with utopian promises of posthuman transcendence as ageless disembodied tele-presences without nationality, gender, creed, or color. Thus, we have become living ghosts in the machine.[2] The World Wide Web seems to be huge ghost town because human presence there is rarely synchronized temporally and never spatially. I won't find another person on the Web, I'll only find the traces they left behind. In other words, we might say that Softcity is not really about contemporaneity but about commensurability; it's not that people are *sharing* virtual space but that they all have *access* to that space—and they do it each individually, alone. There may be a hundred, a thousand, a million people accessing the Internet at any given moment, but I don't experience their presence, only evidence that they were once there. Even with so-called synchronous chat systems and instant message services, I can only infer their existence from whatever texts (and occasional images) they leave behind. Thus, Softcity is a city of software-driven ephemerality, its fantastic edifices built from the bits and bytes of data retrieved from remote mainframe servers. Its inhabitants exist as disembodied nomadic identities with no simultaneity of presence, only a collective solitude.

Virtual Ethnomusicology

Since the summer of 1997, I have been involved in conducting online ethnographic field research. It involved learning particular musical and linguistic skills, interviewing composers, visiting numerous research sites, collecting relevant texts and audio recordings, and observing various kinds of music-related activities. Indeed, I used classical field method in this research: participation-observation, interviews, documentation, etc. All in all, the fieldwork was extremely time consuming and intellectually demanding. It involved studying some rather esoteric knowledge, developing new skills in making music, and challenging my understanding of musical performance. The experience was very different from my earlier research in a mountainous region of Central Java, where I spent years studying rural musical traditions. My Internet fieldwork involved long periods of solitary work in front of a computer monitor while, in Java, I was often out and about, engaged in face-to-face encounters with people, all of us immersed in the hot, tropical environment. My online experiences seemed to be made up entirely of viewing images and listening to sounds and music, of reading and writing texts, of absorbing abstract coding information. My memories of Java, however, are filled with visceral experiences: driving my motorcycle on dusty and dangerous two-lane highways filled with chaotic traffic, sitting cross-legged at all-night performances, playing music with professional musicians, eating rice and fried foods, enduring intense bouts of flu and diarrhea, suffering through periods of profound loneliness, and so on—all far more vivid than the more intellectually demanding tasks of learning the esoteric aspects of Javanese culture (language, history, religious beliefs and practices, musical concepts and structures, and so on).

My Internet fieldwork was a solitary experience that, at the same time, seemed to be intensely social. Although I conducted my research without ever leaving my home, my work involved immersing myself in an entirely new musical community and engaging with informants that lived far away. I dealt with people that I would never actually meet face to face, people whose presence I could only infer from their textual messages. I often asked myself whether what I was doing was *really* fieldwork, since I never had to go anywhere physically, never made demands on my body, never endured the tangible hazards that field researchers routinely face. I simply spent many late nights in front of my computer, connected to the Internet, "traveling" the far corners of cyberspace. I visited homepages originating in countries throughout America, Europe, Australia, Asia, and the Middle East. And, using real-time electronic "chat" systems and email, I communicated with many musicians and music experts who were equally scattered all over the world. I often stayed connected the entire night, virtually traveling from one homepage to another, deeply engaged in the social network of my farflung friends. My sessions on the computer involved reading and typing messages to electronic musicians and fans of music, gazing at images and animated applets, and listening to random songs that would play automatically at some websites. Some of these sites were installed with plug-ins (small applications, or applets) that allowed me to listen to a song without necessarily first downloading it and then playing it back on my own software; others offered excerpts of songs using streaming audio programs. Throughout the night I often checked my email, since many of my composer friends or fellow enthusiasts would send me their most recent pieces for discussion.

Synchronous message systems or chat networks allowed me to interview informants and make new friends. One, a young composer from Israel who called himself TrackZ, became my tutor, and he taught me a great deal about using some of the software involved in electronic music composition. Occasionally I felt it necessary to set my synchronous message program to the "hide" mode—to conceal my online status—since idle textual chatting sometimes became a nuisance, with the number of people wishing to chat with me unmanageable and distracting. Even my teacher, TrackZ, often wanted to communicate when it wasn't convenient for me. This also was somewhat like my experience in Central Java, where I was often visited by strangers from neighboring villages, curious about the foreigner living in their area. In both cases, I could not control my encounters with the Other and had to learn to adapt to the immediacy of each situation as it arose. Additionally, I experienced disorienting moments of temporal suspension when I downloaded large files of music, long moments of isolation and boredom. Everything seemed to pause, in the speeded up temporality of cyberspace, for an interminable period. I had similar feelings of separation in Java during Ramadan, when performances and many social activities were halted for a month of fasting and introspection. I tried to fill these times with reading and writing, but the days seemed to stretch into eternity. In retrospect, it seems astonishing that, on the Internet, such seemingly endless periods could actually be measured in minutes rather than days or weeks.

Just as in Indonesia, I sometimes found myself utterly overwhelmed by the massive amount of information I had to learn and frustrated at my inability to quickly grasp the concepts my teachers viewed as fundamental. Yet the overall experience was deeply satisfying, because many of the artists and fans I met in my work have grown to become my friends, and the knowledge I received has enhanced my understanding of music. While I sat alone in front of that flat computer monitor, it took on depth in my imagination, fleshed out with people actively engaged in relationships as real as any I have encountered in the world of flesh and blood. Like my fieldwork experience in Central Java, I gained access to forms of musical knowledge and skills that have changed me forever.

At the same time, however, I do not want to imply that my fieldwork experience in Java is equivalent to my online experience. In Java, my physicality—that is, being a tall white man—had always mediated my relationships with Indonesian friends and acquaintances. The Javanese word *landa* (usually pronounced something like "londo") refers to white foreigners like myself, and it reflects a further mediating difference, one that implicates me in the colonial history of Java (*landa* is derived from *wong Belanda,* or "Hollander") as well as the economic and political imbalances between the first and the third worlds. This became more pronounced during my last three-month visit in the summer of 1998, when the Indonesian economy fell into chaos and riots flared up in small towns and villages throughout the region. While mutual understanding and friendships might be mitigating factors in my relationships in Java, no amount of language ability, musical knowledge and skills, or cultural sensitivity could ever erase profound differences of circumstance, worldview, and embodied experience.

Fieldwork on the Internet is also about difference, but along entirely new lines. Difference here is not inevitably built around race, class, or gender, but may be constructed by each online presence according to individual needs and desires. Identity online is largely constructed through texts, created out of one's own words and the words of others. However, we also know that the Internet is still a privileged world, open only to those that have access to sophisticated computer and media technologies. As someone once told me, spoofing a recent TV commercial: on the Internet, no one may know you're a dog, but, chances are, you're most likely a white middle-class male. While the Internet may hold utopian possibilities for the future, its current realities are still grounded in fundamental economic and political contingencies. Nevertheless, the Internet gives rise to new social formations, even while those formations are still limited to those privileged people that have access to cyberspace.

As the title of this essay indicates, I refer to the Internet as "Softcity." As a discursive strategy, I am intentionally shifting the perspective from offline presence to online telepresence, placing the Internet in the center of experience and reality at the margins. Thus, I see the Internet as a metaphorical urban center located in the middle of embodied reality—what might be construed as Softcity's rural outlying territories. The rest of this essay is about a small but socially complex community existing within the larger matrix of Softcity, one that emerged uniquely as a result of computer technology and coheres entirely through Internet communications. It is a social collectivity that goes far beyond the textually based discussion groups and email lists that now form a familiar part of the online media landscape and are widely analyzed in scholarly literature on late-twentieth-century technoculture. This music community, known as the *mod scene* (referring to the digital music *modules* that the scene members create and exchange), has led me to believe that the social implications of computing and media technology are much more far reaching than we have perhaps imagined and that the beginnings of radically new grassroots organizations are already emerging on the Internet.

In this predominantly descriptive examination of an online community, I want to touch on several issues raised by Barbara Kirshenblatt-Gimblett (1996: 23) in an article entitled "The Electronic Vernacular." More specifically, I will address two questions drawn from her essay:

(1) What is the nature of "presence" in a disembodied medium such as the Internet, and what are the implications for thinking about the "real" and about materiality? (2) How are locality and community established in a medium dedicated to the seamless flow of data through a network of nodes that are addresses but not places?

The aim of this study is to argue that at least some online communities, like the mod scene, are as real as communities offline. Communities are, after all, based on social relationships, not necessarily on physical proximity. What makes the mod scene particularly interesting is its orientation toward music production and dissemination. Unlike most other communities, the mod scene involves much more than the exchange of email messages or postings in a discussion group. Its members exchange music and music-related software as they share their common passionate interest in electronic music. The Internet, then, provides a context in which social interaction and group formation can take place—it forms, as some say, the culture of simulation. In the following section, I will attempt to locate online virtuality within the larger discourse of postmodernism and the supposed triumph of simulation over reality.

Postmodernism and the Culture of Simulation

In her recent book *Life on the Screen,* Sherry Turkle states that "we are moving from a modernist culture of calculation into a postmodernist culture of simulation" (Turkle 1997: 20). She argues that the "modernist computational aesthetic" (1997: 18) revolved around calculation, one-way processes, and hierarchical relationships. Computers and computation reflected the grand narrative of scientific empiricism: causal relationships, reductive reasoning, and an absolute division between lofty science and popular culture. In the postmodern culture of simulation, this narrative has given way to complexity, interactivity, and decentralized relationships. The GUI (graphical user interface)—along with hypertext media and the Internet—has changed the way we interact with our computers, allowing us to navigate within simulated environments and to manipulate metaphorical objects. As an increasingly important part of our lived experiences, the computer and media technologies are changing the way we think about the real and representations of the real.

Turkle's argument of a culture of simulation is based to a large extent on Frederic Jameson's well-known essay (1983) on postmodernism. It might be useful here to list some of his points, particularly those that are reactions against high modernist aesthetic sensibilities: blurred categorical boundaries (between highbrow and lowbrow culture, between the Scientific and the Popular), stylistic pastiche ("cut 'n' paste" creativity, imitation without reference),[3] an end to privileged subjectivity (the decline of the heroic [white male] narrative and the advent of multiple perspectives), and, most importantly, the conquest of representation over reality. Since, as Jameson points out, postmodernity emerged out of late-twentieth-century multinational capitalism, the culture of simulation is informed by an aesthetics of commodity consumerism. Jameson suggests that the penetration of advertising, television, and the media throughout society, the accelerated rhythms of fashion and style, and the development of rapid transportation and communication systems all contribute toward an aesthetic sensibility in which reality becomes transformed into images and time fragmented into a series of perpetual presents (Jameson 1983: 125).[4]

It is important to recognize that there are various understandings of postmodernity, and that these are differentiated mainly along several axes relating to particular sets of social practices and trends in intellectual thought. Early on, Hal Foster (1983) discussed two main ideologies of postmodernity. The "postmodernism of reaction" is described as neoconservative, rejecting modernism yet preserving the status quo, while the "postmodernism of resistance" seeks to critique and deconstruct modernism yet resist the status quo, thus following a more progressive impulse. E. Ann Kaplan (1988) takes the debate further, arguing that discourse around postmodernism can also be differentiated by two main intellectual trends: feminist scholarship and cultural studies. She calls the former "utopian" postmodernism and describes it as a movement away from traditional and oppressive cultural binaries and hierarchies. Kaplan refers to the latter as "co-opted" or commercial postmodernism, "linked to the new stage of multinational, multi-conglomerate consumer capitalism, and to all the new technologies this stage has spawned." Following this particular vein, co-opted postmodernity might be understood as the complete transformation of the subject through consumer capitalism and the media industries. Indeed, this kind of postmodernism ties in neatly with Turkle's culture of simulation, but it also suggests that human experience is radically different than what it once was: "Technologies, marketing and consumption have created a new, unidimensional universe from which there is no escape and inside which no critical position is possible. There is no 'outside,' no space from which to mount a critical perspective. We inhabit, on this account, a world where the television screen has become the only reality, where the human body and the televisual machine are all but indistinguishable" (Kaplan 1988: 4–5).

According to this logic, we cannot stand outside of this world because, as Beaudrillard (1994: 2) points out, "the era of simulation is inaugurated by a liquidation of all referentials—worse: with their artificial resurrection in the systems of signs . . ." The real and the simulated have thus become one and the same. This suggests, then, a nihilistic perspective in which the social meanings and practices that emerge out of this culture of simulation are illusory and false. From this dystopian viewpoint, postmodernity sets the stage for a darker, seamier side of Western imperialism, where human consciousness itself is now open to colonization and oppression: "The end of modernity ushers in the all-embracing totality of postmodernism. In Other peoples' reality, Other ways of knowing and being, Other identities, postmodernism has discovered new spaces to conquer and subdue. Here 'progress,' 'modernisation' and instrumental rationality are replaced with relativism, real human beings are filtered through electronic screens to render them into virtual images—all the better to exploit them and butcher them without feeling real emotions" (Sardar 1996: 16).

Polemics aside, such arguments suggest that postmodernism is grounded in Western cultural and political expansionism while also constructed out of uniquely American, neo-avantgardist aesthetic sensibilities (Huyssen 1986). We might again raise the question that Dana Polan (1988: 46) had already asked over a decade ago: "to what extent can a description of a postmodernism *in* our culture be extended into a description of our culture *as a whole*?" In another critique, Mike Davis (1988: 80) points out that postmodernism is a totalistic concept that "tends to homogenize the details of the contemporary landscape, to subsume under a master concept too many contradictory phenomena which, though undoubtedly visible in the same chronological moment, are none the less separated in their true temporalities." Finally, this leads us to ask whether postmodernity is a viable or even possible alternative to the grand narrative of modernity. In a recent article, Dilip Parameshwar Gaonkar argues that rather than a totalizing and self-enclosed postmodernism, there might be many different kinds of modernities, relevant to particular circumstances: "One might be justified in pronouncing the end of modernity in a narrow and special sense, as Jean-Francois Lyotard does. But to announce the general end of modernity even as an epoch, much less as an attitude or an ethos, seems premature, if not patently ethnocentric, at a time when non-Western people everywhere begin to engage critically with their own hybrid modernities" (Gaonkar 1999: 13).

Arturo Escobar's "Welcome to Cyberia" makes the compelling argument that, while the culture of simulation (what he calls "cyberculture") might orient itself toward constituting a new order, it "originates in a well-known social and cultural matrix, that of modernity" (Escobar 1996: 112). In other words, while the Internet may be rooted in familiar terrain, it still holds the promise of new cultural narratives and social formations. It remains to be seen whether these will be understood as an extension of modernity, inherent to a larger postmodern condition, or simply one of many alternative modernities. Clearly, in the lofty realm of cultural theory, literary criticism, and the high arts, postmodernism offers ways of understanding the various aesthetic responses to modernism, yet it seems to fall short in explaining the grinding realities of everyday life. If we want to make the argument that postmodernity represents a fundamental shift in subjective experience, then we need to examine specifically how that experience has changed. And if subjective experience is now indeed largely constituted through simulated forms, we also need to ask whether the subject can find meaning in such forms.

We might say that cultural forms emerge out of the materials at hand: if representation exists in the absence of the real thing, then it is representation, or simulation, that constitutes the stuff of culture. As Turkle notes (1997: 23), we are becoming "increasingly comfortable with substituting representations of reality for the real." Indeed, we communicate with one another by telephone or email. The music we listen to is recorded in a studio and played back on our stereo systems, and much of what we see of the world is brought to us through broadcast or videotaped television. In other words, an increasing amount of our day-to-day experiences is mediated rather than live. New technologies can also enhance mediation, creating virtual and hyperreal experiences that might be almost indistinguishable from the real or impossible to find in real-world contexts. Sometimes called the "Disney effect," such simulated environments and social interactions are becoming a major part of our lives. Try talking to a person when you call a utilities company or an airlines. Indeed, a telephone conversation with a real human being may seem profoundly personal compared to the email and voice messages that we usually interact with. On the other hand, claims of one form of mediated interaction being more or less socially intimate than other forms may simply have no relevance: a friendly email exchange with a financial administrator may be far more gratifying than, say, a face-to-face encounter with a surly bank teller. Thus, it might be better to say that we are experiencing radically new forms of social interaction as a result of media technologies—but the *quality* of these social interactions, however mediated they might be, still depends on the embodied humans that give rise to them.

Whether arising out of the postmodern condition or simply constituting a reconfiguration of modernity, new media technologies have altered our relationship with the world around us so radically that the real and the simulated seem to be indistinguishable. Yet despite having such a profound impact on our daily lives and, indeed, our understandings of the world, these technologies are rarely seen as being part of culture. It is too easy to forget that the realm of science and technology is, to use Geertzian phraseology, as much of a cultural system as the realms of art or religion. Turkle might argue that we live in a culture of simulation, but we also live in a culture of hardware, made up of computers, telephones, fax machines, televisions, stereo systems, and so on—a world of technological devices with sharp edges, machined parts, and wiring—and all filling physical space, all breaking down over time, all requiring considerable money to maintain and replace.

In an earlier work, I pointed out the importance of research examining *technoculture* and its relation to music.[5] With this term I point to an understanding of technology not simply as the social and personal intrusion of scientific hardware into "authentic" human experience but as a cultural phenomenon that permeates and informs almost every aspect of human existence—including forms of musical knowledge and practice. More specifically, I am interested in social groups and behaviors characterized by creative strategies of technological adaptation, avoidance, subversion, or resistance. This study will continue by addressing the following questions: If Internet communication is based on what some might consider the *illusion* of presence and others call "telepresence," can it nevertheless support contemporaneous social collectivity? Can individuals logged on to the Internet be at home alone in front of their computers and, at the same time, also participate meaningfully in the formation of communities constituted by other solitary individuals scattered throughout the world? If so, are such "virtual communities" different from communities in the embodied world? Are they extensions of, substitutions for, or alternatives to offline real-world communities? How do the technical realities of online communications (that is, presence based on electronic access, texts, and mediation rather than physical proximity, context, and situation) affect individual and group identity?

In the following pages, I want to describe in detail a set of social relationships mediated through computer-based technology, relationships that emerged entirely out of the Internet. Drawing from Turkle's idea of a "culture of simulation," my ultimate purpose is to determine whether the set of social relationships described below constitutes a "virtual" collectivity, since it is based wholly on Internet communications—that is, whether such online relationships can be described as meaningful and constituting a true community.

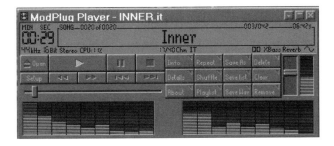

Fig. 1. MODPlug Player (freeware program available at <http://www.modplug.com>).

Electronic Music Modules: Mods

My research has been on an online community devoted to a particular form of music production, generally known as *mod*—short for *digital music module* and no relation to the British counterculture of the 1960s. Mods are music files in binary form created through music editing programs known as "mod trackers" and played back for listening using software called "mod players" (see figs. 1 and 2). Mod music files contain digitally recorded samples as well as coding for sequencing the samples in playback.[6] The "trackers" that are used to create these files provide composers with the means to control and manipulate sound samples in almost limitless ways to produce music. What is unique about these editing programs is that, in the hands of sophisticated composers, the resultant product rivals studio-recorded music in production quality. Creating music using such editing programs is simply called "tracking"—a term perhaps more appropriate than "composing" since it involves a kind of music notation played not by humans but by computers. In other words, *tracking* means "coding music." However, this is not to say that tracking is simply doing empty computer programming. It is a highly creative and skillful activity that perhaps involves a much closer contact with

Fig. 2. MOD4WIN Player (shareware program available on the Internet).

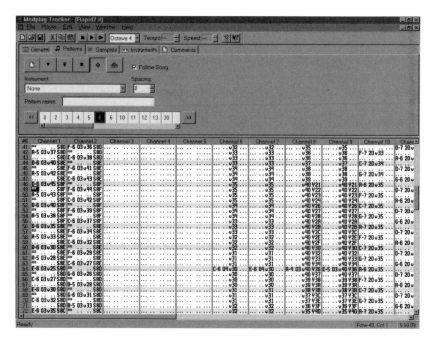

Fig. 3. Mod tracker, coding page (MODPlug Tracker, available at <http://www.modplug.com>).

musical sound than conventional composition, because every aspect of each sonic event is coded (see fig. 3), from pitch and duration to exact volume, panning, and laying in numerous effects (such as echo, tremolo, fades, and so forth). One might say that mod musicians are, at the same time, the composers and the performers of their music. Additionally, they are their own studio sound engineers, producing high quality digital "recordings" with virtual mixing boards that have as many as sixty-four channels. In this way, mod music is subversive, much like punk garage bands with their DIY sensibilities, because it undercuts the division of labor in conventional music production and distribution, as I will discuss in greater detail below.

The Mod Scene

Once the mod is finished, it is released to the larger Internet community of mod enthusiasts, called "the scene." The composer uploads the newly completed composition to one or more of several sites where mods are archived, making it available to his or her audience, who will download the file to their own computers (see fig. 4). Literally thousands of mod files can be found on the Internet, if one knows where to look for them. Although most compositions emulate high-tech popular dance musics like house, jungle, ambient, and so forth, the mod scene is not by any means limited to a particular style or genre of music. One can also find rock, heavy metal, new age, remixes of old and current radio hits, TV and movie themes, blues, jazz, and even classical music in mod format.

Mod websites are important centers for the exchange of information among community members. They not only provide elaborately catalogued archives of mod music files, with detailed charts of new and past hits, but also allow mod fans to publicly comment on any song. These sites also provide news of the scene, including announcements of "compos" (mod composition contests) and related music events. As community centers, such websites often offer bulletin boards and chat services for members to communicate with one another more directly, encouraging the free flow of ideas and information.

By encoding textual information within each mod file, composers also maintain contact with their audiences and with one another by including their email addresses, greetings to fans and other composers, and virtual signatures (often in the form of textual graphics). The text is displayed for the listener during playback. It is either scrolled in the title window of the

Fig. 4. Mod website (Trax in Space at <www.traxinspace.com>).

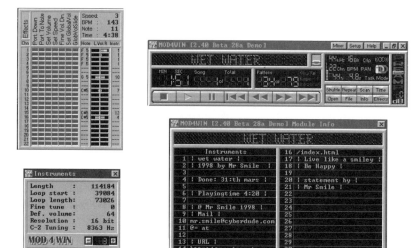

Fig. 5. Mod player, with informational windows (MOD4WIN).

player by clicking a button or displayed in a separate window, along with other song information, by clicking another button (see fig. 5).

In their communications, mod composers affect the language and style of the rave and dance club scene. They rarely use their own names to identify themselves, taking on futuristic or otherwise fantastic aliases like Timelord, bionic, equinøx, kosmic, mysterium, and so forth. Often the names themselves are textualized graphics that play with conventions of spelling and capitalization.

To have any kind of status as an active member of the mod community, or "the scene," one must either be a mod composer, a programmer of mod-related software, or the owner of a website that features mod music. In other words, social status is directly linked to social action. Even fans are able to contribute to the scene: they can make mods available to the wider public by having a "plug-in" mod player on their personal webpage that automatically plays a composition when the site is visited. Most of these personal webpages also provide links to other mod websites and sometimes include small archives of featured or favorite mods. Like the music they advocate, such homepages are ranked according to ongoing surveys or the number of hits (total or daily) by several of the central mod sites. Homepages are reviewed by scene members at the major websites according to their originality and creativity as well as their impact on the scene (the number of hits they draw).

While most mods are composed by individuals, many are the result of collaborative efforts. In either case, and perhaps also for economic reasons, composers often form alliances like bands and set up elaborate websites for their groups that both promote and distribute their music. These alliances are known as "crews," and they may also include programmers and graphic artists (if the composers themselves do not have these skills). The composers, graphic artists (sometimes called "graphicians"), and programmers in the mod scene are predominantly young men, varying in age from the late teens to the early thirties. They are generally middle-class White Europeans and Americans, college educated and possessing some basic computing knowledge.[7] Some have access to fast, high-end computers and broadbandwidth websites as a result of their jobs in the computer industry, where they work as programmers and coders. Most, however, work on conventional PCs in their own homes, perhaps upgrading the soundcard or adding RAM or other hardware, depending on their financial resources and computing knowledge.

As a community, the mod scene has a relatively short but rich history. It emerged in the early to mid-1980s (as legend has it) with teenage computer hackers "cracking" (deciphering) the software codes of early Atari, Commodore, and Amiga game programs. As a kind of computer graffiti, hackers would then code short intros of modified music and graphics into the cracked games and pass them around like war trophies. The intros grew increasingly larger and more sophisticated, eventually becoming autonomous entities. The interests of these young hackers slowly turned from cracking games to PC programming, creating "demos" or multimedia programs that had no other purpose than to demonstrate virtuosic coding abilities. Created entirely for aesthetic pleasure, demos have been described in the online *Wired Magazine* (Green 1995) as the "last bastion of passionate enthusiast-only programming." This resulted in what is now commonly known as the "demo scene," in which teams of young computer hackers compete against one another to produce the most elaborate and sophisticated multimedia programs using the least amount of disk space. Ideally, a good demo is one that has interesting music and graphics but that can be saved onto a 3.5-inch floppy disk. Mod tracking arose out of the demo scene, because it uses relatively little disk space but produces near studio-quality music. While creating demos remains a popular activity today among a small but active group of programmers, it has been largely subsumed by the mod scene. What is known as "demo music" has come to be recognized by most community members as simply a style of composition.

A vestige of the early demo scene is a category of mod music known as "chiptunes," in which composers create fullblown mods using only computer-created sounds, often simply modified sine or square waves and the like. To qualify as a chiptune, the mod must be quite small (about fifty kilobytes or less) and use only "hand-drawn" tones (that is, tones created by the composer) instead of sampled sounds. These mods, while usually having the distinctive bleep and beep quality of transistor-generated tones, are often astonishingly creative and rich in expressive nuances. This kind of mod music remains a viable option for composers, perhaps because it poses particular challenges and limitations while providing a subversive alternative to the bigger and more elaborate compositions that use increasingly larger samples of real-world sounds and are created with sophisticated new tracking programs widely distributed throughout the Internet.

The social order of the mod scene is formed from a hierarchy of graphic artists, musicians, programmers, music experts, and fans. At the top are the seasoned demo artists who have been around since the early days of Amiga and Commodore computer hacking. They serve as the community elders and dictate the aesthetic standards of mod composing. In the middle are many experienced younger mod composers, many of whom are dedicated specifically to composing music and are not necessarily interested in creating demos. Novice composers and serious fans make up the largest but lowest level of the community. They generally have little voice in serious discussions of aesthetics and technical matters but make their presence felt through chat networks and email by asking questions or expressing enthusiasm over a particular mod or mod composer.

The musical aesthetics that developed out of the peculiar history of the mod and demo scene still has a strong hold over the community today. New composers, in particular, are reminded of the rigorous tracking standards through the sometimes devastatingly harsh reviews of mods that are published regularly on the Internet. The aesthetics is, in its extreme manifestation, based on a kind of geek adolescent techno-machismo—music coding is damn hard and not suited for the technologically handicapped. Some community elders have lamented (in various web documents) the decline of musical coding standards, a result of increasingly user-friendly tracking software that allows anybody to create a "lamer," i.e., a poorly or sloppily coded piece of music. Yet generally, apprentice composers are welcomed into the scene and naive attempts at tracking are viewed with considerable tolerance. While some hardcore tracking enthusiasts may emphasize highly polished technological production and sophisticated coding techniques over musical substance, most experienced mod composers have an astoundingly sophisticated knowledge, or at least a profound intuitive understanding, of music theory and compositional technique, producing mods of outstanding quality. Ultimately, however, it is the popularity of the composer's mods that determine his or her status within the community.

All active members of the mod scene communicate with one another regularly by email, through Usenet discussion groups, and on bulletin boards, as well as over real-time chat systems. Frequent communication is particularly important in the mod scene for establishing and maintaining friendships as well as promoting one's own music and/or webpage. Like any community in the offline world, the mod scene is dependent on personal networks of communications among its various members. These personal networks constitute smaller collectivities of friendships and alliances, in which individuals are in direct contact with other individuals (even while they may never meet as embodied people in the offline world). Indeed, very few members are known personally by everyone throughout the scene. In a musical community based almost entirely on the disembodied exchange of information, status is entirely dependent on name recognition: the more widely a person's name is known, the greater prestige he or she enjoys—unless, of course, one gains notoriety for certain transgressions against the scene (discussed further below).

Websites may also reflect a form of social action in the mod scene. To maintain a large archive of mods and to make mod programs available to the public require considerable resources. Some site owners use their own computers as servers (that is, as host computers from which the public may retrieve files), but most obtain storage space on commercial servers. Generally, this means renting disk space for data storage and paying a monthly fee for a certain amount of bandwidth (i.e., public access to the site). Major sites might be staffed with volunteers carrying out specific jobs, such as mod reviews, graphic art, news reports (on the mod scene as well as music generally), research, and so on. Additionally, major mod sites are the center of scene activity and must be maintained on a daily basis. Important sites like Trax in Space, MODPlug Central, or United Trackers reflect a profound commitment to the scene, and those members associated with them enjoy considerable prestige throughout the mod community.

Thus, the mod scene is made up of a social hierarchy based primarily on prestige and authority. Prestige for composers is gained from having a large following of fans and name recognition, while authority arises out of a thorough knowledge of tracking and computer programming esoterica (and, to a lesser degree, music theory and compositional technique). Prestige can also be obtained through other forms of commitment to the scene, most commonly through software development or the establishment and maintenance of a major website. Although those at the highest levels in the social hierarchy (mostly seasoned composers and computer programmers) have little real control over community members, they do command respect for their knowledge and experience.

Imaginary Communities in Virtual Places

The idea of community is predicated on a collective sense of common interests and purpose. Members of a community are bound together through comradeship and a desire to seek one another out. In the past, communities were perhaps based, at least in their conception, on proximity—members lived near one another. In most cases, as Benedict Anderson has shown (1983: 6), even small communities are imagined communities, since it is unlikely that any member will ever know all of his or her fellow members, yet (as Anderson puts it) "in the minds of each lives the image of their communion." All members know some of their fellow members and, as a result, define their sense of belonging to the larger group in terms of these specific relationships.

Earlier, I quoted Barbara Kirschenblatt-Gimblett's argument that the Internet is made up of IP addresses but not places. I believe, however, that *the Internet is all about place;* we can perhaps best understand it as a kind of imaginary universe without real space yet absolutely filled with a multitude of places. When we connect our computers to the World Wide Web, we suspend disbelief and embark on a metaphorical voyage. Web-related language is already filled with metaphors of journey and place: we *go* to sites, we *visit* webpages, we *surf* or *cruise* the Internet, we *travel* on the information highway to certain *home*pages, and so on. Most webpages are, in fact, often graphically oriented around themes related to place: homes, offices, castles, rooms, spaceships, and so forth. In some ways, they almost become interchangeable with the physical sites they represent—such as certain government or religious webpages that stand as virtual monuments to their real world counterparts.

The imagined community of the mod scene is certainly based on common purpose and a strong sense of place among its members. Their purpose is to create, distribute, and listen to mods. More importantly, however, their sense of place is defined specifically in terms of the Internet rather than the real world. *There is no mod scene in the physical world; there are only isolated mod composers and fans.* Indeed, the mod scene adds further meaning to the notion of transnationalism. Many scene alliances are made up of composers living literally thousands of miles apart from one another. A group's webpage becomes the locus for the activities of its members and a place in which they share their common purpose (see fig. 6). It institutionalizes the group, providing its geographically scattered members

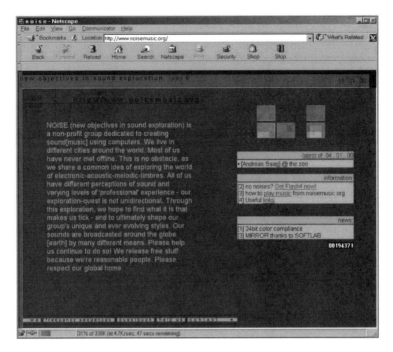

Fig. 6. Website for NOISE (New Objectives In Sound Exploration), a transnational mod crew.

with the means to present themselves to the world as a unified whole. Thus, while members may come together only in the disembodied telepresence of the World Wide Web, their home page provides them with a kind virtual corporeality that is compelling to their fans. Finally, the various webpages of the mod scene are linked with one another through the Mod Web Ring, an Internet application that allows Web surfers to jump from one mod homepage to another by simply clicking a button.

I want to explore the social structure of the mod scene a bit further by arguing that it is closely tied to a kind of exchange economy that resembles, in some ways, a system of commerce. However, what is exchanged is not money and goods; rather, data and communications form the basis of the economy of the mod scene. This exchange system can perhaps be best understood in terms of what we might call the *virtual materiality* of graphical interface–based computing.

Digital Mediation and Simulated Materiality

The computer has been described as a kind of "language machine" (Biggs 1996: 320), but this seems to me rather vague and perhaps even misleading from a cultural perspective. Certainly, in the strictest linguistic sense, a computer is indeed a language machine, because it processes symbolic information and, in the understanding of most people, is mainly concerned with communication. While some computer experts might be able to "read" program languages, much of the information that flows from the hard drive or the Internet to the computer processor remains invisible as well as incomprehensible to most of us.[8] Computer data, as such, means little to the user. Yet when it is processed to become images, texts, or sounds, the same data is now saturated with social relevance. I want to argue here that, if we are to consider the broader sociocultural implications of computing technology, we need to examine how people actually *use* this technology on a day-to-day basis and, as a result, how they derive *meaning* from it. With such a view in mind, we can define three main kinds of computer files most commonly encountered on the Internet: (1) executable files or programs, i.e. files that do things or act on other files; (2) binary data files, i.e. files that are acted upon by executable files; and (3) text files, i.e. files that mainly contain meaningful texts in the traditional linguistic sense. Obviously, these three categories may overlap, but they provide a convenient point of departure for distinguishing what kind of data we use on our computers and how we make such data socially relevant.

For most of us, program files are seen as icons in Windows or Mac environments. When we click on the icon, it becomes a kind of virtual device that can do things for us and often reads and manipulates data files. For example, clicking on the Netscape or Internet Explorer icon initiates the computer modem and a so-called browser (itself represented as a kind of stylized window) to download data found on the Internet. As we explore the World Wide Web, the browser we use processes various kinds of data that become, for the user, meaningful texts, graphic images, or sounds. The various executable programs involved in such processing are known as plug-ins, and most remain invisible, embedded within the browser. Some executable files appear as graphic images, resembling their corresponding real-world objects. For example, VDO (a digital video software program) looks vaguely like a television set, and Beatnik (a music program) is depicted as a stylized boombox. From the perspective of the user, such programs take on a reality of their own—they become virtual devices that allow us access to particular kinds of media on the Internet. Thus, the advent of GUI (graphical user interface) is, aside from the technology itself, one of the most important innovations in the computing industry, because it has changed the way people use computers. As Sherry Turkle (1995: 23) notes, "these new interfaces modeled a way of understanding that depended on getting to know a computer through interacting with it, as one might get to know a person or explore a town." In other words, GUI brought a kind of materiality that simulated the physical experience of handling objects (like an audio component) or moving through space (like traveling through the Internet).

The practitioners and followers of the mod community have been using the Internet from its earliest days both to maintain a common group identity and to disseminate compositions and various related software programs. That is to say, the community members not only exchange information but also engage in trading "products." Since, broadly speaking, this simply involves bytes of computer data, it is important make a distinction here: by information, I mean textualized forms of knowledge—that is, they are engaged in textually based communication (such as email, chat systems, discussion groups, and so forth); by products, I mean programs, images, and data files that, when they are processed, have functions analogous to three-dimensional objects we can literally hold in our hands. For example, mod program files, such as trackers and players, are analogous to audio equipment like cassette tape decks, mixing boards, synthesizers, or CD players (see figs. 1–6). One might view these as a kind of *virtual* media technology. As software, they provide access to certain media forms, but they are themselves entirely dependent on computer hardware as a primary media technology. Similarly, music compositions in binary files, such as mod files, are analogous to audio cassette tapes or compact disks. More important, in addition to being playable on particular software programs, they are also collectible, like recordings on disk or cassette tape. Fans collect and exchange the mods of favorite composers or musical styles. I have my own elaborate directory system based on categories of genre and style and further subdivided into composers—my collection now spans several hundred megabytes. Additionally, composers also collect and exchange digital samples. For example, a mod composer may post a request for acoustic guitar samples on the <alt.music.mod> discussion group. Other composers respond by attaching sample files to their reply posts. In any case, this exchange of virtual products is what distinguishes the mod scene from most other online communities. Past studies of such online communities have focused on textual material (discussion groups, lists, and chat systems), viewing them almost exclusively in terms of public discourse. This makes sense in those cases, since online discourse (i.e., the textual dialogue) is the only evidence one might have of communal activity. In the mod scene, however, the distribution and exchange of virtual products, particularly mods, is absolutely crucial to the vitality of the community.

Thus the mod community is dependent on the use and exchange of programs, data files, and textual files. Active interest in the community is continuously maintained by a steady supply of new compositions, uploaded daily to key archive sites. However, the question remains: what provides composers with the incentive to produce new works? What's in it for them? Everything related to the mod scene is free. All the software programs are either freeware or fully functional shareware, and there is no fee to download mod files from the many available webpages and public archives. In one sense, the mod scene is utopian, because it has a kind of economy based entirely on social prestige. After participating in the scene for several months, I came to realize that the most important form of currency in the mod scene is an email message from a listener telling you that your music is awesome. This is why almost every single mod on the Internet is identified with the composer's handle (alias) and an email address, sometimes including a message by the musician begging for comments. Several websites provide reviews of mods with featured songs and artists. Others provide rankings of each song based on how many times it is downloaded—not only total downloads but the number of daily or weekly downloads. For example, administrators of the Trax in Space site use complicated formulas for determining top-ten or top–one hundred songs based on the number of downloads per day, week, or month. Composers seek recognition for their creative work, and, while Billboard-like charts might indicate the popularity of their music, many simply want feedback from their fans. Some composers create mods following a rather narrowly defined musical style, such as death metal or acid jazz, and they will not gain the popularity that the more mainstream composers are able to achieve. For these composers, listener response is particularly important. All in all, websites that provide mod files for downloading, or services like bulletin boards and mod reviews, are extremely active and, most important, they form the infrastructure for the prestige system on which the scene is based.

Simulation and the Creative Process

As I stated earlier, we live in a culture of simulation, in which the representation of reality seems to have triumphed over reality itself. However, the mod scene perhaps exists within yet another order of simulation, since most other so-called virtual communities are extensions of real world counterparts. In its own peculiar form of virtuality, the mod scene is a social simulacrum—it has no real world counterpart. It emerged within and as a result of computer technology and the Internet. Like other virtual communities, the mod scene depends on the Internet for communication, through which texts are exchanged. What makes it unique is that its products—the music files, the trackers, the mod players and other associated software programs that form the basis of exchange among its disembodied members—and services (provided on major webpages for scene members) are also entirely dependent on the Internet. The music modules as well as the related software might constitute what Jean Baudrillard (1983) calls "true simulacra": third-order simulations that have no (physical) originals and are themselves based on other simulations.

Creativity in electronic music is based on what Arthur Kroker (1993) views as digital recombinancy, where identifiable elements of one expressive form are reconfigured to become another, entirely different, form: "Western music goes recombinant. It is suddenly uplinked into a starlight horizon of a thousand billion data bytes, becoming a malleable object. Not a sound-object, that's too hard-edged, formalistic, and mechanical, but more like liquid sound, where noise melts down into a fluid, viscous material, endlessly combinatorial, following an indefinite curvature of violent velocity" (Kroker 1993: 58).

Kroker's phraseology aside, music recombinancy redefines the compositional process. Instead of writing *for* certain music instruments, as in past Western musical practice, mod composers create music *with* instruments. In other words, mod composers work directly with sound rather than writing instructions (i.e., musical notation) for musicians specializing in particular instruments. This has direct implications not only for the creative process but also for the social organization of music production.

When music is produced in the simulated environment of computer technology, it forces us to reconsider the Cartesian mind-body divide. Traditional Western art music, on the one hand, valorizes virtuosity, which in turn hinges on the close relationship between body and music-making—even while musical creativity is understood to be centered in the mind. Virtuoso musicians are valued for their technical skills in performance in the way that athletes are valued for their speed, flexibility, control, stamina, and power. Furthermore, musicians are evaluated according to their *mastery* over a musical instrument, their *technical control* of it, and their physical ability to make it do extraordinary things. They have an intimate and direct relationship with the sound they produce. And that relationship has always been through the body. On the other hand, Western art music composers have come to be regarded as great musical thinkers. Being a composer might be less glamorous than being a performer, but composers are seen as the fount of creativity, while performers are merely interpreters and technicians: composition results from the creative mind while performance is executed by the interpretive body. Thus, the Cartesian divide is manifest even in the social division of musical labor.

In electronic music such as that produced in digital modules, the mind and body divisions are blurred: the composer determines all aspects of the music, including how it is to be executed down to the most minute detail. In this way, the composer both creates and interprets music with the computer. Yet Western notions of musicality, virtuosity, talent, and so forth have all been bound up in an ideology of human agency, individualism, and personal autonomy. Since electronic music is not performed by a virtuoso but rendered by a software program or a digital sequencer, what are the implications for the understanding of musical talent, ability, and virtuosity? Is the electronic music programmer an artist or simply a computer technician? Where does musicality lie? Is it in the mind or the body of the programmer? Or somewhere between? Or nowhere?

In traditional art music the composer scores out ideas to create a piece that is performed by a singer, instrumentalist, or ensemble. Similarly, in electronic music the composer codes instructions for software programs that, in turn, render those codes into music. Both scoring and coding music, broadly considered, are forms of inscription. A conventional music score, however, is a kind of text that is (re)interpreted by the performer to become music. In this way, the score becomes a textual record of an ephemeral phenomenon, but before the fact. That is, when music is rendered into notation, it becomes textualized in a way similar to language—performers "read" the score in expressive ways, like actors "read" the script to a play. Conventional music notation is thus often implicated in the simple-minded but persistent analogy drawn between music and language. Western music "seems" to be a kind of language (a "universal" language, some have said), perhaps because it can be rendered into text. While musicologists today might deny the analogy, Western music scholarship is still largely founded on musical texts that are, in effect, analyzed much like language texts. Through notation, music theoreticians are able to determine the grammar, the syntax, and even the semantics of musical composition and, thus, perpetuate the music-as-language myth. Most important, however, Western music institutions are founded on the notion of musical literacy—that is, being able to "read" music notation—maintaining a kind of priesthood based not only on music as expressive sound but also music as text.

On the other hand, electronic music code might also be considered a kind of text, but it's not written to be read by humans. It forms a set of instructions to the computer to carry out particular commands in precise order and detail. Like the old hole-punched paper rolls of player pianos, music coding is written by humans but "read" by machines. In this way, music coding is more like an audio recording than a text. Consider this: a recorded performance is always fixed in relation to the medium on which it is inscribed. The same holds for coded music. With audio recordings, however, the performance always *precedes* inscription, while in mod music the performance *follows* (i.e., results from) the inscription (the coding). This means that mod composers do not need to understand "music theory," nor are they by necessity musically literate (in the traditional sense of being able to read notation). A composition (or any part of it) can be immediately "played back" by the computer (using a tracker software program) so that the composer can listen to any passage throughout the compositional process. In this way, composition becomes a more intuitive form of musical expression, somewhat akin to improvisation in jazz or blues. I am not trying to say this necessarily results in intrinsically better music but that it provides more possibilities for "untrained" composers to experiment with musical sound and structure.

Coding music with the use of digital samples also implicates the way listeners understand particular sounds. Sound samples no longer necessarily have a parasitic relationship with real-world sounds. That is, certain musical instruments may remain in the collective imagination only as sounds, not as physical objects. This is already beginning to happen with particular instruments, such as the Hammond B3 organ or the Moog synthesizer. Mod composers are generally quite familiar with these instruments, even if most have never played or even seen them. They might know them only from the sampled tones they have obtained from commercial sound collections and fellow mod composers (or commercial audio recordings). For many mod fans, these sounds may simply be free-floating sonic symbols without a material referent. In any case, thinking about this leads us to ponder the temporality of recorded musical sound: the audio recording fixes performance in time, specifically in the past (after all, the recording is by definition an archival document). Musical coding, too, fixes the performance in time, but in the present.

Musical Creativity and Ownership 1: Ripping

Authenticity, appropriation, and originality all remain slippery concepts in the mod community. Mod composers pay close attention to one another's works and to the music industry generally. In this way, ideas circulate freely into and around the scene. Distinguishing between imitation and innovation requires an understanding of the creative limits of mod composition. Mod scene members draw a fine line between artistic emulation and slavish imitation, between creative borrowing (or quoting) and outright theft, and between stylistic affiliation and mindless fashion. Compositions are judged according to different criteria ranging from the overall artistry and technical execution to details of specific sound elements. The most difficult criteria in evaluating mods, however, has to do with authenticity and originality.

One way mod composers demonstrate their skills in tracking is by creating remixes (or simply "mixes") of other compositions. Remixing, or rendering a new version of a known piece in its entirety, is a common practice. Composers may remix almost any popular piece of music. For example, the widely known theme to the movie *Bladerunner* by Vangelis has been rendered by several mod composers. Some versions attempt to sound as much like the original as possible, while others simply evoke the original theme. In either case, the title is retained and Vangelis is credited as the original composer. Generally, remixing is considered a form of homage to a particularly outstanding piece of music or a tribute to a renowned artist. Remixing mod music is also widespread in the scene.[9] For example, composers often render more than one version of their own compositions. This practice is not unique to the mod scene, of course, and occurs commonly in dance club music and related forms (such as ambient, jungle, etc.—all broadly designated "electronica").[10] Occasionally, one mod composer will create a remix of a piece by another mod composer as a form of tribute. However, only scene insiders would likely understand the significance of these remixes, since very few mods break through into the larger world of popular music. In any case, remixing is a risky enterprise, since the new version may be unfavorably judged against the older, more established version.

While remixing is an interesting aspect of the mod scene, the practice known as "ripping" is ultimately more important. Defined broadly, ripping refers to appropriation in the Ricouerian sense of taking something foreign and making it one's own. In its most innocuous form, ripping is what many mod composers do all the time: they appropriate sound samples from other mod files. While this is allowable—and even encouraged—composers will contact or at least acknowledge the owners of the samples they rip. Samples range from extremely brief sounds (such as percussive "one shots") to melodic or rhythmic motifs and sometimes even whole melodic phrases or harmonies. Because of this, the practice of ripping samples remains a key topic of debate, and discussions focus on whether musical creation begins with the sound sample itself (recording sounds and modifying them) or in the sequencing of a sample. For example, the following quote from the composer known as Ming reflects a common sentiment in the scene (posted on the MODPlug Central discussion forum): "Re-using samples from a mod is IMO not a bad thing. Instead, I encourage people to use my samples, it makes me proud to see that I have created something useful for someone else's work . . . Do not take the opportunity away from this free society of ours to do that. The samples are our instruments, the patterns are the music, and it's the music that we shall not steal . . . Let's define what is stealth [stealing] and what is not. I claim sample ripping is definitely NOT stealth, and that free use of samples should be encouraged" (Ming, July 7, 1999).

Many mod composers use only their own samples (that is, sounds they themselves have digitally recorded and modified), while some use samples ripped from other sources (synthesizers, computer games, audio recordings, etc.), with or without modification. Composers can also go to sites devoted to archiving sound samples and other mod-related resources, such as MAZ (a well-known site to mod composers). These same samples are often then ripped by other composers and in this way move from one mod composition to another. When samples travel so far and wide, they become public property, free-floating, anonymous sound bytes on the Internet that are available to any and all composers. Indeed, the concept of ripping here suggests the ideology of the mod scene—music available for free to all people. In such cases, therefore, ripping has purely technical connotations: removing a sound sample from a source and using it in a mod composition. Technically, one could "rip" a sample from one's own mod to use in another mod. Indeed, mod composers sometimes use software programs to obtain digital samples from computer games—these are, in fact, called "rippers."

On the other hand, the concept of ripping may also have negative associations. When the original creator of musical material is not acknowledged, transgressions associated with ripping can range from simple discourtesy (i.e., using someone's samples or musical ideas without acknowledgment) to outright theft (i.e., plagiarizing an entire composition or section of a composition). While most composers welcome the appropriation of their samples, even without acknowledgment, some feel that it is a form of theft and are careful to declare their ownership in the comments to the music or include a copyright symbol (©) with the list of samples.[11] The use of ripped samples also impinges on the evaluations of mods. Reviewers not only judge the timbral quality of the samples (clicking or distortion, etc.) in a mod but also their uniqueness. For example, novice composers are often ridiculed for using samples that had became hackneyed as a result of their popularity and, thus, are encouraged to come up with unique sounds (or at least find less widely heard samples) for their music.

Recently, however, an important issue in the scene has become plagiarism. Since the number of mod composers has grown significantly over the past several years, music theft and plagiarism has been an increasingly common problem in the scene and remains an ongoing topic of discussion and heated debate. One case of mod plagiarism involved an artist known as Melcom, who discovered to his delight that one of his songs, called "Another World," was featured on a large website that archived MP3 music.[12] However, he was outraged when he learned that another person had taken credit for the composition without changing anything in it, not even the title. An article posted on the MODPlug Central website reporting the theft (June 1999, by Kim, aka Mister X, site owner) poses an unusual dilemma for mod composers. If mods are freely distributed and, as in this case, plagiarism is not for financial gain, what legal recourse does the composer have?

This situation raises an interesting question: exactly what rights does Melcom have? The website in question is a non-commercial venture, and does not profit from his work. Also, the "Digital Property" laws, at least in the US, are still rather vague. Let's talk about it. MODPlug Central moves that, if there is an interest, a collective of MOD musicians should be formed to provide legal counsel in the case of an artists work being used for commercial purposes without their permission. Although we walk a fine line by releasing music for free on the Internet, the holder of the copyright on wholly original works is still the author. If this site was using Melcom's song in a commercial venture, he would have every right to sue them for copyright violation. Problem being, does Melcom have the legal connections and the money to pursue such a lawsuit?[13]

The offending piece has been removed from the website and the plagiarizing composer banned forever. In another case, an artist calling himself DJ Carbon posted several compositions to a website and even released a CD of his works, several plagiarized from various highly regarded mod composers. What is particularly interesting in this and other cases like it is that many mod composers themselves are well-trained computer programmers, and some are even gifted computer hackers. (I mentioned earlier that the mod scene is rooted in computer game cracking, known as the demo scene.) Several highly skillful hackers in the scene were able to trace DJ Carbon's real name, personal website, place of employment, and other confidential information about him. DJ Carbon's punishment was humiliation and banishment from the scene. Since social standing is dependent on social action in the mod scene, DJ Carbon was publicly denounced and all evidence of his activities (his own compositions and his reviews of other compositions) were removed from websites. In effect, DJ Carbon ceased to exist in the scene—the mod community collectively sentenced him to virtual banishment. It is unlikely that he can return to the scene in another virtual incarnation (except, perhaps, as a so-called "lurker"), since scene members know who he is in real life.

Mod scene members discussed the various sides and possible remedies for curbing mod plagiarism. One thread, on the MODPlug Central website bulletin board, explored the idea of establishing an Internet watchdog organization to investigate and remedy copyright abuse in the mod scene:

ICOMM Discussion Group (first posting)

This Bulletin Board is here to discuss the idea of ICOMM, the "International Consortium Of MOD Musicians," a group which could provide legal counsel and other protective services to MOD Musicians to keep their music free—free from charges and free from being ripped off.

Please post all of your ideas here, and let's discuss this idea. Is it necessary? Would you personally support such an organization? Let's hear what you have to say! (Mister X, webmaster, June 28, 1999)

Ripper Stockade

Whether ICOMM gets off the ground or not, I'd like to see a high-profile site dedicated to making confirmed rippers infamous, including posting the full details of what they did, links to their e-mail address and web page, and asking for people to come forward with information on those and other confirmed rippers. I've personally helped to blow the lid off of four different rippers, yet only one of them ever even bothered to apologize or to come clean, even in the face of convincing evidence. (That one ripper actually admitted to ripping songs that nobody had accused him of ripping yet, so I believed the sincerity of his apologies.) I figure if they don't want to come clean, then why should we allow them to maintain any dignity in the scene at all? Ripping seems to be on the rise again, and if we don't fight it now, it'll get out of hand. (Novus, June 30, 1999)

Be wary . . .

The ICOMM seems like a good idea at first, but we should really think about what this would mean. I would like to first point out that the initial idea was regarding commercial ripping, not just stupid-fuck ego-boosting "i made this" bullshit. But then—really is there that much of a problem right now with people using scene music for commercial ventures? If, say, a production video game DOES use scene music, and it is an obvious rip, it really is not hard for the composer to prove it in court . . . (Michael, June 30, 1999)

You read my mind . . .

I had just been thinking about this same sort of situation a while ago; what is there to protect us from being ripped off? Especially since I am a lesser-known artist, it is definitely in my best interests to have someone looking out for me. I think it is a brilliant idea to have protection.

However, on the flip side of the coin, we must be careful to avoid the self-serving greed-mongering practices that govern the record industry today. I'm all for anti-plagiarism protection, but let's be careful not to undo the open and sharing atmosphere in the MOD community; the best thing about MODs, after all, is that they're free to make and free to check out. (Craig, July 02, 1999)

A place to start.

[...] Remember, the best weapons a community such as this has are word of mouth and solidarity. Yes, the net is a big place, but there are a lot of people watching it, and as long as they speak out the people who try and pull crap won't get away with it. Having a place where people can speak out seems to be the bottom line of this project. [...] (Whiggy, July 28, 1999)

Policing mod authorship and distribution, as the discussion above shows, may protect composers from copyright abuse, but it also can have a chilling effect on the subversive character of the scene and its ideology of free music and free software. While mod composers do indeed enjoy copyright protection of their creations (according to current U.S. Code),[14] it is unlikely that they can pursue redress when little or no money is involved. An important aspect of the scene, pointed out by Whiggy, is community solidarity and the rapid dissemination of news among its members. Yet the strongest weapon against plagiarism is suggested by Novus in his post: publicly humiliate transgressors. Since the mod scene is based on social prestige rather than financial gain, public humiliation is a powerful method for discouraging musical theft, at least among mod composers.

Prestige may also be the reason why plagiarism among mod composers is considered a particularly serious transgression by scene members. What infuriated them in the DJ Carbon case was not issues of compensation, since mod composers rarely, if ever, make money from their creations. Rather, plagiarism is a serious crime in the scene because it robs trackers of their hard-earned status in the community—their prestige. Indeed, DJ Carbon's transgression was that he stole the thing that trackers want most: fan recognition. Because of this, as one posting argues, plagiarism can be construed as a direct threat to the scene itself: "Keep in mind though that often times the only 'pay' a tracker ever gets is acknowledgement from his or her fans. Without this acknowledgement, very few people would keep tracking, and the scene would die. And the reason why ripping is a problem is that they drain away this acknowledgement from the person who rightfully deserves it" (MODPlug Central discussion forum, posted by Novus, June 30, 1999).

I have shown how ripping is a broad concept within the scene. When composers borrow digital samples from one another, ripping is seen as a positive aspect of the creative process. After all, tracking was founded on the subversive practice of ripping sounds from game software and music recordings (as well as television and film soundtracks, audio effects, dialogues, etc.), using the rationalization that such brief sound elements should be free for public use. Indeed, for most scene members, samples are simply considered publicly available sounds that can be rendered into music only through the creative process of tracking—i.e., music coding. It is in the appropriation of coding that ripping becomes a serious transgression. While some trackers may identify certain samples as part of their signature sound (usually if they have created or modified a truly unique sound), most regard only their music *coding* as intellectual property. This became apparent in the vigorous protest one tracker aimed at another for using some of his "patterns" (or sections of coding).

Musical Creativity and Ownership 2: MP3 Software

While plagiarism remains a serious threat to the scene, it is controllable by scene members themselves. A greater threat, some scene members have argued, is the advent of MP3 compression software. To some extent, MP3 undermines the option of ripping as well as denies artists the ability to examine one another's coding techniques. MP3 is a digital compression format that allows audio files to be reduced to about one-tenth of their size yet retain considerable sound quality. Most computer users are more or less familiar with digital sound as "wave" files (usually with the suffix <.wav>). These might be likened to uncompressed bitmapped image files (often using the suffix <.bmp>). Similar to images, the greater the detail, or resolution of the file is, the larger the file. Compositions digitally recorded with CD-quality resolution tend to take up considerable disk space in uncompressed wave-file format. However, software compression technologies like MP3 (using files with a suffix such as <.mp3>) reduce the audio file size dramatically while retaining most of the overall audio quality, thus allowing digitally recorded music to be easily distributed on the Internet.[15] However, for mod music, this means playing the song on the computer and recording it as a wave file that is then compressed (i.e., encoded) into the MP3 format. As an MP3 file, the song no longer contains any music coding and as a result, cannot be played back on mod player or tracker software. It is now a true recording rather than a set of computer instructions with audio samples—other artists cannot examine the composer's coding techniques. On the other hand, some mod composers have argued that MP3 protects them from plagiarism, because their creative techniques, as well as their samples, become hidden and far more difficult to rip. Furthermore, once a song is rendered into MP3 format it becomes "fixed" just like any other audio recording, thus discouraging unauthorized versions by other artists. The files can be copied, but they cannot be easily altered.

Ironically, many mod artists and fans find that MP3 technology takes mods one step closer to commercialization. For them, when mods are converted into MP3 format, they become like any other recorded music—they are no longer mods. Some have even accused the major sites like Trax in Space of "selling out" to the popular music industry, because they allow artists to submit songs in almost any digital format, including MP3. The irony is that MP3 and similar technologies are also considered a major threat to the music industry, as Reebee Garofalo (1999) points out. MP3 takes the record companies out of the equation, linking artists and fans more directly through the Internet. Furthermore, distribution over the Internet is difficult, if not impossible, to regulate: MP3 files can be copied just like any other computer files. In response, the RIAA (Recording Industry Association of America), in cooperation with major record companies, is proposing an open standard that would encode each digital recording with a "watermark" that identifies its owner and origin in order to discourage illegal copying and distribution (Garofalo 1999: 250). Other related software technologies are on the horizon or have already arrived, like Napster (a program that facilitates the exchange of MP3 files on the Internet), further subverting the hold of the large music corporations, but it remains to be seen what this may mean for the mod community.

Conclusions: Techno-performativity and Virtual Community

Our understandings of selfhood and community have been radically altered by media and communications technology. In the discourse of postmodernist logic, the autonomous subject has been displaced—or, better yet, fragmented into many selves. When we go online, the computer extends our identity into a virtual world of disembodied presence and, at the same time, allows us to take on other identities. We lurk in or engage with online lists and usenet groups, constructing different versions of ourselves on the fly. The computer, in this way, allows for a new kind of performativity, an actualization of multiple and perhaps idealized selves through text and image. Communities, too, emerge out of the convergence of the many disembodied selves on the Internet. Freed from the constraints of real time and real space, so-called virtual communities are not bound by proximity or face-to-face contact. What makes an online community cohere is common interest and a suspension of disbelief—a faith in the community's realness by its members.

The idea of "virtual community" has been the focus of discussion in a great deal of literature regarding computers and the Internet.[16] Too often, however, "community" is described in simplistic or entirely utopian terms. On the one hand, Howard Rheingold (1993: 5) defines virtual communities as "social aggregations that emerge from the Net when enough people carry on . . . public discussions long enough, with sufficient feeling, to form webs of personal relationships in cyberspace." Others writers compare the virtual communities of the Internet to the Habermasian ideal of "public sphere."[17] Here, the virtual community has utopian possibilities; it is a kind of place where access is guaranteed to all citizens, where they can gather, organize themselves, and hold informed and reasoned debate. In the rarefied textual virtuality of literary criticism and cultural studies, scholars don't investigate whether communities actually exist on the Internet but instead argue over the *concept* of virtual community. In other words, for many theorists, if it does not fulfill our utopian dreams of the public sphere, then it isn't a community at all but simply a myth created by the corporate world "to inspire and privilege commodified desires. . ." (Lockard 1997: 225).

One might also argue that not all people have access to the Internet—indeed, most people in the world do not. Furthermore, it is rarely informed and reasoned debate that occurs online. Some may believe that what happens on the Internet isn't really (or virtually) community, as Joseph Lockard (1997: 224) argues, but simply a *desire* for community. Clearly, the users of Microsoft Windows 98 do not constitute a community, despite the advertising hype. Yet this begs several questions: what *does* constitute community? Is communication enough to create community—that is, can community exist as a virtual, immaterial entity made up of disembodied telepresences? Or does it require material, concrete components in proximity (houses, buildings, parks, embodied humans, and so forth)?

I would argue that all communities are based less on material and embodied proximity (humans sharing physical space) than on a collective sense of identity, of feeling that one belongs and is committed to a particular group. And the group coheres through the common interests, ideals, and goals of its membership. It may fall far short of any utopian ideal, but the cohesion of its collective identity is stronger than the centrifugal pull of its individual members—all of them real, embodied people. The Internet lays the groundwork for community by providing access to sustained communication, informational resources, and, most important, a common locus for members to gather (even while that locus is virtual). Internet communities have emerged, despite temporal and spatial displacements, because they are formed entirely out of social relationships that are very real to members—relationships emerging out of communication, exchange (of information, ideas, even goods and capital), common interests and purpose, and mutual commitment. As Nessim Watson (1999: 120) argues, "we should begin thinking of community as a product not of shared space, but of shared *relationships* among people."

Communities on the Internet are not "virtual," they are real—as real as the offline communities we belong to as embodied humans.[18] Even in the material offline world, a community does not come into existence simply because of the physical proximity of its members—we do not see our zip codes as necessarily identifying who we are (although they might indicate our socio-economic circumstances). A community is defined by the social relationships that form its underpinnings, and these are its reality. Such a perspective recalls the now widely known notion of "imagined" community as set forth by Benedict Anderson (1983: 15): "Communities are to be distinguished, not by their falsity/genuineness, but by the style in which they are imagined." Thus, discussions of whether communities do or do not exist on the Internet, or of whether they are real or virtual, should be focused instead on the nature and quality of the relationships among members. In other words, rather than question whether this or that online collective is or is not a community, we should perhaps ask what these social networks *do* and *mean* for their members.

We might also extend Anderson's point that communities are imagined in particular *styles* by arguing that "styles" refer to the ways in which groups take on particular identities and how these identities are translated in social action. Following this reasoning, the concept of community could thus be considered the unique manner in which a network of relationships is conceived by its members and represented to the wider world as the group's identity. That is, we might understand community as a collective and ongoing performative practice of group representation (to itself and to others). Drawing from Webb Keane's recent study, representations are both entities made up of specific formal properties and kinds of practice distinct but inseparable from the full range of human activity (1997: 8). For Keane, representational forms and practices get at the core of understanding human interaction. They are "embedded within and implicit to the demands of interaction" and provide evidence of "how people constitute themselves and act as social subjects" (1997: 224).

Keane's ideas suggest possibilities for developing a discourse on how Internet communities such as the mod scene constitute themselves, and how members find meaning for such communities. If we also consider Turkle's ideas regarding cultures of simulation, we might then argue that the mod scene is a community constructed entirely of representational (i.e., simulated) forms and practices, drawing together symbolic structures and processes such as hierarchies of prestige and status (where fans, newbies, established composers, technical experts, etc. organize themselves into functional collectivities), informational systems of exchange (in which "goods" themselves are simulated forms based on computer data), rituals of inclusion and exclusion (manifested through member websites, Internet forums, discussion groups, etc.), and social drama (narratives of subverting the popular music industry, resolving conflicts among members, adjudicating cases of theft or plagiarism, etc.). Thus, it is the *context* of online communities that might be virtual (or electronic, or cyber, or whatever), not the sets of social relationships such collectivities engender. What is radical about the mod scene is not the fact that it constitutes a community but the way its members *use* computer technology to build, maintain, and represent real social relationships.

The mod scene suggests an almost utopian scenario: music commodification without mass production and consumerism without the exchange of capital (at least in terms of financial assets). In Baudrillardian terms, mod music files are "are conceived from the point of view of their very reproducibility." The wrinkle, however, is that mods are not conceived to be mass-produced commodities, even as they are mass-mediated through the World Wide Web. Commodification only occurs when the music file is copied and downloaded by fans ("consumers"). These fans often contribute to the distribution of mods by making copies and sending them to friends and acquaintances via email, discussion groups, or Internet chat systems. Bear in mind that copying mod files is not the same as making cassette or CD copies of storebought recordings, since the source recording might constitute a kind of original (holding a kind of prestige that the copy does not have). Because mods are digital data files, the copies are, for all intents and purposes, the same as the original (and there is no greater prestige associated with owning the "original" file). Furthermore, the concept of commodity is problematic here because, while the Internet provides a means for mass-mediation, consumption takes place without the exchange of money. Even the *means* of production and consumption are third-order simulacra: the necessary audio "technology" needed for mods (the mod trackers and players) is available in the form of software that can be downloaded from the Internet, also for free.[19] Thus, the political economy of the mod scene is one based on new divisions of labor, new concepts of capital (prestige, status, name recognition, etc.), and new configurations of social relations (websites as community centers, discussion groups as public forums, etc.).

In Jacques Attali's utopian view of the future of music (1992: 133), cultural production and consumption merge with the ascent of new social and aesthetic orders: "Today, a new music is on the rise, one that can neither be expressed nor understood using the old tools, a music produced elsewhere and otherwise. It is not that music or the world have become incomprehensible: the concept of comprehension itself has changed; there has been a shift in the locus of the perception of things."[20]

In a way, Attali's idealistic rhetoric of a new musical order seems to describe, to some extent, the mod scene. It is indeed a music, to use Attali's words, that requires "new tools" to be expressed and understood, and it is indeed produced elsewhere and otherwise. These "new tools" have emerged out of the ingenious combination of personal computing, the Internet, and the unique software technologies created by mod scene members to compose, distribute, and listen to their music. In Attali's utopian scenario, music becomes a participatory enterprise in which consumers are involved in the creative process itself; distinctions between high and low culture disappear (or at least blur); economic networks of use and exchange dissolve; and, finally, composing is no longer a specialized profession, aimed at the production of a musical object, but an ongoing communal activity that remains perpetually unfinished, undertaken for the sheer pleasure of social interaction—what Christopher Small (1998: 13) calls *musicking*:

> The act of musicking establishes in the place where it is happening a set of relationships, and it is in those relationships that the meaning of the act lies. They are to be found not only between those organized sounds which are conventionally thought of as being the stuff of musical meaning but also between the people who are taking part, in whatever capacity, in the performance; and they model, or stand as metaphor for, ideal relationships as the participants in the performance imagine them to be: relationships between person and person, between individual and society, between humanity and the natural world and even perhaps the supernatural world.

Indeed, following Small's argument, mod music is as much about the mod scene and the relationships it entails as it is about the abstract pleasure of listening to patterns of sounds. We might listen to a mod song and dismiss it as a banal, derivative, or simply amateurish imitation of the popular music styles we hear on the radio or find on commercial recordings—and, indeed, many (if not most) mods are musically conservative—but we would be missing the point. Small's notion of musicking is similar to an argument that John Blacking made more than twenty-five years ago: "musical things are not always strictly musical" (1973: 25). As I stated earlier in this essay, the mod scene is somewhat akin to garage rock and the DIY sensibilities of punk. We might view the mod scene as a musical subculture within the larger popular music scene as well as the commercial computer software industry. Following Hebdige (1979), a subculture is based on the materials of the parent (or dominant) culture, reconfigured to become meaningful to its rebellious participants. In the mod scene, similar to punk sensibilities, aesthetics of abstract beauty do not necessarily prevail over subcultural ideologies of self-expression, resistance, subversion, personal or group empowerment, and so forth. The pleasure of listening to mod music is deeply enmeshed in the complex social relationships that the mod scene engenders, in its often conflicted ideologies of resistance and subversion, and in its quirky techno-geek values. In heroic terms, we might argue that the mod scene is a community of young computer and popular music enthusiasts trying to undermine the hegemony of commodity capitalism and Western technoculture. Yet "resistance" here is based not on rejecting but on embracing mass media and computer technology, choosing and reconfiguring those elements that serve the aesthetic sensibilities, ideologies, and logic of the mod community. Whether they are successful or not in resisting the music industry, whether their community is utopian or not, real or virtual, mod scene members have nevertheless carved out a social place for themselves—and they have carved it out in the non-space of the Internet.

It is clear that the Internet provides fascinating possibilities for social networks, especially for innovative new uses of current technology and for collective and collaborative artistic creativity. The mod scene demonstrates that it is the communal enterprise that impels new music—composers respond to one another (and to their fans) directly and immediately. Yet these communities are also far more than some people exchanging ideas by email. The Internet provides a place for individuals to gather and, as a collective, to generate emergent (sub)cultures, complex prestige systems, elaborate commodity exchange networks, and structured governing bureaucracies. When we take such online social collectivities seriously and acknowledge the reality they can accrue (thinking of reality as a kind of capital) in the social relationships they engender, we might then finally understand what constitutes community, whether we theorize it as real or imagined.

Notes

1. This article is based in part on Lysloff 2003. Many thanks to Deborah Wong for her many valuable suggestions. I am also grateful to my virtual friend Kim Kraft (known as Mister X in the mod scene) for his astute observations, and to my tracking teacher, TrackZ, for his boundless patience.

2. Katherine Hayles (1999) argues that we live in the era of the posthuman, a time when information has lost its body. According to Hayles (1999: 2–3), the concept of the posthuman carries with it several assumptions: 1) privileging informational pattern over material instantiation so that embodiment in a biological substrate is seen as an accident of history rather than an inevitability of life; 2) viewing the mind as an epiphenomenon, the result of informational processing (rather than as the primary source of processing); 3) believing the body to be the original prosthesis that we all learn to manipulate, with the possibility of other prostheses; and 4) reconfiguring the concept of self so that it can be seamlessly articulated with intelligent machines.

3. Jameson calls this "blank parody" (1983: 114 and 1991: 17). Arthur Kroker describes this condition as "recombinant culture," in which "every fragment of the cultural archive, and subjectivity with it, can be resequenced at will, reconstituted into an endless combinatorial of media effects" (Kroker 1993: 66).

4. See the title chapter to Jameson 1991 for a more fleshed-out version of his discussion of postmodernity. See also Hutcheon 1989 for a detailed examination of the postmodern condition.

5. See Lysloff 1997.

6. Fans state that mod music is superior to the MIDI (Musical Instrument Digital Interface) music available on most Windows-based PCs, because mods are entirely software driven and do not require special equipment beyond a PC with a soundcard. MIDI data files contain only coding that instructs the computer to play certain pitches in specific ways, while mods contain both coding and actual sound samples in digital form. Thus, playback of MIDI files (data files ending with .mid or .rmi) requires relatively expensive equipment (and software) for acceptable sound reproduction, but mod songs can be played back with satisfactory results on almost any PC today that has a soundcard and the mod player software (obtained for free on the Internet). Imbedded sound samples allow the composer to have greater control over his or her music and make mod files accessible to anyone that owns a PC and is connected to the Internet. The composers can sample sounds themselves and even incorporate texted vocal lines in their music, or they can create and use synthesized sounds as well as appropriate sounds from various recorded sources, including compact disks and cassette tapes, computer games, and even other mod files.

7. This kind of information is found on several of the scene's main websites, such as Trax in Space (<http://www.traxinspace.com>) and United Trackers (<http://united-trackers.org>), provided voluntarily by members. Personal homepages and crew sites are also important sources for scene demographic data.

8. In any case, many programs are encrypted and have source codes unavailable to the general public.

9. In the staged music of rock, such forms of tribute is known as covers.

10. This practice might also be likened to what Dick Hebdige (1987: 12–14) calls "versioning."

11. Most mod tracker and player programs can display the composer's comments as well as a list of the samples (or "instruments") used in a piece.

12. MP3 is a recently developed and still somewhat controversial form of digital sound data compression in which commercial music files are significantly reduced in size yet retain considerable sound quality in playback. This new technology makes it easier to distribute commercial music through the Internet and has opened new possibilities for piracy.

13. Quoted exactly as written, without editorial changes.

14. See under Title 17 ("Copyrights"), Chapter 1, Section 102 (Subject matter of copyright: In general):

(a) Copyright protection subsists, in accordance with this title, in original works of authorship fixed in any tangible medium of expression, now known or later developed, from which they can be perceived, reproduced, or otherwise communicated, either directly or with the aid of a machine or device. Works of authorship include the following categories:

(1) literary works;
(2) musical works, including any accompanying words;
(3) dramatic works, including any accompanying music;
(4) pantomimes and choreographic works;
(5) pictorial, graphic, and sculptural works;

(6) motion pictures and other audiovisual works;
(7) sound recordings; and
(8) architectural works.

The U.S. Code is published online through U.S. House of Representatives, Office of the Law Revision Counsel (at <http://law2.house.gov/download.htm>).

15. For more on MP3 as well as earlier technologies that have had an impact on the popular music industry, see Reebee Garofalo's overview of the music industry during the twentieth century (Garofalo 1999).

16. See, for example, Shields 1996, Rheingold 1993, and Porter 1996.

17. See Habermas 1996.

18. This point was brought to light by Nessim Watson in an article entitled "Why We Argue about Virtual Community" (see Watson 1998).

19. While the hardware needed for the software is still relatively expensive (in real monetary terms), many mod fans use computers available in schools and libraries, at work, or in other public places.

20. This same passage is posted on the homepage of one of the major websites in the mod scene: <www.castlex.com>.

References

BOOKS AND ARTICLES

Attali, Jacques. 1992. *Noise: The Political Economy of Music.* Minneapolis: University of Minnesota Press.

Baudrillard, Jean. 1987. *The Ecstacy of Communication.* Bernard and Caroline Schutze, transl.; Sylvère Lotringer, ed. New York: Semiotext(e).

———. 1983. *Simulations.* Paul Foss, Paul Patton, and Philip Beitchman, transl. New York: Semiotext(e).

Benjamin, Walter. 1969. "The Work of Art in the Age of Mechanical Reproduction." In *Illumination.* Edited with an introduction by Hannah Arendt. Translated by Harry Zohn. New York: Schocken Books.

Biggs, Simon. 1996. "Multimedia, CD-ROMO, and the Net." In *Clicking In: Hot Links to a Digital Culture.* Lynn Hershman Leeson, ed. Seattle: Bay Press.

Blacking, John. 1973. *How Musical is Man?* Seattle: University of Washington Press.

Davis, Mike. 1988. "Urban Renaissance and the Spirit of Postmodernism." In *Postmodernism and Its Discontents.* E. Ann Caplan, ed. New York: Verso.

Garofalo, Reebee. 1999. "From Music Publishing to MP3: Music and Industry in the Twentieth Century." *American Music* 17 (3): 318–53.

Green, Dave. 1995. "Demo or Die!" *Wired Magazine* 3.07 (<http://www.wired.com>).

Habermas, Jurgen. 1996. *The Structural Transformation of the Public Sphere.* Cambridge: MIT Press.

———. 1983. "Modernity—An Incomplete Project." In *The Anti-Aesthetic: Essays on Postmodern Culture.* Hal Foster, ed. Seattle: Bay Press.

Hayles, N. Katherine. 1999. *How We Became Posthuman.* Chicago: University of Chicago Press.

Hebdige, Dick. 1987. *Cut 'n' Mix.* New York: Methuen and Co.

———. 1979. *Subculture: The Meaning of Style.* New York and London: Routledge. First published by Methuen and Co.

Hutcheon, Linda. 1989. *The Politics of Postmodernism.* New York: Routledge.

Huyssen, Andreas. 1986. *After the Great Divide.* Bloomington: Indiana University Press.

Jameson, Frederic. 1991. *Postmodernism, or, The Cultural Logic of Late Capitalism.* Durham: Duke University Press.

——. 1983. "Postmodernism and Consumer Society." In *The Anti-Aesthetic: Essays on Postmodern Culture.* Hal Foster, ed. Seattle: Bay Press.

Jones, Steven G., ed. 1998. *Cybersociety 2.0.* New Media Culture Series. Thousand Oaks, Calif. and London: Sage Publications, Ltd.

——. 1997 *Virtual Culture.* Thousand Oaks, Calif. and London: Sage Publications, Ltd.

Keane, Webb. 1997. *Signs of Recognition.* Berkeley: University of California Press.

Keisler, Sara, ed. 1997. *Culture of the Internet.* Mahwah, N.J.: Lawrence Erlbaum.

Kroker, Arthur. 1993. *Spasm.* New York: St. Martin's Press.

Kirshenblatt-Gimblett, Barbara. 1996. "The Electronic Vernacular." In *Connected: Engagements with Media.* George E. Marcus, ed. Chicago: University of Chicago Press. Late Editions: Cultural Studies for the End of the Century 3, pp. 21–65.

Lysloff, René T. A. 1997. "Mozart in Mirrorshades: Music, Technology, and Ethnomusicological Anxiety." *Ethnomusicology* 41 (2): 206–19.

——. 2003. "Musical Community on the Internet: An Online Ethnography. *Cultural Anthropology* 18 (2): 23–64.

Lockard, Joseph. 1997. "Progressive Politics, Electronic Individualism, and the Myth of Virtual Community." In *Internet Culture.* David Porter. ed. New York: Routledge.

Ludlow, Peter, ed. 1996. *High Noon on the Electronic Frontier.* Cambridge: MIT Press

McLaughlin, Margaret L., Kerry K. Osborne, and Nicole B. Ellison. 1997. "Virtual Community in a Telepresence Environment." In Jones 1997, 147–68.

Mitchell, William J. 1996. *City of Bits.* Cambridge: MIT Press.

Polan, Dana. 1988. "Postmodernism and Cultural Analysis Today." In *Postmodernism and Its Discontents.* E. Ann Kaplan, ed. New York: Verso.

Porter, David, ed. 1996. *Internet Culture.* New York: Routledge.

Rheingold, Howard. 1994. *The Virtual Community.* New York: HarperPerennial.

Rushkoff, Douglas. 1999. *Playing the Future.* New York: Riverhead Books.

Sardar, Ziauddin. 1996. "alt.civilizations.faq: Cyberspace as the Darker Side of the West." In *Cyberfutures: Culture and Politics on the Information Superhighway.* Ziauddin Sardar and Jerome R. Ravetz, eds. New York: New York University Press.

Schafer, R. Murray. 1994. *The Soundscape: Our Sonic Environment and the Tuning of the World.* Rochester, Vt.: Destiny Books. (First published in 1977 as *The Tuning of the World.* New York: Knopf.)

Shields, Rob, ed. 1996. *Cultures of Internet.* London: Sage.

Turkle, Sherry. 1997. *Life on the Screen.* New York: Touchstone.

Virilio, Paul. 1991. *The Lost Dimension.* New York: Semiotext(e).

WEBOGRAPHY

United Trackers, http://www.united-trackers.org
Trax in Space, http://www.traxinspace.com
MODPlug Central | Resources, http://www.castleX.com
Altered Perception, http://www.alteredperception.com
IMM (formerly the Internet Music Monitor), http://surf.to/the-imm
Rocky's Mod Mansion, http://www.welcome.to/themodmansion
The Mod Archive (includes list of known mod plagiarists, songs and originals, and comments including details of one thief), http://www.modarchive.com/
The Digital Music Archive, http://www.s3m.com/dma/

Tracker Instruments Repository (resources: sound samples), http://sbis.komi.ru/ti-r/
Chiptune Central (archive of mod chiptunes), http://pweb.de.uu.net/msconex/
Black Hole (mod archive), http://bhole.cjb.net/
MAZ Sound Tools, http://www.maz-sound.com/
exStreamities, http://massakr.port5.com/exstream.htm

CHAPTER THREE

A Riddle Wrapped in a Mystery
Transnational Music Sampling and Enigma's "Return to Innocence"

Timothy D. Taylor

Globalization as we currently discuss it and theorize it cannot be conceived of without taking into consideration digital technologies that have sped up the movement of information. It is for this reason that Manuel Castells prefers to label this era both global and informational, not simply one or the other.[1] This essay, therefore, takes off from Castells to examine the ways that music moves around the world, and the consequences that this movement can sometimes have, with respect to a specific case of the German band Enigma and their appropriation of a song by aboriginal musicians in Taiwan.

Globalization/"Glocalization"

Much has been made in the last few years of the new globalized, transnational world that we all live in, a world with flows of "technoscapes," "ideoscapes," "ethnoscapes," "mediascapes," and "finanscapes" that Arjun Appadurai has so influentially labeled and theorized.[2] To this list I would add, or tease out, another, an "infoscape," which to some extent is the atmosphere in which the others exist, made possible by the computer, the Internet, and other digital technologies.[3]

With the rise of these recent "–scapes," however, and terms such as transnational or global economy, it has become important to wonder just how new this "new" global economy is, for claims about the new global economy are almost never historically informed. There is a good deal of evidence that, in terms of overseas investment, we aren't really any more global than we were at the height of the imperial era early in the twentieth century. Doug Henwood has written that in terms of exports as a share of the gross domestic product, the United Kingdom—the biggest imperial power among developed nations—was only a little more globalized in 1992 than in 1913; Mexico exported more than twice as much in 1929 as in 1992; and today, the United Kingdom exports almost twice as much as Japan, which most people think is the biggest exporter. The U.S. economy is more internationalized now than it was at the turn of the century, but it nonetheless exports far less than the United Kingdom and Japan and is in fact closer to Mexico in these terms.[4]

If we step further back in history we affirm that people—and thus their cultures, and more specifically, for my purposes here, their musics—have always interacted. Historian Jerry H. Bentley asserts that "cross-cultural encounters have been a regular feature of world history since the earliest days of the human species' existence."[5] Asian, African, and European peoples regularly traveled and interacted, he argues, via trade routes that crossed the Eurasian landmass; religions such as Buddhism, Christianity, and Islam influenced people far from their points of origin.[6]

Bentley distinguishes three main periods of travel and intercultural exchange, beginning with the Roman and Han empires (he begins in this moment because of the scarcity of historical sources before). He first identifies the era of the ancient silk roads, which he dates at roughly 400 to 200 B.C., as the first major period of intercultural contact. The next major period began around the sixth century; crosscultural exchange was fostered by the foundation of large imperial states such as the Tang, Abbasid, and Carolingian empires and relied on the cooperation of nomadic peoples, who provided transportation links between settled regions. In this period, there was also more frequent sea travel across the Indian Ocean. This second era blended into the third, the last pre-Columbian one, from roughly 1000 to 1350 A.D.; long-distance trade increased dramatically over both sea and land and was marked by the rise of nomadic peoples, namely the Turks and the Mongols, into political power and expansion. The bubonic plague in the later fourteenth century disrupted trade until the fifteenth, leading to a fourth and more studied colonial expansion of European powers.[7]

Immanuel Wallerstein has written of a period a little later than this, the expansion of European empires after Columbus's "discovery," and he coined the term "modern world-system" to describe the establishment of regular contact around most parts of the world. For Wallerstein, modernity itself *is* this rise of capitalism and world trade that began in the sixteenth century. This expansion, he tells us, wasn't just a geographic (that is, colonial) expansion but also an economic one, accompanied by demographic growth, increased agricultural productivity, and what he calls the first industrial revolution. It was also, he notes, the period in which regular trade between Europe and the rest of the inhabited world was established.[8]

At one level, then, while some of the foregoing may resonate with today's headlines, it's still old news. Today's globalization is less something new than a continuation of global processes that have been in place since the late fifteenth century and were themselves preceded by precapitalist forms of crosscultural exchange. To think in binary terms—as if we are now in a moment of "globalization" that renders the past a monolithic moment of "pre-globalization"—doesn't get us very far.

And yet, of course, some things are different today. Today's globalization, as people in the so-called developed countries are experiencing it, would not be happening without digital and other technologies; the exchange of information is faster, information travels further, and there's thus more of it. The main difference, though, isn't merely the speed of dissemination, or even the seeming glut of forms and signs, but rather the fact that there are more and more signs from elsewhere coming to the developed countries. What we are in the midst of today isn't simply a globalization in which forms flow everywhere but rather a moment in which forms from elsewhere are coming to the West with increasing frequency; it was this increase in recordings from other places to European and American metropoles that prompted the invention of the term "world music" over a decade ago.[9] As Stuart Hall writes, "our lives have been transformed by the struggle of the margins to come into representation."[10] It is thus a bit Euro- and Americentric to think that the world is newly globalized, since Western forms have been globalized for decades, through colonialism, imperialism, and the movement, as Wallerstein writes, from economic cores to peripheries (and the subsequent extraction of materials to the core).

So why is the term "globalization" so frequently used if it describes processes that have been ongoing since the beginning of recorded history? Because it obfuscates, as Doug Henwood, Timothy Brennan, and others have argued.[11] The hype surrounding the new global economy covers up to a certain extent the fact that capitalism is as exploitative as it ever was—perhaps more so—and is constantly seeking new people around the world to use as cheap labor, which, according to Wallerstein, has been the impetus behind global expansion for centuries.[12] The term also helps preserve an old binary opposition that is increasingly waning, the binary between "global" and "local" that has been much theorized lately.[13]

Perhaps a better term than globalization is "glocalization," a word that originated in Japanese business in the late 1980s and was quickly picked up by American business.[14] Glocalization emphasizes the extent to which the local and the global are no longer distinct—indeed, never were—but are inextricably intertwined, with one infiltrating and implicating the other. Indeed, it may now be difficult or impossible to speak of one or the other.[15] Older forms and problems of globalization continue but are increasingly compromised, challenged, and augmented by this newer phenomenon of glocalization.

Beginnings

Now let me turn to the specific musical case of this article. In May 1988, the cultural ministries of the French and Taiwanese governments brought roughly thirty Taiwanese residents of different ethnic groups to France to give some concerts. These musicians ultimately performed in Switzerland, France, Germany, the Netherlands, and Italy for a month, earning a stipend of fifteen dollars per day. Unbeknownst to the musicians, some of their concerts were recorded, and the following year, the Ministère de la Culture et de la Francophonie/Alliance Française issued a CD called *Polyphonies vocales des aborigènes de Taïwan* that contained some music from these concerts. A Taiwanese ethnomusicologist, Hsu Ying-Chou, says that the musicians signed a contract before the tour and that the Chinese government approved the French recording.[16] Pierre Bois, of the Maison des Cultures du Monde, told me that it was his understanding that the musicians did in fact know they were being recorded, and that the Chinese Folk Arts Foundation was in regular contact with the musicians, whom he assumed were paid for an earlier recording.[17] (This CD also included music recorded a decade earlier by a Taiwanese ethnomusicologist and was accompanied by liner notes by two Taiwanese ethnomusicologists, one who had made the original field recordings and another who had issued them in Taiwan.)

In the meantime, Michael Cretu, a Rumanian émigré to Germany (also known as "Curly M.C.") and his band Enigma were busy scoring a colossal international hit with their album *MCMXC A.D.* This recording came out of nowhere to sell seven million copies worldwide, which made it the most successful German production abroad ever; the single from the album, "Sadeness Pt. 1" (after the Marquis de Sade) became the fastest-selling single in German recording history. *MCMXC A.D.* went to number one in several European countries, including the U.K., and peaked at number six with a run of 150 weeks on the Billboard 200 chart in the United States. The gimmick (or combination of gimmicks) that proved so salable was

Cretu's mixing of sampled Gregorian chants with dance beats, which resulted in a kind of sped-up New Age litany.[18]

A few years later, in 1992, Michael Cretu sat in the studio he built with the proceeds from *MCMXC A.D.* on the island of Ibiza off the coast of Spain. Cretu, in the words of one fan, took "3 years [to] work his way through hundreds of CD's of native song, sampling, cataloguing and synchronising many sounds before he began his songwriting process."[19] Cretu himself said, "I'm always looking for traces of old and forgotten cultures and I'm listening to hundreds of records and tapes."[20] Stumbling upon *Polyphonies vocales des aborigènes de Taïwan*, Cretu found what he wanted in the first track, called "Jubilant Drinking Song." Cretu's publishing company, Mambo Musik, paid 30,000 francs (about $5,300) to license the vocals from the French Maison des Cultures du Monde; half of this money went to the Chinese Folk Arts Foundation. The resulting single—the most successful single from Enigma's second album, *The Cross of Changes*—is called "Return to Innocence," and it went to number two in Europe, number three in United Kingdom, and number four in the United States. *The Cross of Changes* went to number two in Europe, number 1 in the United Kingdom, number nine in the United States, and number two in Australia in 1993. Because of Enigma's earlier success, 1.4 million advance orders were made for this album, which ultimately sold five million copies and was on *Billboard's* Top 100 chart for thirty-two straight weeks.

Two years later, Kuo Ying-nan, a seventy-six-year-old betel nut farmer and musician in Taiwan of Ami ethnic ancestry, received a phone call from a friend in Taipei. "'Hey! Your voice is on the radio!' And sure enough," said Kuo, "it was me."[21] "I was really surprised," he said elsewhere, "but I recognized our voices immediately."[22] Kuo and his wife, Kuo Shin-chu, were two of the musicians who had toured Europe in 1988, and they were also on the earlier recordings collected on the *Polyphonies vocales* recording.[23]

Cut to Atlanta, Georgia, in 1996, where the International Olympic Committee was selecting music to showcase at the games their city was to host that summer. They commissioned ex–Grateful Dead drummer Mickey Hart, a leader in the "world music" genre, and winner of the first Grammy award for that music in 1991, and Hart duly composed "A Call to Nations," a work that featured many different kinds of drumming as well as Tibetan Buddhist chanting and other sounds, asserting the notion that we're all one world. Other composers were commissioned, and, relevant to this discussion, previously recorded works were also made official songs of the Olympics. Enigma's "Return to Innocence" was named one of these. It thus appeared on a collection featuring official Olympics music and was used by CNN and NBC in advertisements for their coverage of the Olympics, though I have been unable to locate this album.[24] (Some press reports say that this Olympics exposure is how Kuo learned of the use of his voice.)[25] Gill Blake, assistant producer of the project that produced the promotional video for the Olympics, wrote, "We listened to several pieces we felt had something spiritual and timeless about them. It was then purely a matter of making a subjective choice. . . . In addition, 'Return to Innocence' seemed to work in conjunction with the ideas expressed in the video of fair play, peace, unity, etc."[26]

On July 1, 1996, just before the beginning of the Olympics, Magic Stone Music, a record label in Taipei, issued a press release that said they were representing the Kuos in a lawsuit against EMI (presumably as the parent company of Virgin, Enigma's record company), and that they were also producing a new album by them, an album of their traditional music mixed with pop sounds.[27] On July 26, 1996, it was announced that the president of the International Olympic Committee, Juan Antonio Samaranch, had decided to send an official thank-you to the Ami couple, following a report to the committee by Wu Ching-Kuo of Taiwan.[28] The Kuos' attorney claimed that the original use of their voices was illegal, and thus all subsequent uses were also.[29] Their attorney also said that this was not just an intellectual property case, but that the Kuos' human rights had been violated: the musicians "think EMI is ignoring the human rights of the Ami people."[30] "Minority peoples around the world have been treated unfairly over and over in this way," Magic Stone said in their press conference. "In the 17th century, people cheated the aborigines out of their land, but why are the basic rights of aboriginal peoples still being ignored today?"[31]

At some point (a date has not been mentioned), Enigma was reported to have sent a check for $2,000 (another report said 15,000 francs, which is almost $3,000) to Hsu Tsang Houei, who had made the original field recordings in 1978.[32] Professor Hsu deposited the check in an Ami community trust fund. Some reports said that this money was sent to Kuo himself. One account said that Enigma suggested the possibility of further collaborations with Kuo. The Taiwanese government said that the higher figure was paid by the French Maison des Cultures du Monde, which was responding to a letter from Hsu Tsang Houei, and that the money was paid to the Chinese Folk Arts Foundation, which had brought the singers from Taiwan to Europe in the first place. To date, however, the money appears to have remained in the hands of the foundation and has not been paid out to the Kuos or anyone else.[33]

As far as I can tell, this threatened lawsuit went nowhere for nearly two years. I sent a few faxes to Magic Stone inquiring about its status but received no reply; in the last of these I volunteered my professional services as a musicologist, but still nothing. Finally, in March 1998, two press reports clarified matters. This lawsuit had indeed stalled, because the Kuos' "representatives" were told by the (presumable) defendants in the suit that it could cost about $1 million to bring suit. Attorneys willing to take on the suit pro bono could not be found until, finally, a Chinese-American intellectual property lawyer agreed to take the case.[34]

This lawyer, Emil Chang of Oppenheimer, Wolff, and Donnelly in San Jose, California, posted a plea to the Internet newsgroup <alt.music.-enigma-dcd-etc> in June 1998 headed "HELP STOP EXPLOITING ABORIGINAL CULTURES! ABORIGINES SUE FOR JUSTICE AND RECOGNITION: JUSTICE FOR THE KUOS!"[35] Chang included more explicit information about the suit that demonstrates the chain of ownership in today's multinational music world, for the suit was filed against a variety of music production and recording companies, including EMI and Sony. The basis of the suit was copyright infringement and "failure to attribute the plaintiffs as the original creators and performers of their work."

In the middle of the suit, Emil Chang left the firm, and the case was taken over by E. Patrick Ellisen, who told me early in 1999 that the judge was anxious that the case be settled out of court before the scheduled court date of midsummer 1999.[36] But mediation in the spring of 1999 failed to produce results, and Ellisen then expected to go to court. The failure of this mediation meant that another lawsuit was filed, against various licensees of EMI, since "Return to Innocence" appeared on many compilation albums as well as in films, television programs, and television advertisements. Ellisen's office also considered another lawsuit, against EMI in France and Maison des Cultures du Monde, that was not filed.[37] Ellisen and his staff faced an uphill battle, for most traditional music is not copyrighted, so it is easy for defendants in such cases to rely on that fact or claim that any usage of it constitutes fair use. For this reason, Robin Lee, director of Taiwan's Association of Recording Copyright Owners, said that Kuo had no legal case: "The original authors of traditional folk chants have long been dead. And since performers are not authors, they have no copyrights."[38] Lee was wrong, though: it isn't true that folk music can't be copyrighted. It has become standard for folk musicians to list the music as traditional, but the "arrangement" of it as copyrighted, as in "All music traditional, arranged by X." The defendants' attorneys also claimed that the Kuos knew that they were signing away rights to the concert recordings made in France in 1988, as Pierre Bois of the Maison des Cultures du Monde maintained.

Finally, in June of 1999, the parties reached an out-of-court settlement, most of which is confidential. What is known is that the Kuos will be given written credit on all future releases of the "Return to Innocence" song and would receive a platinum copy of *The Cross of Changes* album. Additionally, the Kuos would be able to establish a foundation to preserve their group's culture, particularly its music, an act that Ellisen says was "not to be construed as implying there was any money" in the settlement terms. For its part, Virgin Records America thanked the Kuos "for the important contribution that their arrangement and performance of the vocal chant 'Jubilant Drinking Song' made to the song 'Return to Innocence.'"[39] The careful use of the word "arrangement" here indicates that Virgin never altered its position on the Kuos' music—it is an "arrangement," that is, a version of a work for which they do not hold the copyright.

While the lawsuit was in progress, an established Taiwanese pop band called Xin Baodao Kangle Dui (or New Formosa Band) released an album on which they sing in two local dialects: Minnan Holo, also known as Taiwanese Minnan, and Hakka Kejia.[40] The first track is described by a Taiwanese fan as an "Enigma-like song reminding a person most strongly of 'Return to Innocence.' The only thing is that it's done in a mix of Kejia and Taiwanese. It definitely bewildered me the first time I played it."[41] Think of this: a Taiwanese group singing music in local languages in the style of Enigma's song which had extensively sampled music by an Ami couple singing in their native language. New Formosa Band's song was compiled and remixed on a later release and advertised as a dance tune with world music rhythms, entitled, in English, "Song for Joyous Gathering." The band also added a new member on this compilation album: an Aboriginal musician from Taiwan. Bobby Chen, a pop star in Taiwan, has recorded yet another version of this song.[42]

Twists

That's the story as clearly and as simply as I can now put it, though there are some interesting twists. Enigma's fans responded to the claims by the Kuos that they had not been consulted or recompensed, and I am going to turn now to discussing fans' reactions to this case, for they, too, are no less a part of this "infoscape" surrounding the case of the Kuos and Enigma. The press release mentioned above provoked some angry responses from fans on the Enigma Internet mailing list; a few posters were concerned about the incident, but for the most part Enigma's fans were angry that someone was, in effect, questioning their hero's creativity. The most vociferous (and most loudly agreed with) statement was this:

> Wow, foreign greed, tis but strange since most greed comes from the States. Now I've raved about this before, but I'm sad to say "screw this guy". He took

his cut, and now that his voice is famous people are getting him to cash in on it. There should definately be a statute of limitations on stuff like this, espically if the suits come _after_ the song is a big hit. I shall participate in my very small "one man boycott" (OHHH AHHH :)). And be sure not to help these people profit in any way. But not that it matters to anyone.

Still, why didn't they sue 3 years ago eh? Ya gotta wonder . . .

PS Enigma is still the best _where-ever_ and _how-ever_ they get the samples!!!

And so forth. The gist of this and most subsequent posts was that the Kuos had been paid (though, as I indicated, it isn't clear if they actually received any money) and that they had no right to ask for anything else.

Another post by this same user, slightly mollified by a calm call for fairness, wrote back, saying that

anything he [Kuo] deserves should not have anything to do with Enigma and/or its management. Let the people they bought it from deal with this guy. Also, you have to wonder if it had been some other band and/or the song made little money would anyone care? The only thing left that would make this perfectly _American_ is if this guy claimed some sort of racism or something. eheh :) Seriously though, I think the original party who sold to Enigma should have to be responsible if anyone is going to take the fall. I mean if this original anthro guy made this recording and such then it is kinda public domain stuff. Enigma basically paid the society for their "efforts" and that's about all that was neccessary. Now if the guy's original recording had a bunch of dance beats and other vocals then we'd have a problem ;)

This user's view seems to be that the original recording was of raw material in the public domain, but if the original recording had been refined by the addition of "dance beats and other vocals," that would indicate that their music had been produced in a studio and thus copyrighted.

After this flurry of responses, lawyer Emil Chang's later posting, quoted above, generated some rather nasty responses; most Enigma fans (the vast majority of whom subscribe to the mailing list and do not frequent the newsgroup that Chang posted to) were not sympathetic. Most argued that, even though the Kuos contributed something, "Return to Innocence" simply did not exist before Michael Cretu worked his magic. One person wrote, "As so far as 'Recognition' goes, an almost 80-year old Taiwanese singer is not credited on each and every copy of Enigma2's album because Michael Cretu **is** the creator of RTI ["Return to Innocence"]. Period."

It is clear that Enigma's fans are heavily invested in their highly romantic perceptions of Michael Cretu's genius; they view Cretu as a supremely gifted maker of meanings and speak of him in heroic terms. Their denunciations of, and impatience for, the Kuos' lawsuit makes this clear: they don't like Cretu's claim to genius and originality questioned at all. Or, if they admit that Cretu took somebody else's music, he is described as refining it, turning some raw material into art.[43] Here's one post to the Enigma mailing list during discussions of the Kuos' lawsuit: "OK, so Cretu probably realized that he could afford (and it would be well worth) a hell of a lot more than $2000 for the recording he made. But look; who else do you know who can take a two thousand dollar recording and make it into a multimillion dollar recording? Do you see that Andy guy who sang part of the chorus complaining? Let's not forget that even though the dude from Taiwan has a great voice, it was Cretu['s] creativity that made the real music happen."

Clearly invoking Romantic ideas of the genius as a person driven to work, and working in isolation, the Enigma FAQ on the Internet describes Cretu working alone in the studio at all hours, sorting through hundreds of CDs to sample: "He is a self-confessed night owl, and also a workaholic, this being seen by the fact that the production phase for *The Cross of Changes* took 7 months with the computer log of his sound bridge often stating that recording sessions from 10pm to 11am occurred. During this whole period he rarely saw the sun."[44]

One of the ways Westerners appropriate other music is to construct the original makers of that appropriated music as anonymous. Anthropologist Sally Price was told by a French art dealer that "If the artist isn't anonymous, the artist isn't primitive."[45] Cretu positions Kuo as anonymous and timeless in order to advocate his "return to innocence," a return to a spiritualized past. But when the makers of the original sounds talk back, Cretu's originality is called into question.

Kuo and his wife are assumed to be "primitives," but they're inconveniently privy to the rest of the world via the various "–scapes" mentioned earlier. At the same time, though, there's a refusal to recognize this. The Kuos' music is constructed as "pure," primitive, and thus infantilized, by Enigma, but by attempting to get credit and remuneration Kuo is behaving too much as a contemporary, worldly person: the subaltern speaks.

Enigma contributes to perceptions of their originality and the "primitive" and/or ancient nature of their music iconographically. The cover art on the single is faux "native" art, the Persian mystic poet Jalal ad-Din Muhammad Din ar-Rumi (1207–1273) is quoted, and more. Enigma also uses a typeface on the cover of the "Return to Innocence" single that looks as though it were made by a manual typewriter, as though Enigma is just a small band who make and sell their own recordings, inviting a degree of credibility with listeners.

One last wrinkle concerns the reticence of Michael Cretu and the people behind Enigma; they claim, through their unnamed manager, that they want to avoid cheap imitations of their music, that is, people who take the samples in an attempt to make music like Enigma's, and so they rarely disclose where their samples come from, unless, as we saw, they are forced to.[46] The keeper of the Enigma FAQ on the Internet, Gavin Stok, met with Enigma's manager, who works in Mambo Musik based in Munich, and asked him about Enigma's sampling problems. The manager claimed that license agreements state that they don't have to credit some samples. It became clear in the course of this interview that Mambo Musik was more worried about other musicians who track down Enigma's samples in order to make cheap imitations of Enigma. Here are Stok's words: "Their major concern is of commercial rip-off artists who steal the samples and try to quickly release a song to 'cash in' on the popularity of the first single from a new album. Evidence of this was apparent with the release of *MCMXC A.D.* and Mambo does not want to see it happen again."[47]

This is quite an interesting statement. On the one hand, Mambo Musik adheres to the letter of the law, listing sampled musicians only when required to do so; on the other hand, by not crediting other musicians, they are thus making it much harder for people to find those samples. In practices such as these, Mambo is asserting a kind of de facto ownership over Enigma's samples in these cases. Simon Frith writes that "Samplers have adopted the long established pop rule of thumb about 'folk music'—a song is in public domain if its author is unlikely to sue you. And so sample records make extensive use of sounds lifted from obscure old tracks and from so-called (far away) 'world' music; lifted, it seems, without needing clearance."[48]

Without the benefit of an attribution in the liner notes (except in the first European pressing of the album), several people came forward with very different statements about the origins of the sampled music in "Return to Innocence." Ellie Weinert wrote in *Billboard* that "the archaic-sounding vocals on 'Return to Innocence' are not sung in any particular language but represent a sequence of vowels."[49] A later *Billboard* article referred to the "Indonesian voices" on the album.[50] An online review by a Norwegian Enigma fan, keeper of one of the Enigma webpages on the Internet, said that "this track cleverly blends the joik (Lapp chant) with modern rhythms and song structure. The joik is used as the chorus. This track gives me a feeling of pleasure and happiness, and some of the reason is that Enigma has turned to the ancient Nordic musical culture, the Lapps living in the northern parts of Norway, Sweden, Finland and Russia."[51] (I should point out that the preferred term for "Lapp" is now Saami.) A Finnish Internet user also thought that the music was joik.

But the song is perhaps most frequently heard as Native American. The video that accompanied the song uses images of "Indians in some tropical jungle," as one fan writes.[52] In a class discussion, a student presented this sampled music as Native American, and it has been used as a Native American song on television and in films. One Enigma fan, who claimed to be "part Native American," heard the song as Native American. And "Return to Innocence" appears in the Jonathan Taylor Thomas/Chevy Chase film *Man of the House*, a 1995 Disney release about a boy and a man attempting to bond while in the Indian Guides together.

This scramble for attribution provides one example of what this glocalized/informationalized world is bringing. Information may be moving about, but it is not always true or accurate. The Internet is essentially a giant word-of-mouth network, which means that ascertaining "truth" can be problematic, difficult, or impossible. Anyone who paid attention to the Pierre Salinger fiasco (in which the former White House press secretary claimed that a U.S. Navy missile brought down TWA flight 800 over Long Island, based on "proof" obtained on the Internet) knows what I mean. Salinger clearly approached Internet-disseminated texts as journalistic sources (i.e., more conventional texts), only to discover that he had made a rather large mistake.[53] For ethnographers and fellow travelers like myself, this is less of a problem, since we are interested less in "truth" than in (re)presentations of truth. But we are also information gatherers, and more than once I have felt stymied by the absence of information—or, just as frequently, the welter of contradictory information—about a particular musician, recording, incident, or what have you.

This is not to say that ours is the Misinformation Age but that the rapid movement of information around the planet does result in mistakes and, sometimes, bizarre forms of relativization. As an example of the latter, take Microsoft's Encarta encyclopedia CD-ROM. The most recent edition (as of this writing in 1999) was issued in nine versions, each of which was aimed at a particular regional/linguistic market. In the American version, Alexander Graham Bell invented the telephone; in the Italian version, it was invented by a little-known candlemaker named Antonio Meucci five years earlier. In the American Encarta, Thomas Alva Edison invented the light bulb at the same time as the Briton Joseph Swan, who is credited in the British Encarta as having been first. While such a strategy identifies Microsoft as more of a marketing corporation than one concerned with knowledge, its Encarta staff insists it is attempting to be responsive to different viewpoints. "We're not changing facts," says the editor of the U.K. edition, "we're changing emphasis."[54] But because these changes in "emphasis" can travel beyond their intended audience, others can learn of them, as the spate of publication about this new Encarta illustrates.[55]

Music

Now let's move to a discussion of Enigma's music itself. The song on the original recording of Ami music that Enigma sampled is the first of "Two Weeding and Paddyfield Songs" (called by the couples' lawyer, Emil Chang, "Jubilant Drinking Song," and on a later album, "Elders Drinking Song"—the title changes with the use) with the following text:

> Friends, we need this hard work, we the people of the land
> Let us not despise it!
> Friends, we will undertake this difficult task with joy,
> So that we may live off the fruits of our labours.
> Friends, have no fear of the difficulties, nor the burning sun,
> For we are only doing our duty![56]

Enigma doesn't manipulate the Ami song at all, save for the addition of a little reverberation. The fact that Enigma leaves this music largely unchanged points to their usage of it as a kind of artifact, not as something to be ripped apart and scattered throughout their track, as has happened in other cases. The Ami music we hear in "Return to Innocence" is clearly used not as "material" or as "local color" but rather as a largely intact sign of the ethnic/exotic unspoiled by technology or even modernity. This use of identifiably "ethnic" music samples is part of a growing trend in popular musics, and dance musics in particular, even giving rise to a new genre name: "ethnotechno."[57] Some of Enigma's fans credit Cretu with spawning this new genre with "Return to Innocence."

Enigma's song is four minutes sixteen seconds in length, and samples the Kuos' voices for over two minutes of this time, as shown in table 1. Cretu effectively takes the Kuos' pentatonic melody and undergirds it with lush, synthesized, diatonic harmonies. The song concludes with a wash in the dominant seventh that takes us to the tonic in minor, which then segues directly into the next track, entitled "I Love You . . . I'll Kill You." The stop-and-start quality of the original Ami music is echoed by the drum track in the Enigma song, an effect that compels the listener's attention. It also announces the song as one without a practical function, that is, to allow for dancing; you couldn't dance to it easily with the drum track starting and stopping as it does. Cretu is making another kind of point with this song, moving away from dance and physical pleasure toward something more introspective. The beat isn't fodder for discotheque music here; it is recoded by Cretu as something primal and timeless, in keeping with the partly "spiritual" orientation of the album, and the band's style and image more generally.

What Enigma fans seem to like about the song is its homogeneity, its consistency, and its refusal to make the Kuos' music markedly distinct from

TABLE 1. Enigma: "Return to Innocence," Lyrics and Events

Time	Lyrics/Event
0:00	[Kuo Ying-nan sampled voice]
0:26	Love—Devotion
	Feeling—Emotion
0:49	Don't be afraid to be weak
	Don't be too proud to be strong
	Just look into your heart, my friend
	That will be the return to yourself
	The return to innocence.
1:09	[Kuo Ying-nan sampled voice]
1:29	The return to innocence.
1:31	If you want, then start to laugh
	If you must, then start to cry
	Be yourself, don't hide
	Just believe in destiny.
1:42	Don't care what people say
	Just follow your own way
	Don't give up and use the chance
	To return to innocence.
1:53	[Kuo Ying-nan sampled voice]
2:08	[Kuo Hsin-Chu joins sample]
2:26	[drum track stops]
2:28	*Spoken:*
	That's not the beginning of the end
	That's the return to yourself
	The return to innocence.
2:37	[drum track starts]
2:59	[both Kuos sampled]
3:53	[drum track stops]
3:56	*Spoken:*
	That's the return to innocence.
4:02	[V^7, modulation to minor]
[4:16/0:00]	[beginning of next track]

the music by Enigma. Enigma experts and fans have commented on the simple formal structure of the song—the Norwegian webpage owner mentioned above calls it their most traditional song—it is a verse-chorus arrangement, with the Ami music serving as the chorus. Some fans like this simplicity, but others find it too simple, too traditional. Other fans believe it to be conservative because the vocals are too intelligible—in other words, too much like what they derisively call "pop music."

The similar version by the New Formosa Band, however, sounds like pop music. It's called "Song of Joyous Gathering," makes use of an Amis folk song from an indigenous group in Taiwan, and is sung or spoken in several dialects, a folk dialect as well as the Holo Minnan and Hakka Kejia dialects. The mention in verse 5 of the Chinese mainland refers to those Chinese who followed Chiang Kai-shek in 1949; the song is generally about setting aside differences. Table 2 includes the lyrics with musical events.

The main similarities with Enigma's song concern the use of the folk music as a chorus, though the New Formosa Band sings the chorus themselves; it isn't sampled. The message of unity is made partly in this way, but in other ways as well. The song is almost a study in the possibilities of harmonious, egalitarian combinations and juxtapositions. First, the two singers alternate between the Minnan and Kejia dialects; second, at 3:12 and 5:03 the folk song is sung first in Minnan and then in the folk dialect, making the earlier idea of juxtaposing languages even more localized; third, the folk song is mixed with the music of verse three; and fourth, the folk song itself is harmonized in thirds for the first two bars (see fig. 1).

The New Formosa Band has recently released a collection that introduces a new member of the band. The track on the collection that they highlight is the one I have just analyzed, but this time in a new guise. It's a remix of the earlier version, but, despite that, it's described as a dance tune with world music rhythms, and "world music" is written in English. (There is virtually no English anywhere else, save production and copyright information and

Fig. 1. New Formosa Band: "Song of Joyous Gathering," folk melody.

TABLE 2. New Formosa Band: "Song of Joyous Gathering," Lyrics/Events

Time	Verse	Dialect	Lyrics/Event
0:00			[synth fades in]
0:17		[folk]	spoken words [probably Amis folk song words]
0:52			[drum track starts]
1:01	1		[Amis folk song]
1:21	2	Holo Minnan	There will sometimes come softly within your heart the sound of a beautiful melody A very gentle melody Like the white clouds leaning against a mountain slope A lover's dream of his beloved
1:41	3	Hakka Kejia	Just like how you care for your children, use that love to treasure this island of water, mountains, and the clear four seasons We are all one family
1:51	4	Holo Minnan	Remember to let loose and move your footsteps Happy times cannot be delayed [repeat]
2:01			[fragments of folk song]
2:11			[repeat folk song]
2:31	5	Hakka Kejia	No matter if you're Taiwanese, from the mainland China, an indigenous person, or a Hakka Kejia, pray to Heaven that it will protect this land and the people Safe for all generations
2:41	6	Holo Minnan	No matter if you're Taiwanese, from the mainland China, an indigenous person, or a Hakka Kejia, we will dance tonight Songs do not distinguish between you and me
2:52	7	Hakka Kejia	[same as Verse 3; sung over folk song]

3:01	8	Holo Minnan	[sung over folk song]
			  Let us sing together
			  enchanting melodies
			  Tonight we are fated to be together as one family
			  [repeat]

TABLE 2. *(continued)*

Time	Verse	Dialect	Lyrics/Event
3:12		Folk	[Amis folk song]
		Holo Minnan/	[Holo Minnan]: Let's not fight anymore [folk dialect]
		Folk	[Holo Minnan]: Be my partner [folk dialect]
3:11			[folk song (men only)]
3:21			[folk song with female singer]
3:41			[drum track stops; birdcalls; synth portamenti]
4:01			[drum track starts]
4:22		Holo Minnan	We'll sing together, enchanting melodies Happy time cannot be delayed
4:34	9	Hakka Kejia	Hearing the songs coming from the mountains, I hold in my tears That call from far away times reminds us not to forget the words of our ancestors
4:44	10	Holo Minnan	[same as verse 6]
4:54	11	Hakka Kejia	Just like how our dear parents are always by our sides, giving us their warmth and love, sacrificing day and night, doing everything to protect us Be my partner, this is our homeland
5:03			[Holo Minnan]: Let's not fight anymore [folk dialect] [Holo Minnan]: Be my partner [folk dialect]
5:24			[Amis folk song as original]
5:34			[folk song harmonized with thirds]
5:44			[folk song combined with verse 3]
6:04			[folk song repeated]
6:13			[folk song harmonized]
6:26			[folk song; fades]
6:43			[End]

I would like to thank Yi-Fawn Lee and Rebecca Fan for help with the translations.

descriptions of two other songs, one as "Acid Jazz" and the other as "Techno.") This remix version cuts a few parts of the original, but the main difference is the addition of a drum track that sounds much more hip hop–influenced than the earlier version.

The primary significance of the New Formosa Band's music, for my purposes here, is that it demonstrates the ways in which musics are increasingly caught up in the global flow of sounds, images, ideas, and ideologies made possible by digital technology. Even though this Taiwanese band doesn't expressly address Enigma's appropriation of Ami music, they nonetheless critique the blockbuster German band both by asserting a native perspective and working with an indigenous performer, and by scrupulously sharing the spotlight in the song.

Since the onset of this controversy, Difang (Kuo Ying-nan's Ami name) released the promised recording on Magic Stone in 1998. *Circle of Life* (the title is in English) shows Kuo on the cover (see fig. 2) and features many traditional songs sung by the Kuos. This album topped the charts in both Taiwan and Japan.

The songs on *Circle of Life* were mixed in the studio with drum machines and synthesizers and sound much like Enigma's "Return to Innocence." But the difference, of course, is that all of this was done with the Kuos' knowledge, permission, and cooperation. And as a way of further critiquing Enigma's treatment of them, the penultimate track is a version of the song that Enigma sampled that is far less intrusive. The final track is the original version of their song without any added studio sounds at all.

The Ami music in *Circle of Life* was recorded in a studio in Taipei. The resulting tapes were then sent to the Belgian musician and producer Dan Lacksman, who had produced the album *Deep Forest*. The resulting band of the same name is Enigma's main competitor in the realm of ethnotechno/ New Age pop.[58] Lacksman also released his own ethnotechno album in 1996, entitled *Pangea*. He was reportedly recruited via the Internet and has said that *Circle of Life* represents a crossover between traditional and contemporary music.[59] Lacksman's contribution helps explain the contemporary electronic sound of the album, which may also explain its popularity in Japan, where it was one of the topselling world music albums.

But Lacksman's presence on *Circle of Life* also helps illustrate the circle of musical sounds possible under new regimes of glocalization. Lacksman, who credits only one sample on all of *Pangea* (presumably because that's the only one the copyright holders were likely to hear), occupies a different structural relationship within the music industry for *Circle of Life* but at the same time lends his prestige to it, for the words "Deep Forest" (in English) appear on a cardboard slip on the cover.

Fig. 2. Difang: *Circle of Life,* cover.

That cover was reportedly designed by a Japanese magazine that had sent people to visit Kuo for an interview: "They were so touched by the Ami singers' songs that they volunteered to do the photography and design work for the record."[60] Kuo thus appears alone on the cover—a true world music star—surrounded by a sea of green grass that situates him and his music in the realm of the "natural," thus justifying Lacksman's refinement of this natural musical resource. Photographs inside, and even the picture on the CD itself, continue this theme. The Kuos and the other singers are photographed in their natural habitat, completely exoticized.

This self-exoticization is abetted by the liner notes, which begin in the form of a fable (in fact, the first line is "This is a fable").[61]

> There was a great eagle, circling in the sky for several generations. Its eyes gazed relentlessly on a flock of people on the ground who had, for many generations, been worshiping the eagle. They night and day, ceaselessly, sung the legends of the eagle. Because of this the eagle was immortal.
> But with the passage of time, little by little, the great eagle was no longer able to hear the singing of these ballads. They were being replaced by a catastrophic flood

of love stories and wild, violent cryings. The large eagle lost its direction of flight. From then on, the eagle disappeared.

Fortunately, there are still people on the ground who remember those songs, those brave legends—the honesty and purity passed down for generations. Like prophets, they continued to sing and pass them on. But the ignorant have viewed it as a new sound, causing at first apprehension and fear. Later, they blindly followed, plagiarizing. The value of those singing (and passing on) the songs was overlooked.[62]

The notes then continue to tell Kuo Ying-nan's story and also relate the story of Enigma's appropriation of their song, though only obliquely mentioning the lawsuit by referring to "an explosion of international controversy over the authorship and rights of native peoples' music."

While it may seem as though the opening of the liner notes perpetuates an old notion of the natural primitive, writing the notes thus has another, clever, aspect, for this style allows Enigma to be accused of plagiarism, a charge that, in the absence of a settlement of the lawsuit, could not be straightforwardly made at the time of the release of the recording. The musical-rhetorical strategy employed by Cretu in "Return to Innocence" to convey feelings of mysticism and timelessness has been used against him to advance the Kuos' and other Ami viewpoints.

The Enigma and Kuo story is illustrative of older "globalizations" competing with this newer "glocalization," facilitated mainly by new digital technologies of communication and the dissemination of (mis)information in the "infoscape."

Enigma's success with "Return to Innocence" and the entire *Cross of Changes* album has awakened the music industry to the potential of "indigenous cultures," to which royalties are almost never paid. This, to recall Wallerstein, is definitely cheap labor. Roger Lee, senior marketing director of Sony Music Taiwan, persuaded Sony's huge worldwide hit band, Deep Forest, to sample music from the Ami, as they did on their 1995 album *Boheme*, which won the Grammy award for best world music album in 1995.[63] (This album is also popular with Enigma fans, judging by the response on the Enigma Internet mailing list). According to Lee, "This era has revealed the infinite business potential of indigenous culture. We shouldn't just passively go with the flow of the predominant cultural mechanism. Mainstream needs to be countered by non-mainstream, and any non-mainstream influence may turn out to be tomorrow's mainstream."[64] The kind of appropriation Lee is advocating has roots, of course, in much older appropriations, as pointed out above, and such a statement points to the kinds of old and continuing problems currently occurring under the rubric of globalization, or, for that matter, postmodernism.

The term globalization can hide old forms of exploitation dressed up in contemporary business language like Roger Lee's. Capitalism in this global/informational economy is finding new ways of splitting sonic signifiers from their signifieds and from their makers in a process Steven Feld has called "schizophonia." This newer phenomenon of "glocalization" helps us understand the ways that there may be at the same time new forms of resistance to this process.[65] Digitized sounds move to the centers in ways they didn't before, but, for the first time, the original makers of these musical signs are finding ways of bringing them back home again.

Notes

An earlier version of this essay appears in Timothy D. Taylor, *Strange Sounds: Music, Technology and Culture* (New York: Routledge, 2001).

1. Manuel Castells, *The Rise of the Network Society*, vol. 1 of *The Information Age: Economy, Society and Culture* (Cambridge, Mass.: Blackwell, 1996), 66.
2. Arjun Appadurai, *Modernity at Large: Cultural Dimensions of Globalization*, Public Worlds, vol. 1 (Minneapolis: University of Minnesota Press, 1996).
3. This term is derived from Paul Virilio's idea of the "infosphere," a kind of information-scape that he believes will assume biological proportions in the near future (Virilio, *Open Sky*, trans. Julie Rose [London and New York: Verso, 1997]), 84.
4. Doug Henwood, "Post What?" *Monthly Review* 48 (September 1996), 6–7.
5. Jerry H. Bentley, *Old World Encounters: Cross-Cultural Contacts and Exchanges in Pre-Modern Times* (New York: Oxford University Press, 1993).
6. Ibid., 5.
7. Ibid., 26–27.
8. Immanuel Wallerstein, *The Modern World-System: Capitalist Agriculture and the Origins of the European World-Economy in the Sixteenth Century*, Studies in Social Discontinuity (New York: Academic Press, 1974), 102.
9. For more on this point, see Timothy D. Taylor, *Global Pop: World Music, World Markets* (New York: Routledge, 1997).
10. Stuart Hall, "The Local and the Global: Globalization and Ethnicity," in *Culture, Globalization and the World-System*, ed. Anthony D. King, Current Debates in Art History, vol. 3 (Binghamton, N.Y.: Department of Art and Art History, State University of New York at Binghamton, 1991), 34.
11. Timothy Brennan, *At Home in the World: Cosmopolitanism Now*, Convergences: Inventories of the Present (Cambridge, Mass.: Harvard University Press, 1997).
12. Immanuel Wallerstein, *Historical Capitalism* (London: Verso, 1983), 39. Some have argued that global expansion has been driven not by the search for cheap labor but new markets (see, most famously, V. I. Lenin, *Imperialism, the Highest Stage of Capitalism: A Popular Outline* [New York: International, 1939]).
13. See, for just two examples, Jocelyne Guilbault, "On Redefining the 'Local' through World Music," *World of Music* 32 (1993): 33–47; and Rob Wilson and Wimal Dissanyake, eds., *Global/Local: Cultural Production and the Transnational Imaginary*, Asia-Pacific: Culture, Politics, and Society (Durham, N.C.: Duke University Press, 1996).
14. Roland Robertson, "Globalisation or Glocalisation?" *Journal of International Communication* 1 (1994): 33. See also Roland Robertson, "Glocalization:

Time-Space and Homogeneity-Heterogeneity," in *Global Modernities,* ed. Mike Featherstone et al., Theory, Culture and Society (London and Thousand Oaks, Calif.: Sage, 1995). Virilio, *Open Sky,* also uses the term. For just two examples of the term in business discourse, see Christopher Conte, "A Special News Report on People and Their Jobs in Offices, Fields and Factories," *Wall Street Journal,* May 21, 1991, sec. A, p. 1; and Martha H. Peak, "Developing an International Style of Management," *Management Review,* February 1991, 32–35; a recent scholarly article that considers the term is Marwan M. Kraidy, "The Global, the Local, and the Hybrid: A Native Ethnography of Glocalization," *Critical Studies in Mass Communication* 16 (December 1999): 456–76. For other uses of the term, as well as alternatives to it, see Philip Hayward, "Cultural Tectonics," *Convergence* 6 (spring 2000): 39–47. Lastly, see Timothy D. Taylor, "World Music in Television Ads," *American Music* 18 (summer 2000): 162–92, for a discussion of American and European business attitudes toward the global and the local.

15. For a discussion of the false dichotomy of "global" and "local," see Charles Piot, *Remotely Global: Village Modernity in West Africa* (Chicago: University of Chicago Press, 1999). I would like to thank Louise Meintjes for this reference.

16. "IOC President to Thank Ami Singers," *Free China Journal,* <http://ww3.sinanet.com/heartbeat/fcj/0726news/16_E.html>. This URL is no longer active.

17. Pierre Bois, email communication, March 29, 1999. He reiterates this rather defensively in a later email, April 7, 1999.

18. By "New Age" I'm referring to a middle-class rejection of mainstream religions and a turn to other forms of spirituality. See Wouter J. Hanegraaff, *New Age Religion and Western Culture: Esotericism in the Mirror of Secular Thought,* Studies in the History of Religions, vol. 72 (Leiden: Brill Academic Publishers, 1996); Paul Heelas, *The New Age Movement: The Celebration of the Self and the Sacralization of Modernity* (Cambridge, Mass.: Blackwell, 1996); and Deborah Root, "Conquest, Appropriation, and Cultural Difference," chap. 3 in *Cannibal Culture: Art, Appropriation, and the Commodification of Difference* (Boulder: Westview Press, 1996). I should note that there has been a good deal of discussion on the Enigma Internet mailing list concerning the New Age category; some liked it, but the majority didn't.

19. Enigma FAQ, <http://www.spikes.com/enigma/faq/faq5.htm>. All quotations from the Internet appear with their original spelling and punctuation unless otherwise indicated.

20. Enigma Live Chat Event on the Internet, December 13, 1996.

21. Renata Huang, "Golden Oldie," *Far Eastern Economic Review,* November 2, 1995, 62.

22. Ashley Esarey, "An Ami Couple Seeks Recognition for Their Music," <http://www.sinica.edu/tw/tit/special/0996_Innocence.html>. This URL is no longer active.

23. There is also another story that Kuo tells: "Two years ago, my granddaughter brought a tape home and played me the song on the Enigma record. That's the first time I heard that Enigma had used my voice. I was very surprised and happy. It felt good to have people using my voice, but I was also surprised because I never sang such a song with all those other sounds, I wondered how it was made" (Lifvon Guo [Kuo Ying-nan], interview by Frank Kohler, *All Things Considered,* National Public Radio, June 11, 1996).

24. "Olympic Music from Taiwan," <http://ourworld.compuserve.com/homepages/smlpp/enigma.htm>. This URL is no longer active. "Return to Innocence" has appeared on many other compilations, however, including *Dance Mix USA,* vol. 2, Quality Music 3902, n.d.; *Dance Mix USA,* vol. 3, Quality Music 6727, 1995; *First Generation: 25 Years of Virgin,* Virgin 46589, 1998; *High on Dance,* Quality Music 6741, 1995; *Loaded,* vol. 1, EMI America 32393, 1995; and *Pure Moods,* Virgin 42186, 1997.

25. "Ami Sounds Scale Olympian Heights," <http://www.gio.gov.tw/info/sinorama/8508/508006e1.html>. This URL is no longer active.

26. Ibid.

27. "From Betelnuts to Billboard Hits," <http://pathfind.com/@@CkDc-MAcAoVE9Uaiw/Asiaweek/96/1122/feat5.html>. This URL is no longer active. The new recording, called *Circle of Life* (Magic Stone MSD030), was released in October of 1998.

Lost in the shuffle in the lawsuits and new recordings are the other singers on this original Ami song, Panay, Afan, and Kacaw, according to "True Feelings from the Bosom of Nature," *Sinorama Magazine,* <http://www.gio.gov.tw/info/sinorama/8508/5080/161e.html>. This URL is no longer active.

28. "IOC President to Thank Ami Singers."

29. "Ami Sounds Scale Olympian Heights."

30. "From Betelnuts to Billboard Hits."

31. "Ami Sounds Scale Olympian Heights."

32. "Olympic Music from Taiwan." This is the figure also reported on the Internet newsgroup <misc.activism.progressive> by someone who seems to be an activist on behalf of the Kuos.

33. "Ami Sounds Scale Olympian Heights."

34. Deborah Kuo, "Taiwan Aborigines Sue Enigma, Music Companies," Central News Agency, March 28, 1998; and "Taiwan Couple Sue Enigma Over Vocals on International Hit," Associated Press, March 27, 1998.

35. "dcd" refers to the Australian band Dead Can Dance.

36. E. Patrick Ellisen, telephone communication, February 16, 1999.

37. E. Patrick Ellisen, telephone communication, April 21, 1999.

38. Quoted in Huang, "Golden Oldie," 62.

39. Brenda Sandburg, "Music to Their Ears," *Recorder,* June 24, 1999, 1. See also Victor Wong, "Taiwan Aboriginal Singers Settle Copyright Lawsuit," *Billboard,* July 31, 1999, 14.

40. The Hakka people were a migratory group who were persecuted at various times by the native peoples in whose territories they settled.

41. From <http://www.mpath.com/~piaw/bunny/shinbao.htm>. This URL is no longer active.

42. It is included on the soundtrack to the film *Chinese Box,* Blue Note 93285, 1998.

43. This is a common discourse by Western musicians and fans about non-Western musics—they are simply raw material for the genius of the Western star. See, for example, Paul Simon's discussion of his work with black South African musicians on *Graceland* in the documentary *Paul Simon: Born at the Right Time* (Warner Reprise Video, 1992), in which he clearly narrates himself as an adventurer/explorer going into the heart of darkest Africa to bring back natural, unrefined materials that he transforms into something valuable and worthy. See also Timothy D. Taylor, *Global Pop,* chap. 1, for a more general discussion of this assumption.

44. Enigma FAQ, <http://www.spikes.com/enigma/>.

45. Quoted in Sally Price, *Primitive Art in Civilized Places* (Chicago: University of Chicago Press, 1989), 100.

46. <http://www.spikes.com/enigma/Mambo1.html>. This URL is no longer active.

47. Ibid. Cretu and Enigma were sued for a sample on their earlier recording. An interview with Cretu published in Norway in 1990 says that "the Gregorian church chant is recorded in Rumania, and [Cretu] answers frenetically affirmative when we wonder if the singing monks have received their share of all the D-marks he gets" (Catharina Jacobsen, "Michael's Mystical Music," *Verdens Gang* (Oslo,

Norway), December 19, 1990 [trans. Joar Grimstvedt], available at <http://www.stud.his.no/~joarg/Enigma/articles/verdensGang1290.html>). But it later transpired that the recording was made in the 1970s by the Munich-based choir Kapelle Antiqua, which recognized its recording of chant sampled on the Enigma album. Even though this recording is in the public domain (and it is not clear why), the group sued, claiming that Cretu had infringed on its "right of personality." The group settled out of court. *Billboard*'s report on the suit says that "it is understood that the bulk of the money paid to Kapelle Antiqua is in recognition of the infringement of its 'right of personality.' Lesser sums have been paid to the record companies Polydor and BMG/Ariola for the unauthorized use of master recordings" (Ellie Weinert, "'Sadeness' Creator Settles Sample Suit," *Billboard,* September 14, 1991, 80).

48. Simon Frith, "Music and Morality," in *Music and Copyright,* ed. Simon Frith, Edinburgh Law and Society Series (Edinburgh: Edinburgh University Press, 1993), 8.

49. Ellie Weinert, "'Changes' in Works for Enigma," *Billboard,* January 8, 1994, 10.

50. Dominic Pride, "Virgin Stays with Proven Marketing for Enigma," *Billboard,* November 23, 1996, 1.

51. Review of *The Cross of Changes,* <http://www.hyperreal.com/music/epsilon/reviews/enigma.cross>. This URL is no longer active.

52. "Olympic Music from Taiwan."

53. See Douglas Rushkoff, "Conspiracy or Crackpot? Cyberlife US," *Manchester Guardian,* November 14, 1996, 13.

54. Paul Fisher, "Connected: Log and Learn Encyclopedia," *Daily Telegraph* (London), December 11, 1999, 10.

55. Here are just some of the newspaper reports on the Encarta discrepancies: Charles Arthur, "How Many People Does It Take to Invent the Lightbulb?" *Independent* (London), June 26, 1999, 13; Kevin J. Delaney, "Microsoft's Encarta Has Different Facts for Different Folks," *Wall Street Journal,* June 25, 1999, A1; and Stephen Moss, "Your History Is Bunk, My History Is Right," *Manchester Guardian,* June 29, 1999, section G2, p. 4.

56. Liner notes to *Polyphonies Vocales des aboriginès de Taïwan,* Inedit, Maison des Cultures du Monde, W 2609 011, 1989.

57. These "ethnotechno" musics generally display a different attitude toward the sampled material, however. It's usually less foregrounded in the mix; the samples tend to be shorter; and the samples are not usually put to New Age ends, as in Enigma's song. See Taylor for a discussion of the differences between these electronic musics and their uses of samples. For more on "ethnotechno," see Erik Goldman, "Ethnotechno: A Sample Twist of Fate," *Rhythm Music,* July 1995, 36–39; Josh Kun, "Too Pure?" *Option,* May/June 1996, 54; and Jon Pareles, "A Small World After All. But Is That Good?" *New York Times,* March 24, 1996, H34.

58. Nancy Guy, email communication, March 29, 1999. See also Guy's "Techno Hunters and Gatherers: Taiwan's Ami tribe, Enigma's 'Return to Innocence' and the Legalities of Cultural Ownership," paper presented at the Society for Ethnomusicology, Bloomington, Indiana, October 22, 1998.

59. Juping Chang, "Ami Group Sings of Bittersweet Life in the Mountains," *Free China Journal,* January 1, 1999, <http://publish.gio.gov.tw/FCJ/fcj.html>. This URL is no longer active.

60. Ibid.

61. I am very grateful to Dale Wilson for providing a translation of the liner notes. For sources on the concept of self-exoticization and self-Orientalism, see Shuhei Hosokawa, "Soy Sauce Music: Harumi Hosono and Japanese Self-Orientalism," in

Widening the Horizon: Exoticism in Post-War Popular Music, ed. Philip Hayward (Sydney: John Libby, 1999); Koichi Iwabuchi, "Complicit Exoticism: Japan and its Other," *Continuum* 8 (1994): 49–82; and Joseph Tobin, *Re-Made in Japan: Everyday Life and Consumer Taste in a Changing Society* (New Haven: Yale University Press, 1992).

62. Liner notes to Difang, *Circle of Life,* Magic Stone Music MSD-030, 1998.

63. *Deep Forest* sampled music from all over Africa. Released in Europe in 1992, it proved to be one of the bestselling albums on college campuses in 1993–94, selling over 1.5 million copies in the United States alone. For more on Deep Forest, see Carrie Borzillo, "Deep Forest Growing in Popularity," *Billboard,* February 19, 1994, 8, and "U.S. Ad Use Adds to Commercial Success of *Deep Forest,*" *Billboard,* June 11, 1994, 44; Steven Feld, "Pygmy Pop: A Genealogy of Schizophonic Mimesis," *Yearbook for Traditional Music* 28 (1997): 1–35, and "A Sweet Lullaby for World Music," *Public Culture* 12 (winter 2000): 145–71; René T. A. Lysloff, "Mozart in Mirrorshades: Ethnomusicology, Technology, and the Politics of Representation," *Ethnomusicology* 41 (spring/summer 1997): 206–219; Andrew Ross, review of *Deep Forest, Artforum* 32 (December 1993): 11–13; the extensive FAQ at <http://www.spikes.com/worldmix/faq.htm>; Hugo Zemp, "The/An Ethnomusicologist and the Record Business," *Yearbook for Traditional Music* 28 (1997): 36–56; and <http://www.music.sony.com/Music/ArtistInfo/DeepForest.html>.

Deep Forest took Lee's advice. Wind Records was established in Taiwan to preserve traditional musics. But it was one of their collections that found its way onto *Boheme,* on which Deep Forest sampled a selection from a CD entitled *The Songs of the Yami Tribe,* vol. 3 of *The Music of Aborigines on Taiwan Island,* Wind Records, TCD-1503, 1993. According to the Deep Forest FAQ, the Ami sample on *Boheme* is "A Recitative for Describing Loneliness," a title so evocative that it would seem hard to pass up (the Deep Forest FAQ is at <http://www.spikes.com/worldmix/faq.htm>). The Yami are a different group than the Ami; it isn't a different transliteration. The program note accompanying the original Yami recording says that "A Recitative for Describing Loneliness" employs a scale uncommon in Yami music and is borrowed from the Ami. But to Roger Lee, it seems that an aborigine is an aborigine: Can I give you Ami? No? Yami? As usual, this appropriative act by Deep Forest becomes converted into a sales tactic. Next to the entry in Wind Records' catalog for Yami recordings, there is a little blurb mentioning that "A Recitative for Describing Loneliness" was "excerpted" on Deep Forest's *Boheme.*

The sampled song is "Marta's Song," Marta being the well-known Hungarian singer Marta Sebestyén. "A Recitative for Describing Loneliness" flits through the background when Sebestyén isn't singing her folk song (beginning at 1:45). A Hungarian/Transylvanian reader of the Enigma Internet mailing list informs me that this is a traditional song about a woman who is lamenting being pregnant and alone.

64. Anita Huang, "Global Music, Inc.," <http://www.gio.gov.tw/info/fcr/8/p42.htm>. This URL is no longer active. See also a more recent report on the growing popularity of music with an "Aboriginal flair" in Taiwan: Victor Wong, "Taiwan's Power Station Brings Aboriginal Flair to What's Music," *Billboard,* July 11, 1998, 48.

65. Steven Feld, "From Schizophonia to Schismogenesis: On the Discourses and Commodification Practices of 'World Music' and 'World Beat,'" in Charles Keil and Steven Feld, *Music Grooves: Essays and Dialogues* (Chicago: University of Chicago Press, 1994).

References

DISCOGRAPHY
Chinese Box, Blue Note 93285, 1998.
Dance Mix USA, vol. 2. Quality Music 3902, n.d.
Dance Mix USA, vol. 3. Quality Music 6727, 1995.
Deep Forest. *Boheme*. 550 Music/Epic BK-67115, 1995.
——. *Deep Forest*. 550 Music/Epic BK-57840, 1992.
Difang. *Circle of Life*. Magic Stone Music MSD030, 1998.
Enigma. *MCMXC A.D.* Capitol 86224, 1991.
Enigma 2. *The Cross of Changes*. Charisma/Virgin 7243 8 39236 2 5, 1993.
First Generation: 25 Years of Virgin. Virgin 46589, 1998.
High on Dance. Quality Music 6741, 1995.
Lacksman, Dan. *Pangea*. Eastwest World 61947-2, 1996.
Loaded, vol. 1. EMI America 32393, 1995.
The Music of Aborigines on Taiwan Island, vol. 3, *The Songs of the Yami Tribe*. Wind Records: TCD-1503.
New Formosa Band. *Best Live & New Remix*. RD-1345, 1996.
Polyphonies Vocales des Aborigines de Taïwan. Inedit, Maison des Cultures du Monde, W 2609 011, 1989.
Pure Moods. Virgin 42186, 1997.

FILMOGRAPHY
Paul Simon: Born at the Right Time. Warner Reprise Video, 1992.

REFERENCES TO SITES ON THE INTERNET
ftp://ftp.best.com/pub/quad/deep.forest/DeepForest-FAQ.txt
http://www.hyperreal.com/music/episilon/reviews/enigma.cross
http://www.mpath.com/~piaw/bunny/shinbao.htm
http://www.music.sony.com/Music/ArtistInfo/DeepForest.html
http://www.spikes.com/enigma/Mambo1.html
http//www.spikes.com/worldmix/faq/

UNPUBLISHED MATERIALS
Bois, Pierre. Personal communications. 26 March 1999 and 7 April 1999.
Ellisen, E. Patrick. Personal communications. 21 April 1999, 16 February 1999, and 22 February 1999.
Guy, Nancy. Personal communication. 29 March 1999.
——. "Techno Hunters and Gatherers: Taiwan's Ami tribe, Enigma's 'Return to Innocence' and the Legalities of Cultural Ownership." Paper presented at the Society for Ethnomusicology, Bloomington, Indiana, 22 October 1998.
Kohler, Frank. Interview with Lifvon Guo [Kuo Ying-nan]. National Public Radio, *All Things Considered*, 11 June 1996.

MATERIALS AVAILABLE ONLY ON THE INTERNET
Enigma Live Chat Event on the Internet, 13 December 1996.

BIBLIOGRAPHY
"Ami Sounds Scale Olympian Heights." Part 1. http://www.gio.gov.tw/info/sinorama/8508/508006e1.html.
"Ami Sounds Scale Olympian Heights." Part 2. http://www.gio.gov.tw./info/sinorama/8508/508006e2.html.
Appadurai, Arjun. *Modernity at Large: Cultural Dimensions of Globalization*. Public

Worlds, edited by Dilip Goankar and Benjamin Lee, no. 1. Minneapolis: University of Minnesota Press, 1996.
Arthur, Charles. "How Many People Does It Take to Invent the Lightbulb?" *Independent* (London), June 26, 1999, 13.
Bentley, Jerry H. *Old World Encounters: Cross-Cultural Contacts and Exchanges in Pre-Modern Times.* New York: Oxford University Press, 1993.
Borzillo, Carrie. "Deep Forest Growing in Popularity." *Billboard,* February 19, 1994, 8.
———. "U.S. Ad Use Adds to Commercial Success of *Deep Forest.*" *Billboard,* June 11, 1994, 44.
Brennan, Timothy. *At Home in the World: Cosmopolitanism Now.* Convergences: Inventories of the Present. Cambridge, Mass.: Harvard University Press, 1997.
Castells, Manuel. *The Rise of Network Society.* Vol. 1 of *The Information Age: Economy, Society and Culture.* Cambridge, Mass.: Blackwell, 1996.
Chang, Juping. "Ami Group Sings of Bittersweet Life in the Mountains." *Free China Journal,* January 1, 1999. http://publish.gio.gov.tw/FCJ/fcj.html.
Conte, Christopher. "A Special News Report on People and Their Jobs in Offices, Fields and Factories." *Wall Street Journal,* May 21, 1991, sec. A, p. 1.
Delaney, Kevin J. "Microsoft's Encarta Has Different Facts for Different Folks." *Wall Street Journal,* June 25, 1999, A1.
Enigma FAQ. http://www.stud.his.no/~joarg/.
Esarey, Ashley. "An Ami Couple Seeks Recognition for Their Music." http://www.sinica.edu/tw/tit/special/0996_Innocence.html.
Feld, Steven. "From Schizophonia to Schismogenesis: On the Discourses and Commodification Practices of 'World Music' and 'World Beat.'" In Charles Keil and Steven Feld, *Music Grooves: Essays and Dialogues.* Chicago: University of Chicago Press, 1994.
———. "Pygmy Pop: A Geneaology of Schizophonic Mimesis." *Yearbook for Traditional Music,* 28 (1997): 1–35.
———. "A Sweet Lullaby for World Music." *Public Culture* 12 (winter 2000): 145–71.
Fisher, Paul. "Connected: Log and Learn Encyclopedia." *Daily Telegraph* (London), December 11, 1999, 10.
Frith, Simon. "Music and Morality." In *Music and Copyright,* edited by Simon Frith. Edinburgh Law and Society Series. Edinburgh: Edinburgh University Press, 1993.
"From Betelnuts to Billboard Hits." http://pathfind.com/@@CkDcMAcAoVE9Uaiw/Asiaweek/96/1122/feat5.html.
Goldman, Erik. "Ethnotechno: A Sample Twist of Fate." *Rhythm Music* (July 1995): 36–39.
Guilbault, Jocelyne. "On Redefining the 'Local' Through World Music." *World of Music* 32 (1993): 33–47.
Hall, Stuart. "The Local and the Global: Globalization and Ethnicity." In *Culture, Globalization and the World-System,* edited by Anthony D. King. Current Debates in Art History, no. 3. Binghamton, N.Y.: Department of Art and Art History, State University of New York at Binghamton, 1991.
Hanegraaff, Wouter J. *New Age Religion and Western Culture: Esotericism in the Mirror of Secular Thought.* Studies in the History of Religions, no. 72. Leiden: Brill Academic Publishers, 1996.
Hayward, Philip. "Cultural Tectonics." *Convergence* 6 (spring 2000): 39–47.
Heelas, Paul. *The New Age Movement: The Celebration of the Self and the Sacralization of Modernity.* Cambridge, Mass.: Blackwell, 1996.
Henwood, Doug. "Post What?" *Monthly Review* 48 (September 1996): 1–11.

Hosokawa, Shuhei. "Soy Sauce Music: Harumi Hosono and Japanese Self-Orientalism." In *Widening the Horizon: Exoticism in Post-War Popular Music*, edited by Philip Hayward. Sydney: John Libby, 1999.

Huang, Anita. "Global Music, Inc." http://www.gio.gov.tw/info/fcr/8/p42.htm. (This URL is no longer active.)

Huang, Renata. "Golden Oldie." *Far Eastern Economic Review* (November 2, 1995): 62.

"IOC President to Thank Ami Singers." *Free China Journal,* http://ww3.sina net.com/heartbeat/fcj/0726news/16_E.html.

Iwabuchi, Koichi. "Complicit Exoticism: Japan and Its Other." *Continuum* 8 (1994): 49–82.

Jacobsen, Catharina. "Michael's Mystical Music." Translated by Joar Grimstvedt. *Verdens Gang* (Oslo, Norway), December 19, 1990, http://www.stud.his.no/~joarg/Enigma/articles/verdensGang1290.html.

Kraidy, Marwan M. "The Global, the Local, and the Hybrid: A Native Ethnography of Glocalization." *Critical Studies in Mass Communication* 16 (December 1999): 456–76.

Kun, Josh. "Too Pure?" *Option* (May/June 1996): 54.

Kuo, Deborah. "Taiwan Aborigines sue Enigma, Music Companies." Central News Agency, March 28, 1998.

Liner notes to Difang, *Circle of Life*. Magic Stone Music MSD-030, 1998.

Liner notes to *Polyphonies Vocales des Aborigines de Taïwan*. Inedit, Maison des Cultures Du Monde, W 2609 011, 1989.

Lysloff, René T. A. "Mozart in Mirrorshades: Ethnomusicology, Technology, and the Politics of Representation." *Ethnomusicology* 41 (spring/summer 1997): 206–219.

Moss, Stephen. "Your History Is Bunk, My History Is Right." *Manchester Guardian,* June 29, 1999, sec. G2, p. 4.

"Olympic Music from Taiwan." http://ourworld.compuserve.com/homepages/smlpp/enigma.htm.

Pareles, Jon. "A Small World After All. But Is That Good?" *New York Times,* March 24, 1996, H34.

Peak, Martha H. "Developing an International Style of Management." *Management Review,* February 1991, 32–35.

Piot, Charles. *Remotely Global: Village Modernity in West Africa*. Chicago: University of Chicago Press, 1999.

Pride, Dominic. "Virgin Stays with Proven Marketing for Engima." *Billboard,* November 23, 1996, 1.

Robertson, Roland. "Globalisation or Glocalisation?" *Journal of International Communication* 1 (1994): 33–52.

———. "Glocalization: Time-Space and Homogeneity-Heterogeneity." In *Global Modernities*, edited by Mike Featherstone, Scott Lash, and Roland Robertson. Theory, Culture and Society. London and Thousand Oaks, Calif.: Sage, 1995.

Root, Deborah. *Cannibal Culture: Art, Appropriation, and the Commodification of Difference*. Boulder: Westview Press, 1996.

Ross, Andrew. Review of *Deep Forest. Artforum* 32 (December 1993): 11.

Rushkoff, Douglas. "Conspiracy or Crackpot? Cyberlife US." *Manchester Guardian,* November 14, 1996, 13.

Sandburg, Brenda. "Voices in the Copyright Wilderness." *Recorder,* December 7, 1998, 1.

Taylor, Timothy D. *Global Pop: World Music, World Markets*. New York: Routledge, 1997.

———. *Strange Sounds: Music, Technology and Culture*. New York: Routledge, 2001.

———. "World Music in Television Ads." *American Music* 18 (summer 2001): 162–92.
"True Feelings from the Bosom of Nature." *Sinorama Magazine,* http://www.gio.gov.tw/info/sinorama/8508/5080/161e.html
Virilio, Paul. *Open Sky*. Translated by Julie Rose. London and New York: Verso, 1997.
Wallerstein, Immanuel. *Historical Capitalism*. London: Verso, 1983.
———. *The Modern World-System: Capitalist Agriculture and the Origins of the European World-Economy in the Sixteenth Century*. Studies in Social Discontinuity, edited by Charles Tilly and Edward Shorter. New York: Academic Press, 1974.
Weinert, Ellie. "'Changes' in Works for Enigma." *Billboard,* January 8, 1994, 10.
———. "'Sadeness' Creator Settles Sample Suit; Will Compensate for Unauthorized Usage." *Billboard,* September 14, 1991, 80.
Wilson, Rob, and Wimal Dissanyake, eds. *Global/Local: Cultural Production and the Transnational Imaginary*. Asia-Pacific: Culture, Politics, and Society. Durham, N.C.: Duke University Press, 1996.
Wong, Victor. "Taiwan's Power Station Brings Aboriginal Flair to What's Music." *Billboard,* July 11, 1998, 48.
Zemp, Hugo. "The/An Ethnomusicologist and the Record Business." *Yearbook for Traditional Music* 28 (1997): 36–56.

CHAPTER FOUR

"*Ethnic Sounds*"
The Economy and Discourse of World Music Sampling

Paul Théberge

One evening some years ago, I received a phone call from a musician—a stranger. He had heard, through a mutual acquaintance, that I had once studied African drumming, and he wanted to know if I could teach him something about the music. I told him, apologetically, that I didn't feel qualified to do so but recommended a couple of good books on the subject. Pressing further, he said that he had heard that there were specific, even secret African rhythms that, when played at the right tempo, could induce a state of trancelike ecstasy in the listener. Did I know them? Could I simply demonstrate some of them to him? I told him that I knew of no such rhythms and quickly ended our conversation.

At the time, I was struck by the impertinence of my interlocutor's questions—secret rhythms, magical powers, "voodoo" trances. Of course I was aware of the mythology surrounding African music, but my own interest had been more "pure" than that. Or had it? What was it in the music that I had responded to, that had spurred my own desire to study it? I still don't have any satisfactory answers to those questions, but I no longer find such inclinations to be quite so impertinent: indeed, I have come to regard the essentializing, exoticizing impulse evident in the questions posed to me in that phone call to be both pertinent and even central to how the music of other cultures, despite our increasing proximity to them, continues to be portrayed by musicians, listeners, critics, and, above all, the recording industry in the Western, post-industrialized world.

In this chapter, I want to explore what appear to be some of the most resilient discourses around world music, and the manner in which they promote and even justify the technological appropriation of it; specifically, I will discuss the fetishization of musical sounds that has taken place since the rise of sampling in popular music. While a number of recent studies, perhaps most notably Timothy Taylor's *Global Pop* (1997), have offered a detailed and theoretically astute account of the transformations presently taking place in global music culture, the commercialization and commodification of sound samples of world music, which must be regarded as an important aspect of this phenomenon, have scarcely been touched upon in the literature. It is this area of fragmented, commodified sounds, and the industry that supplies them, that I want to address.

The perspective on sampling that I will take here differs from that taken in much of the popular music literature, most of which has tended to focus on what is often described as the "disruptive" uses of sampling, especially in various genres of electronic dance music: for example, the spectacular collages of found sound and bits and pieces of music culled from decades-old pop recordings, the uses of sampling as an extension of turntable scratching techniques in hip-hop, and the attendant "crisis" in copyright law provoked by such practices. While these examples are important to understanding the workings of popular culture, the manner in which specific uses can redefine technology, and the nature of corporate economics and power, they do not reflect various other, equally significant uses of sampling technology that have become increasingly common within the contexts of music, film, television, and commercial production.

Within such contexts, samples are valued for their ability to economically reproduce conventional musical instrument sounds and to act as markers, in a general sense, of specific musical genres. In many instances, the samples used have a duration of only a few milliseconds and are digitally processed, looped, and programmed to be played from a keyboard or sequencer. In sample production, much emphasis is placed on the technical quality of the samples—their sound quality and playability—and, more important, the unlimited access to sound material: in the 1990s, virtually every instrument imaginable—from the European classical orchestra to the gamelans of Indonesia—has been sampled and is available for sale.

Producing and Selling Sounds

Popular music since the 1950s has become increasingly dependent upon the technologies of audio production and reproduction, and, partly in response to this overall trend, there has been a corresponding importance attached to "sound"—as both an aesthetic and a commercial category—that is at least equal to that conventionally accorded to melody and lyrics in popular music production and reception. But while marketing terms such as the "Nashville Sound" designate a complex interplay between musical style, instrumentation, and recording techniques (see Ivey 1982), the rise of digital synthesizers and samplers during the 1980s has placed a new kind of emphasis on individual instrument sounds as distinct musical entities and as independent objects of commodity exchange (for a detailed account of this development, see Théberge 1997).

Because of this emergent status in individual sounds as objects of musical style and fashion, the increasing difficulty associated with programming digital synthesizers, and changing perceptions of the user within the musical instrument industry, a small cottage industry dedicated to the supply of new sounds developed during the 1980s (ibid.). As a group of largely independent programmers and developers whose activities lie on the periphery of the electronic musical instrument industry, the scale of their operations hardly warrants the title of "industry" at all. However, with regard to digital sampling, the cost of hiring musicians and professional recording facilities alone made this area of sound development significantly different from the simple creation of sound programs for digital synthesizers (which relies on the samples and/or synthesis resources already present in the instruments).

Initially, the manufacturers of digital samplers took it upon themselves to record and release large sample libraries—first on floppy discs and, later, on CD-ROM—in support of their instruments (many having in-house recording studios of their own). Increasingly, however, manufacturers have become dependent on outside developers who have greater expertise in the ever-changing styles and genres of popular music. The Roland Corporation, for example, began assembling its own archive of sample discs on CD-ROM as early as 1988 and later entered into production, promotion, and/or distribution agreements with no fewer than eighteen third-party developers—in the United States, Germany, the United Kingdom, and elsewhere—who supply sample libraries formatted for its line of digital samplers. By the mid-1990s, over fifty Roland-formatted libraries had been produced, including discs devoted exclusively to guitar sounds, drums, or bass, and collections representing a range of musical styles and sounds, from mainstream rock to techno and from Western orchestras to the vast timbral variety of world music instruments.

Many of the independent sample developers have direct (or indirect) links with either the recording industry or the musical instrument manufacturers. Northstar Productions, for example, was established as part of

Northstar Recording, a recording studio located in Portland, Oregon. Through its original sample work, it has been able to develop a reputation (and a clientele) that is unusual for an otherwise medium-sized, regional recording facility. In contrast, a number of musicians, producers, and sound engineers have used their already established reputations in the recording industry as a means of promoting their own, specialized sample collections, among them jazz musician Miroslav Vitous, recording engineer Bob Clearmountain, and film composer Hans Zimmer. Spectrasonics, one of the most respected sample developers of the mid-1990s, was founded by Eric Persing, the former chief sound designer for Roland who was responsible for the sound programs found in many of that company's most successful digital synthesizers. Coming to the sample business from the retail end of the music industry spectrum, Sweetwater Sound, a large audio and music instrument mail-order house, produces and distributes its own collection of sample discs (along with discs by other developers) primarily for Kurzweil series samplers.

Even with the technical, artistic, and economic advantages that such industry links afford, however, the cost and logistical complexity encountered in recording discs of world music samples is considerably greater than that required for the simple production of samples of, for example, pop drum sounds and patterns. In the case of Spectrasonics' *Heart of Asia* sampling project, the producers teamed up with a commercial music production house, Schtung Music, which has studios in Hong Kong, Singapore, and Los Angeles. Schtung Music, founded in Hong Kong in 1982 by two ex-rockers from New Zealand, Morton Wilson and Andrew Hagen, and later joined by Hong Kong musician Eddie Chung, produces music for Asian television and radio advertising—numbering among its clients corporations such as Singapore Airlines, Sony, Nestle, Esso, and McDonalds—as well as soundtracks for short films. It has also launched its own record label, Schtung Records, dedicated to the recording and promotion of what it calls "Pan-Asian Hip"—a fusion of Eastern traditions with pop, jazz, and ambient music. Schtung's base in the world of radio/TV commercial production and film post-production should not be ignored: given that commercial producers often make use of prerecorded sound effects libraries (as do film soundtrack editors) and even canned music, they would no doubt be predisposed to the idea of developing and using musical instrument sample libraries; indeed, the only real historical precedent for the production and marketing of collections of individual sounds lay within the film and broadcast industries, not the music industry. With Schtung's contacts with local musicians and its technical infrastructure (linked via high-speed ISDN digital phone lines, Schtung claims its studios are "state of the art"), Spectrasonics was able to produce a collection of Asian musical instrument samples that is virtually without parallel in the world of sampling and is reputed to have achieved sales in excess of 10,000 copies (an impressive number, considering that the sample-formatted, CD-ROM version of the two-disc set sells for $399).

Distribution of prerecorded samples, or "soundware," has also evolved into a complex set of industry relations during the past decade. Many of the smaller (and even some of the larger) sample developers are now represented by larger production/distribution agencies, such as Ilio Entertainments and East-West Communications in the United States, Masterbits in Germany, and Time + Space in the United Kingdom (each having foreign offices as well). Most of the producer/distributors specialize in sample discs, but some have also branched out into music software development and, in a few cases, hardware distribution (e.g. sound boards and other computer peripherals). Such moves have brought sample distributors into closer contact with the lucrative computer and multimedia industries. East-West Communications, for example, is the exclusive distributor and primary sound developer for a software product for PCs called "GigaSampler," developed by Nemesys Music Technology, Inc., and based on technologies licensed from Rockwell Semiconductor Systems, the fast-growing multimedia and "infotainment" division of Rockwell International; Rockwell has plans to integrate its proprietary sample access and retrieval technology into a number of future audio and multimedia chipsets. Market alignments of this kind must be seen as part of a larger trend in which some of the most innovative electronic musical instrument manufacturers have become objects of corporate takeovers as the computer industry continues to expand into the area of multimedia: in recent years, both E-mu Systems and the Ensoniq Corporation, two of the most well-known designers of digital samplers and synthesizers in the United States, have been acquired by Creative Labs, Inc., makers of "Sound Blaster," the de facto standard in consumer audio for PCs.

As multimedia uses of sound samples become more prevalent, the cost of prerecorded sound samples will undoubtedly drop, but for present users in the professional and semiprofessional music markets, sample libraries do not come cheap. While a typical audio CD designed for sampling purposes may cost between $69 and $129, most sample users opt for CD-ROM discs on which the samples have been edited, looped, and programmed for the particular sampler that they use, at a price of $199 or more for a single disc and as much as $750 for a multiple-disc set. But many users do not require the hundreds or even thousands of samples that might be found on a CD-ROM, so, in an effort to meet the demands of multimedia users and musicians with a limited budget, some sample distributors have begun to use the Internet as a means of selling individual sounds from their vast holdings. For example, from the East-West Communications website <soundsonline.com>, one can not only order sample collections on CD and CD-ROM but also, for a mere $2.95, download individual samples for use in music, film, multimedia, or jingle production (licenses for use in national advertising campaigns requires a somewhat heftier $29.95 per sample). And like all things in commodity culture, samples are cheaper by the dozen: a 50 percent discount applies when downloading ten or more samples.

While most commentary has tended to focus on the most spectacular uses of sampling by popular musicians, as in the case of Rap and much dance music, it is also important to recognize the full range of commercial sample users and the media contexts in which audiences experience music and sound. Samples of conventional musical instruments and, increasingly, the sounds of world music instruments find their way into a diverse set of uses, from Nintendo to Disney and from popular music to multimedia projects, film soundtracks, and television commercials. Film uses are perhaps especially important: through association with film images and narrative, sounds take on (and contribute to) specific meanings. Indeed, the cliche residue runs deep in popular film: from the brief excerpts of the music of various "Nationalities" found in Erno Rapée's (1924) book of piano and organ music designed for use in early silent cinema to the present-day use of world music samples, Hollywood has always employed music and instrument sounds to evoke foreign lands and cultures. Even when specific nationalities or cultures are not depicted at all, as in animated features such as Disney's *The Lion King*, music is used to connote a sense of place and an overall cultural setting: through the use of African-sounding languages, call-and-response choirs (both "real" choirs and choir samples are heard in the film), and huge sampled drum ensembles, children are introduced to a mythical "Africa" in sonic form. Even in the popular songs written for the film by Elton John, part of synthesist and composer Hans Zimmer's job was to make them sound, in his own words, "more African" (*Roland Users Group* 12 (2), 1994, p. 31).

Over the past decade, soundware developers have served a diverse set of individual and industry interests—a set of interests geared increasingly toward the consumption of the world's musical and sonic riches.

The Role of Discourse 1: Authenticity

At $2.95 per sample, the soundware industry must be considered one of the most extreme manifestations, and perhaps the ultimate commercial outcome, of R. Murray Schafer's notion of "schizophonia"—the splitting of sound from the maker of the sound (1977: 90–91). Once split from its maker, the sound can be appropriated, modified, and made available for sale as a discrete commodity. While schizophonia and commodification can be considered the potential byproducts of virtually any act of sound recording, the fragmentary nature of sound sampling—where individual instrument samples are seldom more than a few seconds in duration and often as brief as a few hundred milliseconds—seems to attract and indeed may require a proliferation of discourses whose first and foremost role is to guarantee the authenticity of the originating sound source, investing it with cultural significance and enhancing the value of the sample in the marketplace in the process.

This play of signification and exchange is especially evident in the case of world music samples where discourses of authenticity are closely linked to those of exoticism, essentialism, and primitivism. But while such discourses are no doubt similar to those found in the recording industry promotion of world music/world beat, they are inflected somewhat differently. On the one hand, much of the promotional emphasis associated with world music is placed on the development of star performers, allowing certain individuals to come to the fore on the international music stage. The ethnic origin and perceived cultural integrity of the artist is taken as the primary guarantee of authentic musical expression. Discourses in world music sampling, on the other hand, tend to downplay the identity of the sampled artists (performers are seldom identified at all in liner notes or promotional material) and are more often condensed around the figure of the musical instrument itself, its sounds regarded as the embodiment of musical culture.

Big Fish Audio's sample disc *Didgeridoo and other Primitive Instruments* is a case in point. The disc contains samples taken from a variety of Australian and Native American traditional instruments, which are presented by the disc's producers as virtually synonymous with the cultures they represent: "each sample contains all the authenticity of its native origins." Furthermore, this reification of recorded instrument sound as authentic cultural artifact is explicitly linked to the notion of primitivism in the disc's title, the didgeridoo being singled out in both title and cover art as embodying the essence of the "primitive" in music. The instrument has been treated in a similar fashion elsewhere: for example, in the early 1990s, the image of an Aboriginal in full body paint and playing a didgeridoo figured prominently in magazine ads for E-mu System's Proteus/3 "World" sample playback module. In these ways, a special status appears to have been conferred upon the didgeridoo as the quintessential icon of both primitivism and authenticity in world music sampling.

However, not all sampling reifies simple instrumental sounds (and images) in this manner. In contrast, some sample CD producers have tended to represent the musical phrase as the essential expression of culture. In its *Heart of Africa* sample collection, Spectrasonics suggests that there is a distinction between instrument samples and phrase samples:

The key to the Heart of Africa library is the combination of both Multisampled Sounds (DISC A), and Phrases (DISC B & Heart of Africa vol. II). The concept is that the Multisamples allow you to create your own melodies, harmonies and rhythms—while the Phrases give you the opportunity to infuse and incorporate authentic live performances (which are difficult or impossible to realize utilizing only the multisamples), into your own music. The Multisamples give the flexibility, while the Phrases bring the authenticity. (*Heart of Africa, Vol. I,* liner notes)

Despite this difference in the conceptual level at which "authenticity" is located, however, the primitive and the exotic are inextricably linked to it. In Spectrasonics' promotional material, it describes *Heart of Africa* as "a fascinating compilation of strange & wonderful instruments, voices, and performances that evoke deep images & primitive emotions," and *Vol. II*, in particular, as "a collection of powerfully primal & untamed tribal ensemble phrases & SFX."

Such generalized appeals to the "primal" and the "primitive" are given greater specificity in advertisements that stress the quasimagical powers to be found in sampled sounds and phrases. For example, Ilio Entertainments' ads for *The Complete Gamelan,* developed by Propeller Island, state "This disc takes you to the edge of the world allowing you to create liquid metallic textures and intricate rhythms which produce a hypnotic, almost hallucinogenic effect on the listener" (*Keyboard,* July 1996, p. 12–13). Thus, the promise implicit in world music sampling is not simply access to new creative materials, duly authenticated, but the fact that the samples carry with them primal powers that can be transferred to the user/consumer—powers that can then be used to induce specific effects on the listener.

In delivering these sounds found only at "the edge of the world," sample manufacturers and distributors invoke a set of narratives drawn from the colonial past. Sample recordists are portrayed variously as travelers, explorers, or pop ethnomusicologists, expending time (as long as two years in one account) and great effort to bring back sonic treasures to the West. In a somewhat more contemporary vein, one reviewer put it thus: "Like pharmacologists spreading out through the jungle in search of potent new drugs, musicians with portable DAT decks are bringing back an amazing assortment of musical experiences that any sampler owner can use" (Jim Aikin, in *Keyboard,* June 1996, 77). Here, the invocation of the jungle and the explicit link to obscure plants and herbs possessing healing powers unknown to Western science create a romantic and almost magical aura around the sampling project.

As romance would have it, even the explorer must occasionally succumb to the power and mystery of primitive cultures. In the second installment of a two-part feature article devoted to world music instruments and playing techniques, Dave Stewart and Barbara Gaskin begin their survey by quoting an extended passage, taken from John Miller Chernoff's *African Rhythm and African Sensibility* (1979), where the ethnomusicologist describes his study of African drumming and his encounter with juju medicine. Commenting on his experience, Stewart and Gaskin state, "Though it may be a little shocking to hear a sober and respectable academic talking in terms of slaughtered chickens and juju medicine, this story serves to illustrate the power of ritual and magic in traditional African music" (Stewart and Gaskin 1996b: 34). Of course, Chernoff's position toward such events was part of a "research strategy" intended as a means of gaining access to a foreign context (Chernoff 1979: 19), but Stewart and Gaskin prefer to use his reflections—and, by implication, the institution of academic musicology—as a way of legitimizing their own attitudes toward the "inner nature of music."

The fact that many sampling "expeditions" are conducted within close proximity to the metropolitan centers of the non-Western world (and, in many cases, to commercial recording studios) is often obscured by these romantic images of the jungle, the desert, and the remote village setting found in most discussions of world music sampling. In the case of Spectrasonics' *Heart of Africa, Vol. II*, the instrumental and vocal phrases were recorded "on location," not in the jungle but in stage performances at a two-week tribal competition in Kenya; here, it is the tribal participants who demonstrated their musical commitment by traveling "many miles by bus over what most would consider impossible road conditions to perform their music" (liner notes). Having revealed the origin of the recordings, the producers take great pains to assure the consumer that the samples are "authentic" in every way, the competition judges cited as the ultimate arbiters of both authenticity and quality in the performances. Wallis and Malm, however, have argued that the music performed at competitions of this kind is often not as "authentic" as it may appear: "Different styles become streamlined. Virtuosity becomes a measure of individual prowess" (1984: 270).

The constant reference to the power and authenticity of traditional music tends to mask the degree to which these "primal" performances are tamed, so to speak, by the technology of digital sampling. Whatever their status as "authentic" performances of traditional repertoire, the recorded performances are subjected to a technological process calculated to render them useful to Western sampling musicians: "In some cases 'live performance' phrases will drift flat or sharp, slow down or speed up during the coarse [*sic*] of the performance. Trying to lock sequenced tracks to such material can be very tricky. In editing these phrases for your use I have attempted, wherever possible, to make them loopable. So even if a phrase drifts slightly in pitch or tempo, most of them should come around to the beginning smoothly" (*Heart of Africa, Vol. II,* liner notes). As in similar sampling CDs, many of the performance loops are catalogued according to their BPM (Beats Per Minute) value, giving remixers and deejays "quick access to the cultural groove" (promotional material for *Ethnic,* Time + Space/Zero-G). The "LoopConnector" system of sample developer Sounds Good goes so far as to impose a set of BPM templates (in tempos of even tens) as well as standardized tonalities throughout their collection of sample discs for "maximum ease in mixing 'n' matching loops and riffs within and between the various titles."

Thus, at the same time that suppliers of prerecorded samples infuse the material with cultural significance and authenticity, they also transform it. Through the techniques of digital sampling—editing, looping, pitch shifting, and time compression, among others—sample developers organize and facilitate easy encounters between a diverse range of musical styles.

The Role of Discourse 2: Appropriation and Identity

Making music with digital samplers and prerecorded samples of world music instruments and phrases is, fundamentally, an art of appropriation. But while samplers and sample discs may facilitate, at a technical level, the acquisition, manipulation, and layering of diverse sonic materials, the discourses of world music sampling play a key role in defining the significance of the act of cultural appropriation, justifying and enabling it at the level of ideology. Furthermore, appropriation is linked, through discourse, to a process in which musical identities are transferred, adopted, or remade.

A fundamental task of this set of world music discourses is to deflect ethical, political, or musical criticism of the cultural appropriation that has taken place. In some instances, writers acknowledge that critical perspectives exist but simply deny their relevance. For example, in a review of the *New World Order: Journey #1* sampling project, one reviewer stated, "If Doerschuk [an associate editor at the magazine] were writing this review, he'd probably raise some provocative question about the cultural exploitation of Third World musicians in postmodern audio collage. He might ask whether, shorn of their context, these beats have any real emotional impact, or whether they're only a cynical, jaded substitute for musicianship.

Since he's not writing it, we won't even raise the issue. We'll just say, 'Hey, happenin' loops!'" (Jim Aikin, in *Keyboard* 21 (2), February 1995, 34.) In raising "the issue" only to summarily dismiss it, the reviewer suggests that aesthetic experience—indeed, simple musical pleasure—should take precedence over all ethical or musical concerns.

Others are more self-conscious with regard to their relationship to those who have been sampled. Their concern, however, often manifests itself only indirectly, again eliding the question of cultural appropriation itself. For example, in the credits to the *Heart of Africa* CD set, sample developer Spectrasonics includes the following note: "This project is dedicated to our brothers and sisters who have suffered so terribly in Rwanda and Liberia. May your faith in our Lord sustain and strengthen you. A portion of each sale of Heart of Africa goes directly to African relief." While such generalized expressions of sympathy and solidarity may be genuine, the altruism evidenced in the desire to share some of the profits generated from the project does not even begin to address (and may even mask) the issue of compensation for those who participated directly in the recordings—in general, musicians who allow themselves to be sampled are given a one-time-only fee and relinquish copyright or any other claim to royalties derived from the sale of the resulting CDs. For their part, they collectively receive little more than a word of thanks in the CD credits: "Special Thanks to: The many gifted musicians and singers who contributed to this project." In contrast, the various individuals who participated in technical, organizational, or other ways to the project are cited by name in the credits, and composers who supplied demo material (constructed out of the sampled contents of the discs) enjoy copyright protection for their work. Thus, a basic asymmetry, both in terms of economic gain and artistic acknowledgement, exists between the makers and those who are the objects of the sampling enterprise.

A second discursive strategy trivializes the act of appropriation, depoliticizing it by rendering it banal. This is accomplished by exploiting and elaborating a common metaphor—that of "taste"—thereby creating an essential equation between musical aesthetics and the gustatory senses. It can be found in the titles of CD-ROM collections—for example, *Ethnic Flavours* and *Spices of India* (both from U.K. developer Time + Space)—and is a common feature of promotional materials, articles, and reviews—"if you'd like to spice up your tracks with a dose of real African flavor, it's hard to see how you could go wrong with Heart of Africa" (CD review, JimAikin, in *Keyboard*, June 1996, 77). Indeed, the metaphor is so pervasive that it has become something of a cliche. However, by characterizing appropriation and the layering of diverse musical fragments in this way, these discourses remove world music sampling from the realm of the extraordinary or the technocratic and place it firmly within the everyday: music as a kind of "fusion cuisine" and sampling artist not as *flaneur* (Mitchell 1993: 314), but rather, as *gourmet*.

A third, and perhaps more significant discursive strategy characterizes appropriation as a form of cultural exchange among equals and the inevitable byproduct of larger social forces, such as the diffusion of modern communications technologies and increased cultural contact in the "Global Village": "Today, more than ever before, cultures are cross-pollinating musical styles. It is sometimes difficult to know what originated where. Many remote communities have unique dance or musical forms which have evolved over hundreds of years. Yet, it is no longer uncommon for young entrepreneurs in once-secluded villages to obtain access to a TV which can be plugged into a car lighter. And so new styles evolve, a rhythm is altered and alters once again" (Dan Portis-Cathers, *Heart of Africa, Vol II,* liner notes). Similar discourses are evident in a variety of mass media, where technology is often represented as a benign force in an increasingly globalized culture—for example, the television ads in which healthy, smiling, dashiki-clad Africans (among others) are depicted as participants in a world of effortless communications via computers and Internet access. But in music the exchange of ideas is often understood as part of a larger process in which cultural identities are also transformed: "In times of peace, bordering communities exchange ideas and traditions mutate. In times of war, conquerors move in to impose change and find themselves being changed by their subjects. Change comes, and with it the beating of new and different drummers. But no matter what changes on the outside, to find true Music, one must always look within. And when you do, you may find yourself looking to Africa, because that's where its heart is" (ibid.). Here, sampling musicians (descendants of a former conqueror) are not only changed by their encounter with African music but appropriate it fully as a part of their own musical identity: by looking "within" for true music, one finds not one's own music but "Africa" in technological form.

The significance of this Global Village discourse lies not simply in the role it plays in facilitating the global music mix and the adoption of new musical identities but, more importantly, in the very manner in which it justifies cultural appropriation: by portraying the recorded subject as active and the sound recordist as passive, it inverts the relationship of power that exists within the process of cultural exchange. This "ideological reflex" absolves sampling musicians from responsibility through an apparent liberation of the sampled subject, in discourse, at the same time that their sounds are technologically reduced to mere raw material, grist for the sampling mill.

The sources of what I refer to here as this "ideological reflex" are complex and located in the more general history of Western science and technology (the expression itself is borrowed from Leiss 1972). One can trace its use, however, throughout the brief history of electronic music in both its avant-garde and popular forms. It has genealogical roots in Varèse's early-twentieth-century notion of "the liberation of sound" and an even more direct link to the present discussion in the 1960s tape collages of Karlheinz Stockhausen:

> I wanted to come closer to an old and ever-recurrent dream: to go a step further towards writing, not "my" music, but a music of the whole world, of all lands and races.... They all wanted to take part in *Telemusik*, often simultaneously and overlapping. I had my hands full keeping open the new and unknown world of electronic sound for such guests: I wanted them to feel "at home" and not "integrated" by some administrative act, but rather, genuinely engaged in an untrammelled spiritual encounter. (*Telemusik*, liner notes, DGG 137012, 1966)

Despite Stockhausen's ingenuous characterization of his work as "an untrammelled spiritual encounter," it is difficult to consider his appropriation and manipulation of a wide variety of recorded traditional musics as anything other than a technocratic "administrative act." His desire to subsume his own musical identity within the "music of the whole world, of all lands and races" obscures his own role in the creation of the work and the role of technology in mediating his relationship with the music of other cultures.

The notion that "they all wanted to take part" not only obscures the "administrative act" that has taken place but, in popular music, an economic one as well. For in the dual process of appropriation and commodification of recorded sounds that takes place in the production, marketing, and use of world music samples, questions of musical identity cannot easily be separated from questions of ownership. In a very material sense, as Steven Feld has argued, "the question 'Whose music?' is submerged, supplanted, and subverted by the assertion 'Our/my music'" (1994: 238); indeed, even in the case of Stockhausen, it is clear that we are always listening to "his" work. And despite the marketing appeals of manufacturers such as E-mu Systems—"As borders dissolve, traditions are shared" (magazine ad for Proteus "World" synthesizer, 1992)—the guarantee that world music samples are copyright-free draws a new "boundary line between participation and collaboration at *ownership*" (Feld 1994: 242; emphasis in the original). Thus, in making music with world music samples, even though the sounds may retain some sense of their original identity, ownership assures that we are always engaged in creating "our" music.

The desire to immerse oneself in the endless variety of recorded world music is, no doubt, one of the more recent byproducts of "schizophonia" as it affects the listener in the age of postmodernism, and is quite different from earlier relationships to the music of other cultures. In the early-twentieth-century music of Bela Bartok, for example, Hungarian peasant music was so completely assimilated that the composer used it as "his musical mother-tongue"; Bartok regarded his work in collecting folk songs on the early recording devices of the day as simply an adjunct to a deeper, more anthropological study of music "as part of a life shared with the peasants" (Bartok 1931/1967). While Bartok's empathy for Hungarian music cannot in itself serve as the sole justification for his appropriation of it, it is nevertheless clear that the fusion of traditional and modernist elements in his music is qualitatively different from the fleeting juxtapositions of collage that have become the preferred idiom of contemporary technoculture (on this point, see Born & Hesmondhalgh 2000: 41). With the vast, decontextualized collections of sampled sound available on CD-ROM, technoculture neither allows for the type of profound encounter experienced by a composer such as Bartok, nor is it required.

In world music sampling, "identity" is a mobile construct, influenced by the ever-shifting surfaces of fashion and personal taste. In this, musicians appear to differ little from other music consumers, where musical taste "is now intimately tied into personal identity; we express ourselves through our deployment of other people's music. And in this respect music is more like clothes than any other art form—not just in the sense of the significance of fashion, but also in the sense that the music we 'wear' is as much shaped by our own desires, our own purposes, our own bodies, as by the intentions or bodies or desires of the people who first made it" (Frith 1996: 237). Sampling technology enhances our ability to deploy an increasingly diverse range of "other people's music" but does so in a manner that is at once fragmentary and exceedingly rich, consisting of individual sounds, timbres, and rhythmic and melodic loops organized and densely layered into a "global mix." And as Frith suggests, this mix has relatively little to do with those who produced the original sounds but everything to do with those who consume them.

Conclusion: Everything Old Is New

> All around the world, there are instruments which have been with us, not just for a couple of decades, but for hundreds, even thousands of years . . . Thanks to the advent of specialized CD-ROM and audio CD sound libraries, we now have access to these exotic, intriguing sounds for the first time, opening up exciting new areas of sonic exploration and composition. (Stewart and Gaskin 1996a: 33–34)

In the commercial worlds of technology, music, and fashion, it is always necessary to create a sense of "the new," the sense of experiencing something "for the first time" even when that experience coincides with something that is simply old or replete with nostalgia. To do so requires, first and foremost, a considerable effort to erase the collective past—a past stretching, in the most general sense, across centuries of musical exchange between various musics of the world and that of both Western classical and popular musics (see Bellman 1998). And, in the specific case of world music sampling, we are expected to forget the fact that pop music from at least the beginning of the twentieth century to the present has been continually infused with, and invigorated by, exotic sounds: from "Hawaiian" guitars to Latin percussion and from the 1950s exotica of Les Baxter and Martin Denny to the sitar tones of 1960s psychedelia (see Hayward 1999). Even the idea of using prefabricated rhythmic loops was anticipated in the tacky, portable rhythm machines associated with the home organs of the 1960s—"Reach for the rhythm selector switch and dial your choice of 18 different rhythmic patterns. Waltz to rock 'n' roll, Bosa nova to Beguine" (Ace-Tone ad, 1967). And long before the name of Peter Gabriel became associated with world music, Joni Mitchell had a hit with "The Jungle Line," employing a Moog synthesizer bassline in conjunction with a recording of the warrior drums of Burundi as a bed track (Mitchell 1975).

But what *is* new, as I have argued throughout this chapter, is the level of commodification and the broad, unfettered access afforded by the combination of samplers and specialized sample compilations—a phenomenon potentially leading to the kind of qualitative changes in music characterized by Steven Feld as a shift from "schizophonia" to "schismogenesis," the progressive differentiation and interaction of world cultures, which is intensified by the economic and industrial interests at play in the global marketplace (1994). That such a development has been accompanied by the most essentializing and exoticizing discourses, on the one hand, and the strategic denial of cultural appropriation, on the other, should come as no surprise, for a more critical perspective on this phenomenon would serve only to constrain the marketplace by calling into question the economic and technical processes of musical globalization and the discourses that sustain the unequal distribution of cultural capital that are so characteristic of them.

References

Bartok, Bela. 1931. "The Influence of Peasant Music on Modern Music." Reprinted in E. Schwartz and B. Childs (eds.), *Contemporary Composers on Contemporary Music,* pp. 72–79. New York: Da Capo Press, 1967.

Bellman, Jonathan (ed.). 1998. *The Exotic in Western Music.* Boston: Northeastern University Press.

Born, Georgina, and David Hesmondhalgh. 2000. *Western Music and Its Others: Difference, Representation, and Appropriation in Music.* Berkeley: University of California Press.

Chernoff, John Miller. 1979. *African Rhythm and African Sensibility.* Chicago: University of Chicago Press.

Feld, Steven. 1994. "From Schizophonia to Schismogenesis: On the Discourses and Commodification Practices of 'World Music' and 'World Beat.'" In C. Keil and S. Feld, *Music Grooves,* pp. 257–89. Chicago: University of Chicago Press.

———. 1996. "Pygmy POP: A Genealogy of Schizophonic Mimesis." *Yearbook for Traditional Music* 28: 1–35.

Frith, Simon. 1996. *Performing Rites: On the Value of Popular Music.* Cambridge: Harvard University Press.

Hayward, Philip (ed.). 1999. *Widening the Horizon: Exoticism in Post-War Popular Music.* Sydney: John Libbey.

Ivey, William. 1982. "Commercialization and Tradition in the Nashville Sound." In W. Ferris and M. L. Hart (eds.), *Folk Music and Modern Sound,* pp. 129–38. Jackson: University Press of Mississippi.

Leiss, William. 1972. *The Domination of Nature.* Boston: Beacon.

Mitchell, Joni. 1975. "The Jungle Line," *The Hissing of Summer Lawns,* Asylum 7E-1051.

Mitchell, Tony. 1993. "World Music and the Popular Music Industry: An Australian View." *Ethnomusicology* 37 (3): 309–38.

Rapée, Erno. 1924. *Motion Picture Moods.* New York: G. Schirmer.

Schafer, R. Murray. 1977. *The Tuning of the World.* New York: Alfred A. Knopf.

Stewart, Dave and Barbara Gaskin. 1996a. "Sample the World, Part I: Sound—The Final Frontier." *Keyboard* 22 (2): 33–53.

———. 1996b. "Sample the World, Part II: The Hit-Makers, or Is That Hit-Takers?" *Keyboard* 22 (4): 33–42.

Taylor, Timothy D. 1997. *Global Pop: World Music, World Markets.* New York: Routledge.

Théberge, Paul. 1997. *Any Sound You Can Imagine: Making Music/Consuming Technology.* Hanover, N.H.: Wesleyan University Press / University Press of New England.

Wallis, Roger and Krister Malm. 1984. *Big Sounds from Small Peoples: The Music Industry in Small Countries.* New York: Pendragon.

CHAPTER FIVE

Technology and the Production of Islamic Space
The Call to Prayer in Singapore

Tong Soon Lee

Introduction

In almost every Islamic community today, the loudspeaker, radio, and television have become essential in the traditional call to prayer, a remarkable juxtaposition of high media technology and conservative religious practice. The loudspeaker simply extended the purpose of the minaret, that towering section of the mosque where the reciter traditionally stood to perform the call to prayer, his voice reaching the surrounding Islamic community. Until recently, this community in Singapore was still located relatively close to the mosque in a homogeneous context. In the early 1970s, however, as a result of urbanization and resettlement programs that accompanied the process of industrialization in Singapore, the amplified call to prayer became a source of conflict in the emerging reinterpretation of social and acoustic spaces. By focusing on the use of the loudspeaker and radio in the Islamic call to prayer in Singapore, I will explore the intricate, and sometimes stormy, relationship between technology and the spatial organization of social life. I want to argue that media technology is not necessarily closely associated with popular culture but can be inextricably bound to so-called "traditional" forms of expressive cultures, in this case, the religious institution of Islam.[1]

Adhan—The Islamic Call to Prayer

The Islamic call to prayer, otherwise known as the *adhan* (or *azan*), is recited five times a day from every mosque to inform Muslims of the prayer times, namely *Subuh* (before dawn), *Zuhur* (noon), *Asar* (late afternoon), *Maghrib* (after sunset), and *Isyak* (evening).[2] For Muslims, the *adhan* is sacred. As a social phenomenon, the *adhan* unifies and regulates the Islamic community by marking the times for prayer and creating a sacred context that obligates specific religious responses. Upon hearing the *adhan*, Muslims are obliged to put aside all mundane affairs and respond to the call physically and spiritually. Indeed, the *adhan* is seen as a microcosm of Islamic beliefs, as it "covers all essentials of the faith" (Fiqh us-Sunnah at-Tahara and as-Salah 1989: 95). Furthermore, the *adhan* symbolizes the presence and blessings of God when it is recited in celebratory events such as births or during calamities.

While the towering minaret of the mosque serves as a physical landmark that signifies the sacred center of the local Islamic community, the call to prayer is a "soundmark" (Schafer 1994: 10) regulating the daily life of each Muslim. In this way, the Islamic community may be identified along acoustic lines, that is, "the area over which the muezzin's voice can be heard as he announces the call to prayer from the minaret" (ibid.: 215).[3] In broader cultural terms, the call to prayer is iconic of the social identity of Muslims.

In Singapore today, the recitation of the *adhan* in every mosque is amplified through the use of loudspeakers. In addition, a prerecorded version of the *adhan* is broadcast over the radio.[4] There are two radio stations in Singapore that broadcast programs in Malay, namely Warna 94.2 FM and Ria 89.7 FM. However, the *adhan* is broadcast only on Warna 94.2 FM, as it caters solely to the Malay/Islamic community.[5] Its broadcasting is entirely in Malay, while Ria 89.7 FM mixes both English and Malay and plays mainly popular music in both languages. At the time of this research, there was no television broadcast of the *adhan* in Singapore, but the Islamic community was able to receive a broadcast over the Malaysian television channels, Malaysia being directly to the north of Singapore.[6] The old mosques in Singapore, particularly those built before 1975, had their loudspeakers placed outward, toward the community surrounding the mosque.[7] After 1975, however, the loudspeakers in what are called the "new generation mosques" (Majlis Ugama Islam Singapura 1991) were redirected toward the interior.[8]

Negotiating Islamic Space

Since 1959, when Singapore was granted self-government in domestic affairs by the British colonial office, the emphasis of the ruling People's

Action Party (PAP) has been political consolidation, industrialization, economic expansion, and urbanization within a general discourse on nation-building (Chua 1995a). Such political motivations engendered the urbanization and resettlement project, formalized in 1967.[9] Since then, "the fate of rural Singapore was sealed," as there were "massive changes in Singapore's cultural and physical landscape" (Sequerah 1995: 186). Attention turned to the construction of new "satellite" towns with public, high-rise apartments constituting the core residential area, surrounded by industrial estates and served by expressways. As a result, rural settlements were gradually demolished to make way for these developments (Wong and Yen 1985; Chua 1995b; Seet 1995; Sequerah 1995).[10]

During the post–World War II period, until Singapore's political independence in 1965, the bulk of the population lived in urban villages and rural *kampungs* (Chua 1995b: 227).[11] The population distribution in the rural areas was largely characterized by ethnic distinctiveness, each forming relatively homogeneous communities.[12] The population in Malay *kampung*s was largely Muslim and predominantly Malay, with Islam regulating "the pulse of *kampung* life" (Seet 1995: 209).[13] Each *kampung* would usually have a mosque or *surau*—a sanctified space used for religious gatherings—headed by a religious leader (known as the *imam*) who led prayer meetings and arbitrated on matters related to the Islamic faith (Seet 1995).

In the urbanization process, the removal of burial grounds and places of worship caused anxiety among all ethnic communities (*Straits Times,* 1 August 1974; *Berita Harian,* 7 June 1974a, 7 June 1974b, 13 June 1974). Muslim leaders from PAP made numerous public addresses to the Islamic population, urging them to accept and adapt to a modern, urban environment. Muslims in Singapore were urged to "discard some of their age-old 'adat' [customs] to make way for progress" (*Straits Times,* 22 December 1973, 24 December 1973, 11 August 1974, 26 October 1974).[14]

One of the most important issues highlighted in their speeches was the assurance that the clearing of religious buildings did not apply to the Islamic community alone but also affected Chinese and Indian temples as well as Christian churches. The following observation, extracted from a newspaper article (*Berita Harian,* 11 June 1974), reveals the sentiments of several Islamic organizations toward this issue, which were expressed during a meeting held on 1 June 1974 (hereafter referred to as "Meeting"), out of which a petition was drawn and sent to the former Prime Minister Lee Kuan Yew[15]: "They were of the opinion [that] efforts were already undertaken [by the government] to destroy the freedom of Islam in our Republic.[16] Muslim leaders emphasized that the removal of religions sites "has only been resorted to when absolutely necessary and unavoidable. In no

Technology and the Production of Islamic Space / 111

way can this be represented as being done only against the Muslim places of worship." Furthermore, they affirmed that the government, "had already publicly announced . . . that in the process of urban renewal and development, adequate sites would be [re]served for a temple, a mosque, and a church in each of the new towns." In addition, the government had provided generous subsidies to the Islamic community for the purchase of sites for the building of mosques in these towns (*Straits Times*, 18 July 1974a).[17]

It was pertinent for Muslim leaders to assure the Islamic community of the impartial attitude of the government toward religious and ethnic issues in a country that emphasizes intercultural harmony. Nevertheless, the relocation of religious sites and the subsequent issue involving the use of loudspeakers for *adhan* recitation evoked larger social issues concerning Malays in Singapore.

Adhan and the Islamic Soundscape

Accompanying the altered landscape, which went from a localized, rural setting marked by ethnic homogeneity to a more urbanized and multiethnic environment, was a transformation in the organization of social space. Muslims found themselves in closer proximity to other ethnic groups and, at the same time, further from one another and from the mosque. Once exclusive to Islamic rural communities, the sacred acoustic environment of the amplified call to prayer was now "inhabited" by non-Muslims. In other words, the urbanization process had brought about the diffusion of closeknit and homogeneous ethnic communities into newly organized, heterogeneous, multi-ethnic communities. Muslims now inhabited highrise apartment complexes, with other ethnic groups and religious followers as close neighbors. These highrise apartments were microcosms of a multicultural environment that the government constructed in the process of building a cohesive nation, with quotas set on the percentage of each ethnic community living in each neighborhood.[18]

In the new context, sound production from traditional practices was sometimes regarded as "intrusive" by community members not involved in those events. The new, urbanized resettlement, therefore, also resulted in new regulations concerning these practices, one of which was the legislation of the anti–noise pollution campaign in August 1974. This campaign affected not only the Islamic call to prayer but also public and religious activities by other ethnic communities. How, then, did the anti–noise pollution campaign affect the Islamic community?

Between May and July 1974, as part of the noise abatement campaign, the government and Islamic organizations in Singapore decided to redirect the loudspeakers of the mosques inward; they had originally faced the exterior of the mosque. Newspaper reports published during this period showed that this decision infuriated a section of the Islamic population, thereby creating conflicts between the government and members of the Islamic community. This conflict arose because of the popular misconception that the government had planned to ban the use of loudspeakers for the call to prayer, thus threatening the very core of Islamic religious practice. A newspaper report (*Straits Times*, 10 June 1974) on the Meeting notes,

[i]n the view of the participants of this meeting, efforts were being made by the authorities to suppress Islam, contain its growth, and restrict the freedom of its followers when they came to know that mosques in this country would no longer by allowed to use microphones.

Some even went so far as to allege that this was an effort to put a stop to the muezzin's call to prayers.

Reversing the directions of the loudspeakers toward the interior of the mosque under the pretext of the noise abatement campaign was not favorably received. A Muslim journalist (*Berita Harian*, 24 May 1974) asked, "why should the . . . loudspeaker . . . be directed into the mosques—wouldn't that be diverting from the reason for azan—to call others to prayer? . . . Can the azan that lasts a minute be considered noise and a disturbance to others?"[19] The Islamic community was urged to understand the necessity of reducing noise pollution in the new environment and to support the implementation of the noise abatement campaign (*Straits Times*, 23 July 1974). More importantly, newspaper reports emphasized that the campaign was not directed solely at the Islamic call to prayer but also applied to various activities of all ethnic communities and social institutions, including "Chinese opera, funeral processions, church bells, Chinese and Indian temples, music during weddings, record shops and places of entertainment . . . the recitation of pledges in schools and school sports" (*Berita Harian*, 14 June 1974b).[20]

The Singapore Muslim Assembly and the Singapore Muslim Action Front were two organizations that filed petitions with the government concerning issues that arose out of the urbanization program, among which was the controversy surrounding the use of loudspeakers in mosques for the call to prayer (*Straits Times*, 18 July 1974a, 18 July 1974b, and 24 July 1974; *Berita Harian*, 2 June 1974 and 14 June 1974b). Indeed, the Singapore Muslim Assembly noted that the Meeting was "a direct result of the controversy surrounding the use of loudspeakers by mosques to call the faithful to prayers" (*Straits Times*, 24 July 1974).

Muslim leaders from the ruling government, however, accused both the Singapore Muslim Assembly and the Singapore Muslim Action Front of "divisive folly" (*Straits Times*, 18 July 1974b). The two Islamic groups were said to have submitted their petitions on Muslim matters (referring to the issues emerging out of the urbanization program) to the Singapore Government to coincide with the Islamic Foreign Ministers' Conference in Kuala Lumpur (Malaysia), held during the period from 21 to 25 June 1974. Without waiting for the government's response, they distributed copies of their petition to the public as well as to the participants at the conference to "discredit Singapore before the foreign delegates" on a "frequently exploited" issue—the "suppression of Malays in Singapore" (*Straits Times*, 18 July 1974a).[21]

The issue of the use of loudspeakers for the Islamic call to prayer reflected larger historical and sociopolitical matters pertaining to the Malays in Singapore. In Singapore, where the majority of the Malay population is Muslim, matters pertaining to the Islamic faith are very often bound up with ethnic concerns. In addition, Singapore was part of the Malaysian Federation from 1963 to 1965, and for historical and political reasons its constitution recognizes Malays as the indigenous population and Malay as the national language.[22] This historical alliance with Malaysia had "given Singapore's minority Malay population a sense of its own interests and political significance as a community in the larger regional picture" (Chua 1995a: 18). Furthermore, as a "Chinese enclave in the Malay sea" (ibid.: 108),[23] the government was concerned with religious harmony and interethnic relationship, particularly between the Chinese—constituting the majority of the country's population and its ruling political party—and Malay. In such a context, the Singapore government was circumstantially impartial in handling the problems that emerged from the urbanization program. It is important to note that this conflict indicates the central, but sometimes ambiguous, place of media technology in the Islamic call to prayer in a country that takes pride in ethnic and religious plurality.

After discussions with the government, Islamic organizations agreed to (1) reduce the amplitude of loudspeakers in existing mosques, where they remained facing outward, (2) redirect loudspeakers toward the interior of new mosques to be built in the future, and (3) broadcast the call to prayer five times a day over the radio, which previously only transmitted the sunset calls. The radio played an important role in resolving this conflict resulting from the use of loudspeakers in the new, urbanized context. Compared to the loudspeaker, the radio is a different form of media technology that was well adapted to the new spatial organization.

Mediating the Islamic Community

In the rural context, it was possible to define the Islamic community in Singapore as, borrowing Murray Schafer's terminology, an "acoustic community" (1994: 214–17). In other words, a community characterized by the acoustic space within which the call to prayer could be heard. In the new, urban environment, however, it became difficult to maintain the Islamic acoustic community, because the environment had become larger and more interspersed with non-Muslims.

In the modern, multi-ethnic context of Singapore, the concept of the Islamic community had suddenly become a matter of fluid boundaries and contested public space. Roger Friedland and Deirdre Boden (1994: 33) emphasize that, "individuals, organizations, and societies construct space and time in the way they do *because* of the meanings they impart to them" (emphasis in original). In this way, we might understand all sounds as cultural phenomena with socially imbued meaning. The sacredness of the call to prayer is an Islamic cultural construct, yet, at the same time, it can be a noise hazard to the majority of non-Muslims in Singapore, just as Chinese street operas might be to the non-Chinese population.

Nevertheless, creating a sacred acoustic space to define community is crucial to an Islamic people struggling to affirm their cultural viability and maintain the social borders distinguishing their community in relation to the larger, non-Islamic environment. In urbanized societies, such borders are fluid, ambiguous, and usually tenuous. Electronic mediation of the call to prayer in the transformed environment provides the means to reclaim the acoustic space that once identified the Islamic community in its rural past. Paraphrasing Arjun Appadurai (1990: 17), who is himself paraphrasing Walter Benjamin, I would call this the work of culture production in an age of electronic mediation.[24] In the rural past, the amplified call to prayer defined a physical space through the acoustic phenomenon that, in turn, defined the Islamic sacred and social space. The radio, however, superseded the loudspeaker in the new urban environment, taking over its role as a tool for culture production and also redefining the concept of community.

Presence/Absence of the Community

Regardless of the physical distances that separate Muslims from the mosque and from each other, they remain a community—an imagined community (Anderson 1991) defined in relation to the radio transmission of the call to prayer. Through the use of radio, the extended and separated profiles of Muslims in the urban environment now form an uninterrupted acoustic space and, as a result, a unified social and religious space. It is the radio, rather than the physical proximity of a mosque, that facilitates the cohesion of the Islamic community, and maintains its identity within the larger, urban context of Singapore. Indeed, the radio is sometimes revered as a symbol of God's presence outside the locale of the mosque.

Friedland and Boden (1994: 23) note that, "the immediacy of *presence* is extended by humans, first through language, [and] now through technology" (emphasis in original). The far-reaching presence, or absence, created by technology implies the existence of extended spatial relationships between the sound producer and its recipient. The amplified call to prayer creates an immediate Islamic presence for a larger, more dispersed community. However, the reorganization of social space in Singapore had made it necessary for Muslims to reinterpret their tools of culture production and adapt to changes in social space.

At the same time, the media itself brought about new spatial and social relationships (Berland 1992). In retrospect, broadcasting the call to prayer over the radio appeared to ease the conflict. The radio is one form of electronic mediation that was well adapted to individuated users, such as the relocated and dispersed Islamic population: it simultaneously separates and reunites listeners in differentiated and expanded space (ibid.: 46). Electronic mediation creates a dialectical relation between *presence* and *absence*: the *absence* of shared physical space among Muslims in Singapore does not affect their religious and sociocultural identity because of the *presence* of shared acoustic space created through radio transmission.

Mediating Culture Production

Schafer (1994: 165) notes that, "all acoustic communications systems have a common aim: to push man's voice farther afield . . . [and] to improve and elaborate the messages sent over those distances." But we can see that, in the case of Singapore, the use of media technology has had larger social implications that concern culture production and the easing of intercultural tensions. It is, therefore, not an exaggeration to note that the reorganization of social space in Singapore during the 1960s and 1970s resulted in an increased dependence on technology to maintain cultural identities. The "new" Islamic context—defined by the electronic broadcast of the *adhan*—facilitates a mediation between technology, space, and social identity, as Islamic identity is "technologically articulated with the changing spatiality of social production" (Berland 1992: 39) in Singapore.

The conflict over the use of loudspeakers in the call to prayer suggests that the technologically aided production of Islamic culture is closely linked to the politics of religious and ethnic expression. The call to prayer does not merely inform Muslims that it is time to pray: it is a statement that says "we are Muslims." As Arjun Appadurai (1990) notes, the process of culture production in changing spatial contexts has become a matter of identity politics.[25] He further suggests that communications technology has relativized spatial dimension in the contemporary world, and in such a context social agents identify and align themselves along different axes.[26] The use of electronic mediation is an example of such an axis that constitutes the heart of identity politics, where Muslims seek to maintain their past in the present—their historical construction of religious and cultural expression through the call to prayer in the larger, secularizing environment. The controversy reflects, as it continues to generate, an awareness of the inextricable relationship between technology and Islamic identity.

Miniaturizing Reception

Performing the call to prayer from the minaret had always assumed the collective, physical proximity of the Islamic community that surrounded the mosque. However, when it is broadcasted over the radio, the call to prayer becomes decentralized: the mosque is no longer the exclusive source from which the call to prayer is recited. In other words, what was previously an inclusive, community-wide tradition has now become decentralized and individualized, reduced to an almost personal, private act of worship.

Compared to the loudspeaker, the radio may be seen as a "miniaturization" (Chow 1993) in the broadcasting of the call to prayer. This altered mode of transmission requires, and is required by, a change in the form of reception and the concept of the "imagined communities." In the spatially reorganized, urban, and multi-ethnic context, Islamic culture production through the call to prayer is now a matter of choice, of consciously creating newly localized and homogeneous acoustic sites through radio transmission, sites that are linked by commonalities of the Islamic culture. Listening to the call to prayer via the radio reunites each member of the Islamic community and creates an abstract communal Islamic space without the encroachment of non-Islamic social spaces. In this way, the individual listener is the context for electronic mediation and, at the same time, a product and process of the technological articulation of Islamic identity in a new spatial context.

Although the personal and almost private act of listening to the call to prayer separates the individual listener from the larger community, "miniaturizing" the call to prayer, to paraphrase Rey Chow (1993: 398), makes its listeners aware of the presence of Muslims as well as non-Muslims in Singapore. This altered state of reception of the call to prayer through the radio—the increased frequency of its broadcasting, and listening in "miniaturized," individualized spaces—offers the Islamic community a means of cultural self-production within the collective, non-Islamic context of Singapore. This new form of reception of the call to prayer over the radio suggests that the "collective," both Muslims and non-Muslims, is not necessarily an "other," but a mundane and imagined part of the Muslim "self" that can be brought to *presence*, or relegated to *absence*, at the switch of a button. The radio provides its listeners with the power and the "ability to control the timing and spacing of human activities and thus the 'locales' of action" (Friedland and Boden 1994: 28). In terms of technology, Muslims are thus able to play an active part in their decision to participate in the imagined Islamic community.

While the government organizes, controls, and provides urban *places* through its resettlement program, the communities of people construct their own *spaces* through their practices of living (Fiske 1992)—in this case, constructing and maintaining an Islamic space through the use of electronic mediation. In this sense, then, "space is a practiced place" (de Certeau 1984: 117). While media technology may be a facet of the larger corporate and culture industry, usually dismissed as an alienating medium in the West, it is reconfigured as a "fully cultural process" (Ross 1991: 3) within the Islamic community in Singapore. In this way, "technology . . . is not simply the social and personal intrusions of big science made manifest; it also permeates and informs almost every aspect of human experiences" (Lysloff 1997: 208).

An important aspect of broadcasting the call to prayer (as well as other forms of religious speeches, *Qur'an* readings, and news of the Islamic community) is the redefinition of the listening context (Berland 1992: 41), which, in turn, enlarges the Islamic community. In Singapore, women do not usually attend prayers in the mosque, while men are obliged to do so. In this case, the radio may be seen as a main (if not the only) vehicle of maintaining the religious and social identity of women (and those unable to go to the mosques) through the broadcasting of the call to prayer. While the mosque serves as a structural space that facilitates group identity largely among men, the radio constructs a larger Islamic community that includes women. In this way, by "drawing new types of listeners" (ibid.), media technology reconfigures the concept of the Islamic community. Technology becomes a tool through which women affirm their status within the Islamic community. In terms of the physical proximity of the mosque, women are generally displaced from the community. However, in an electronically mediated context, women are not only involved in the Islamic community but are positioned equally with men in terms of their reception of the Islamic call to prayer.[27]

A new technological development has further enhanced the significance of the *adhan* in redefining the Islamic community beyond geographical boundaries: the concept of the eMuezzin. Based in Preston, U.K., PatelsCornershop.com allows volunteers to become eMuezzin and to deliver the *adhan* in the form of a text message on mobile phones through the Internet five times a day (www.PatelCornershop.com/muezzin.asp). Until the system can be automated, volunteers are requested to register to become Regional or Mosque eMuezzins, the former responsible for sending prayer messages (as well as other community messages) to a specific region (city, state, or country) and the latter oversees the sending of messages to members of a local mosque or Islamic organization. eMuezzins are allowed access to a special site to initiate the sending of the *adhan* to their respective locales.

The eMuezzin concept is not merely a new mode of delivering the *adhan*. Indeed, as with the radio broadcast of the *adhan* in Singapore, it not only reconfigures the conventional practice of the Islamic call to prayer, but it also significantly transforms and constitutes the habitus traditionally framed by it (Bourdieu 1977). PatelCornershop.com states that "the eMuezzin does not have to be a single individual. A group of people can share the task of an eMuezzin. Women are encouraged to be eMuezzins" (www.PatelCornershop.com/muezzin.asp). Thus, with the emergence of eMuezzin in contemporary technological context, the conventional notion of a *muezzin* as a male religious leader with good vocal skills no longer holds primacy—any Muslim man or woman can "produce" the *adhan* in cyberspace. Viewed in this way, web technology democratizes the practice of the *adhan*. The traditional frameworks that govern the conventional practice of the *adhan* remain intact. Media technology, however, enhances the propensity of the Islamic community to create new habitus, the "structured structures predisposed to function as structuring structures" (Bourdieu 1997: 72). The new habitus generates novel practices of the *adhan*, transcending not only its conventional structures, but redefining its social contexts, and reconfiguring the meanings of "reciting" and "listening" to the Islamic call to prayer.

Conclusion

Past critics of mass media and technology have argued that these electronic forms are part of a larger culture industry that induces social alienation and passive reception, empties meaning from life, and is controlled by a dominating and oppressive power (Adorno and Horkheimer 1993; Postman 1993; Keil and Feld 1994). In Singapore, however, the use of the radio broadcast in the call to prayer demonstrates how a community actively employs media technology to maintain collectivity in a pluralistic society: media technology here affirms religious and cultural identity and is absolutely important in the work of Islamic culture production. We might say that Muslims are "traditionalizing" media technology and defining its social significance. Broadly speaking, modern media technologies have become part of a social process in contemporary cultures in which new uses are defined by different social groups with different needs and interests. In response to critics of media technology, then, I quote Paul Théberge (1993: 152), who asserts that, "the ultimate significance of any technological development is . . . neither singular, immediate nor entirely predictable."

Radio broadcast of the call to prayer creates what Schafer (1994: 90) calls "schizophonia": "the split between an original sound and its electroacoustical transmission or reproduction." Schafer intended the term to be "a nervous word" (ibid.: 91) and indicates his anxiety about the impact of sound technology on the sound environment. However, I have argued that, for the Islamic community in Singapore, the radio has been indispensable in their efforts to maintain their cultural identity.

To my knowledge, recordings of the call to prayer (for purposes other than the radio broadcast) are discouraged in Singapore—perhaps for the same reasons that the *Qur'an* is not translated into other languages—and live recitation is the accepted norm. What this suggests is the importance placed on the ephemerality of live performance: the act of recitation is as sacred as the text itself.[28] Why, then, had it become acceptable for the call to prayer to be broadcast over the radio? Such schizophonic technology seems to run counter to conservative Islamic views of sacred performance. Nevertheless, Islamic leaders in Singapore seemed to be convinced that the benefits of media technology outweighed the costs, whatever they were, and that modernization does not necessarily lead to secularization. It is interesting to note that broadcasting the call to prayer has, for better or worse, produced an aesthetics for "good" versions of *adhan* recitation. An *adhan* reciter whose version is chosen by the Islamic Council of Singapore to be broadcast over the radio is revered as a good and skillful reciter whose interpretation is commended or even imitated by other reciters.

Media technology has played an important role in the situation I have just described, making the abstract principles of "space" immediate and tangible, creating an arena that offers possibilities for conscious acts of negotiation among ethnic communities that are divided in cultural and religious experiences. Where one form of media technology had created conflicts, another eased community disputes and created an environment of multicultural tolerance. It is important to remember that discourses about the meaning of space and value of technology are always socially and culturally constructed, and that these issues are among numerous ever-changing constructs that constantly get articulated, modified, and transformed on multiple fronts, including the field of ethnomusicology.

Notes

1. This essay was first presented at the Fortieth Annual Meeting of the Society for Ethnomusicology, 19–22 October 1995, Los Angeles, California. It is an extension of my master's thesis, *Musical Processes and Their Religious Significance in the Islamic Call to Prayer* (University of Pittsburgh, 1995), written under the supervision of Bell Yung and René T. A. Lysloff. Fieldwork for the thesis was conducted in Singapore from May to July of 1994. I am grateful to René T. A. Lysloff with whom ideas for this essay were initially conceived and for his patient guidance, invaluable criticisms, and insights in later revisions. This version is revised and updated from that published in *Ethnomusicology* 43(1) (1999).

2. The exact prayer times, or *Waktu Sembahyang*, for each time period differs slightly throughout the year. In Singapore, the prayer times are published in a pamphlet by MUIS (Majlis Ugama Islam Singapura), the Islamic Council of Singapore. Prayer times may also be viewed on the website of the Malay television channel, Suria, at http://suria.mediacorptv.com/freebies/prayertime/index.htm.

3. *Muezzin* (or *mu'adhdhin*) refers to the *adhan* reciter, usually a respected male member of a particular Muslim community with good vocal qualities.

4. The broadcast is sometimes used by *muezzin* in their respective mosques as a cue to begin the call to prayer.

5. The majority of the Malay population in Singapore is Muslim, that is, follow the Islamic faith.

6. In 1994, the two main channels were in English and Mandarin; Malay and Tamil programs were broadcast as part of these channels. Today, the Suria and Vasantham Central channels cater specifically to the Malay and South Asian communities respectively.

7. It is difficult to ascertain when loudspeakers began to be used to amplify the call to prayer in Singapore. Recollections of several elderly Muslims suggest that the use of loudspeakers probably began in the early 1950s. Prior to that, the *kentung*, a wooden, cylindrical idiophone usually placed in the mosque, was struck to summon Muslims in the community for prayers. In such situations, the *adhan* would be recited after the congregation arrived.

8. In 1975, a Mosque Building Fund Scheme was established by MUIS. Under this scheme, all Muslims can make monthly contributions towards the building of mosques in the newly developed, urbanized housing estates (Majlis Ugama Islam Singapura 1991).

9. The project was known as the Ring Concept plan, with an emphasis on "the progressive urbanisation of Singapore's landscape" (Sequerah 1995: 186).

10. See Chua 1995b for a critical discussion of how the concept of nostalgia (for the rural past) is appropriated in contemporary Singapore as a form of social critique of the present.

11. "Kampung" (or "kampong") is a Malay word referring to "village." "Kampung" and "village" are often used interchangeably; however, as in this case, "kampung" sometimes suggests a more rural setting in contrast to the village, which connotes a more urbanized context.

12. The three largest ethnic communities in Singapore are the Chinese, Malay, and Indian (predominantly Tamil-speaking); two other prominent communities

are the Eurasians and Peranakans. Chua (1991) delineates two distinct types of villages in Singapore, namely, Chinese and Malay.

13. It is important to note that not all Malays in Singapore are Muslims, and that the Islamic community comprises members of other ethnic groups, such as Indians and Chinese.

14. "Adat" refers to the traditional or customary cultural practices of the Malays.

15. See *Straits Times*, 5 June 1974 and 10 June 1974, and *Berita Harian*, 2 June 1974 and 11 June 1974, and 20 June 1974, for descriptions of the Meeting. These Islamic organizations petitioned against the demolition of religious buildings and burial sites, as well as the noise abatement campaign that regarded the amplified call to prayer as "noise."

16. Extracted and translated from a newspaper article written by Haji Muda Baru, a participant at the gathering who was expressing his views on the Meeting. He was directing his disappointment at MUIS for their apparent apathy towards the removal of burial grounds and religious sites.

17. Extracted from a speech by then Minister of Social Affairs Othman Wok.

18. See Chua 1995a: 124–46, and Lai 1995: 121–34.

19. Extracted and translated from a newspaper article by Mohd Guntor Sadali.

20. See *Berita Harian*, 14 June 1974a, for further discussions. See *Straits Times*, 20 July 1974, 25 July 1974, 26 July 1974, and 6 Aug 1974 for reports on how the noise abatement campaign affected Chinese street opera performance.

21. Extracted from a speech by then Minister of Social Affairs Othman Wok.

22. The lyrics of the National Anthem of Singapore are in Malay.

23. Singapore is situated at the south of peninsular Malaysia and is surrounded by the islands of Indonesia.

24. Appadurai paraphrased Walter Benjamin's "The Work of Art in the Age of Mechanical Reproduction" (1969: 217–52) as "The Work of Reproduction in an Age of Mechanical Art."

25. Appadurai uses the term "-scapes" to describe such spatial fluidity.

26. See also Appadurai 1996.

27. For example, boys are separated from girls in religious teaching classes for children at the Whitechapel Mosque in London's East End (Wazir 2001).

28. Indeed, Benedict Anderson (1991: 12–19) concedes that the cultural systems of distinct religious communities (such as Islam) were imagined largely through the medium of sacred language and script (such as classical Arabic).

References

BOOKS AND JOURNAL ARTICLES

Anderson, Benedict. 1991. *Imagined Communities: Reflections on the Origin and Spread of Nationalism*. London and New York: Verso. Originally published in 1983.

Appadurai, Arjun. 1990. "Disjuncture and Difference in the Global Cultural Economy." *Public Culture* 2 (2): 1–24.

———. 1996. *Modernity at Large: Cultural Dimensions of Globalization*. Minneapolis: University of Minnesota Press.

Adorno, Theodor, and Max Horkheimer. 1993. "The Culture Industry: Enlightenment as Mass Deception." In *The Cultural Studies Reader*, edited by Simon During, 29–43. London and New York: Routledge.

Benjamin, Walter. 1969. "The Work of Art in the Age of Mechanical Reproduction." In *Illuminations*, edited by Hannah Arendt and translated by Harry Zohn, 217–52. New York: Schoken Books. Originally published in 1955, Frankfurt: Suhrkamp Verlag.

Berland, Jody. 1992. "Angels Dancing: Cultural Technologies and the Production of Space." In *Cultural Studies,* edited by Lawrence Grossberg, Cary Nelson, Paula Treichler, 51–55. New York: Routledge.

Bourdieu, Pierre. 1977. *Outline of a Theory of Practice.* Translated by Richard Nice. Cambridge: Cambridge University Press.

Chua, Beng Huat. 1991. "Modernism and the Vernacular: Transformation of Public Spaces and Social Life in Singapore." *Journal of Architectural and Planning Research* 8 (3): 203–21.

———. 1995a. *Communitarian Ideology and Democracy in Singapore.* New York and London: Routledge.

———. 1995b. "That Imagined Space: Nostalgia for Kampungs." In *Portraits of Places: History, Community and Identity in Singapore,* edited by Brenda S. A. Yeoh and Lily Kong, 222–41. Singapore: Times Editions Private Ltd.

Chow, Rey. 1993. "Listening Otherwise, Music Miniaturized: A Different Type of Question about Revolution." In *The Cultural Studies Reader,* edited by Simon During, 382–402. London and New York: Routledge.

de Certeau, Michel. 1984. *The Practice of Everyday Life,* translated by Steven Rendall. Berkeley and Los Angeles: University of California Press.

Fiqh us-Sunnah at-Tahara and as-Salah. 1989. "The Call to Prayer (*Adhan*)." In *As-Sayyid Sabiq* [Purification and Prayer], vol. 1, translated by Muhammed Sa'eed Dabas and Jamal al-Din M. Zarabozo. U.S.A.: American Trust Publication, 95–104.

Fiske, John. 1992. "Cultural Studies and the Culture of Everyday Life." In *Cultural Studies,* edited by Lawrence Grossberg, Cary Nelson, Paula A. Treichler, 154–73. New York and London: Routledge.

Friedland, Roger, and Deirdre Boden. 1994. "NowHere: An Introduction to Space, Time and Modernity." In *NowHere: Space, Time and Modernity,* edited by Roger Friedland and Deirdre Boden, 1–60. Berkeley and Los Angeles: University of California Press.

Keil, Charles, and Steven Feld. 1994. *Music Grooves: Essays and Dialogues.* Chicago and London: University of Chicago Press.

Lai, Ah Eng. 1995. *Meanings of Multiethnicity: A Case-Study of Ethnicity and Ethnic Relations in Singapore.* New York: Oxford University Press.

Lysloff, René T. A. 1997. "Mozart in Mirrorshades: Ethnomusicology, Technology, and the Politics of Representation." *Ethnomusicology* 41 (2): 206–19.

Majlis Ugama Islam Singapura. 1991. *New Generation Mosques and Their Activities Bringing Back the Golden Era of Islam in Singapore.* Singapore: Majlis Ugama Islam Singapura.

Postman, Neil. 1992. *Technopoly: The Surrender of Culture to Technology,* 164–80. New York: Vintage Books.

Ross, Andrew. 1991. *Strange Weather: Culture, Science, and Technology in the Age of Limits.* New York: Verso.

Schafer, R. Murray. 1994. *The Soundscape: Our Sonic Environment and the Tuning of the World.* Rochester, Vermont: Destiny Books. Originally published as *The Tuning of the World* in 1977, New York: Knopf.

Seet. K. K. 1995. "Last Days at Wak Selat: The Demise of a Kampung." In *Portraits of Places: History, Community and Identity in Singapore,* edited by Brenda S. A. Yeoh and Lily Kong, 202–221. Singapore: Times Editions Private Ltd.

Sequerah, Pearl. 1995. "Chong Pang Village: A Bygone Lifestyle." In *Portraits of Places: History, Community and Identity in Singapore,* edited by Brenda S. A. Yeoh and Lily Kong, 180–201. Singapore: Times Editions Private Ltd.

Théberge, Paul. 1993. "Random Access: Music, Technology, Postmodernism." In *The Last Post: Music after Modernism,* edited by Simon Miller, 150–82. Manchester: Manchester University Press.

Turner, Victor W. 1995. *The Ritual Process: Structure and Anti-structure*. New York: Aldine De Gruyter. Originally published in 1969, Chicago: Aldine Publishing Company.

Wong, Aline K., and Stephen H. K. Yen (eds.). 1985. *Housing a Nation: 25 Years of Public Housing in Singapore*. Singapore: Maruzen Asia for the Housing Development Board.

NEWSPAPER ARTICLES: STRAITS TIMES (SINGAPORE)

"Adat That No Longer Applies." 22 Dec. 1973: 7.
"Discard 'Adat' Call: Malay Groups Agree." 24 Dec. 1973: 10.
"Muslim Team Appointed to Send Memo to Lee." 5 Jun. 1974: 23.
"Need for Clarification on Noise Pollution and Call to Prayers: Translation of an Editorial from *Berita Harian*, Jun 7." 10 Jun. 1974: 13.
"Politicians Exploiting Religious Issue—Bid to Smear Government: Othman." 18 Jul. 1974a: 1, 17–18.
"Divisive Folly." 18 Jul. 1974b: 10.
"Clamp on Noise from Wayangs." 20 Jul. 1974: 11.
"Let's Throw Our Weight behind Anti-Noise Move." 23 Jul. 1974: 8.
"We're Not So Foolish: Muslim Groups." 24 Jul. 1974: 11.
"Noise Control: Wayangs Seek Deposit Cut." 25 Jul. 1974: 5.
"Of Funeral Rites and Noisy Wayangs." 26 Jul. 1974: 12.
"Muslims Won't Be Affected by Government Cremation Plan." 1 Aug. 1974: 5.
"Plea for Cut in Wayang Fees." 6 Aug. 1974: 5.
"Mattar: Urban Renewal a Must for Progress." 11 Aug. 1974: 5.
"'Face Problems Squarely' Call to Muslims." 26 Oct. 1974: 13.

NEWSPAPER ARTICLES: BERITA HARIAN (SINGAPORE)

"Adakah tindakan itu sengaja bertujuan hendak menghina? [Is the Action Intended to Be Disrespectful?]" 24 May 1974: 4.
"Badan 15 orang dibentuk untuk uruskan masaalah Islam [A Body of 15 Formed to Handle Islam's Woes]" 2 Jun. 1974: 10.
"Rumah2 ibadat yang terjejas [Houses of Worship Jeopardized]" 7 Jun. 1974a: 1, 8.
"Pembangunan semula: Penerangan perlu diberi [Redevelopment: An Explanation Must Be Given]" 7 Jun. 1974b: 4.
"Perjumpaan badan2 Islam: Sikap MUIS dikesali [Gathering of Islamic Bodies: MUIS Attitude Regretted]" 11 Jun. 1974: 4.
"Resolusi mengenai ganti tapak mesjid sudah dipinda [Resolution on the Replacement of Mosque Site Has Been Amended]" 13 Jun. 1974: 1.
"Jangan lekas marah sebelum pelajari tiap persoalan sedalam-dalamnya [Do Not Be Quick to Anger before Studying in Detail Every Issue]" 14 Jun. 1974a: 4.
"Pertemuan badan2 Islam dan Pemerintah dianjur untuk Pemerintah dianjur untuk atasi masaalah yang mereka hadapi [Meeting of Islamic Bodies and Government to Overcome Problems Confronting Them]" 14 Jun. 1974b: 4.
"Kecewa dengan sikap 'lepas tangan' MUIS terhadap soal bersama [Disappointed with the 'Washing of Hands' Attitude of MUIS Regarding the Issue of Togetherness]" 20 Jun. 1974: 4.

WEBSITES

Wazir, Burhan. 2001. "Call to Prayer: How Bengali Boys in London's East End Are Marrying Tradition with Modern Britain," *The Observer*, 18 March (http://www.observer.co.uk/britainuncovered/story).

PatelsCornerShop.com (http://PatelsCornerShop.com).

CHAPTER SIX

Plugged in at Home
Vietnamese American Technoculture in Orange County

Deborah Wong

It's all so hard to explain! But through the book, and through the CD-ROM, you can understand me.   (Pham Duy in conversation, 1995)

It is no longer possible for a social analysis to dispense with individuals, nor for an analysis of individuals to ignore the spaces through which they are in transit.
(Marc Augé [1995: 120])

This chapter is an extremely localized consideration of one musician and a representation of one piece of music. Ethnographies of the particular offer special rewards as well as a corrective to certain habits in cultural studies. My consideration of the Vietnamese composer Pham Duy[1] in the context of Vietnamese American[2] technoculture is self-consciously placed at the intersection of cultural studies and ethnography, i.e., I work within an expectation of cultural construction amidst the free flow of power, but I believe that the best way to get at its workings is through close, sustained interaction with the people doing it and an obligation to address their chosen self-representations.

I begin with the assumption that technology is a cultural practice and that an examination of technological practices in context is the only way to get at what technology "does." Until recently, far too much of the literature on technology has treated it as something outside of or beyond culture, or has simply valorized technology as a force with specific effects and outcomes (such as cultural gray-out). I assume, on the other hand, that any technology is not only culturally constructed but that its uses are culturally defined as well; the "same" technology can thus have very different applications and/or evocative associations in different societies. A certain body of thought has viewed technology as inherently destructive to "tradition," but I regard this too as a historicized cultural belief system rather than a given; if anything, I err in the other direction, assuming that technology carries with it the potential for democracy and community building.[3]

Current work on communication and technology is increasingly informed by cultural studies and anthropology, and "the hypodermic needle theory" of mass communication and its effects is now routinely cited and discarded. In a discussion of audience studies and its contentious interface with cultural studies and ethnography, communications theorists James Hay, Lawrence Grossberg, and Ellen Wartella (1996: 3) note that "the one general area of consensus across this range of shifting, occasionally contradictory positions was their rejection of the 'hypodermic needle' conception of communication which assumed (some long time ago) that audiences were passive receptors—tablets on which were written media messages." Cultural studies provides effective tools for getting at agency without ruling out the possibility of coercion, rather treating it as one dynamic among many. Undoubtedly, technology and the media have effects, but new theoretical models enable considerations of multidirectional results in which production and reception are no longer constructed as binaries or as mutually exclusive. Constance Penley and Andrew Ross address these matters (and their historical background) with a keen sense for how theories of technology are bifurcated, noting that editing their groundbreaking collection *Technoculture* led them to be "wary, on the one hand, of the disempowering habit of demonizing technology as a satanic mill of domination, and weary, on the other, of postmodernist celebrations of the technological sublime" (1991: xii).

I focus here on an example of localized production that raises interesting questions about reception. In his groundbreaking book on Chinese American karaoke, Casey Man Kong Lum notes that most media studies have focused on texts "such as television programs, popular magazines, and romance novels [that] do not involve their audiences in the process whereby their semiotic resources are produced" (1996: 16). Without ruling out the possibility that audiences can read against a text, Lum argues that "audiences are certainly limited in the extent to which they can negotiate meaning because they interpret on the grounds built by others" (ibid.). While I think one can locate the production of semiotic resources nearly anywhere, it is not helpful to simply level the field in reaction to earlier models that locked making and receiving into particular sites. There are always reasons for locating cultural production in particular places, and the Vietnamese composer Pham Duy has reasons for turning so single-mindedly to local, family-based output.

Television is one of the major forms of technoculture that has undergone extended crosscultural study. In the West, theorists have argued that television causes social isolation and alienation, but the American anthropologist Conrad Kottak has countered that considerable ethnocentrism lies behind studies like the Annenberg School of Communication's project finding that heavy television watching leads to the "cultivation effect," wherein viewers begin to believe that the real world is similar to whatever they see on television (1990: 11). While this may be true for Americans watching American programming, Kottak maintains that television watching in Brazil follows different patterns, whether in the cities or in remote rural areas. He and his research team found, among other things, that many Brazilians watch TV in groups—not alone in their homes—and that, in rural areas, people with TV sets in their homes are expected to open their windows so that neighbors can watch from the street (Pace 1993). Furthermore, television clearly expanded Brazilians' understandings of regional and social differences within their own country as well as their knowledge of the rest of the world; Kottak concludes (1990: 189) that "Brazilian televiewing *expresses and fuels* hunger for contacts and information. . . . Our village studies confirmed that television can (1) stimulate curiosity and a thirst for knowledge (2) increase skills in communicating with outsiders (3) spur participation in larger-scale cultural and socioeconomic systems, and (4) shift loyalties from local to national events." Overall, Kottak finds that "TV is neither necessarily nor fundamentally an isolating, alienating instrument" (1990: 189). If it is isolating and alienating in the United States, then this is a particular cultural response to the technology.

Similarly, ethnomusicologists have addressed particular technomusical artifacts as expressions of localized community issues and concerns. Mark Slobin's work on sheet music as a cultural technology for Jewish immigrants (1982) was an important first step toward conceptualizing the physical products of music cultures as expressions of local conflicts and challenges. More recently, cassette culture has attracted a lot of attention. One line of work treats cassettes as physical texts that can undergo cultural analysis (Wong 1989/1990, Manuel 1993); another follows Slobin's lead and focuses on localized cassette industries for issues of identity work (Sutton 1985, Castelo-Branco 1987). The shifting space between production and consumption has also been redefined. Popular music theorist Steve Jones (1990) addresses how the accessibility of cassette recording puts production into the hands of nearly anyone, thus generating a "cassette underground" of popular music that rejects the values of the music industry. Ethnomusicologist Amy Catlin refers to Hmong American practices of recording young women's courtship songs on cassette (a traditional practice maintained in the diaspora) and circulating them widely—between Rhode Island and California, for instance—and notes that some young women no longer rely on matchmakers but make and distribute the tapes themselves (1992: 50, 1985: 85 for a photograph). Anthropologist Susan Rodgers (1986) considers the role of cassette recordings of music drama for the Batak of northern Sumatra, concluding that the tapes are "cultural texts" in which the Batak both confront and mediate issues of kinship and ethnic identity. More recently still, some ethnomusicologists have begun to consider how cassettes are used, regarding them as sites for social action based in local concerns (see Greene 1995 and 1999 for detailed ethnographies of cassette use in a Tamil Nadu village). In short, the physical byproducts of certain music technocultures—cassettes in particular—have been treated rather differently over time in response to changing theoretical landscapes. Significantly, research has addressed urban popular musics as well as rural "traditional" musics, finding redefinition through local use in both contexts.

Locating Little Saigon

I have been fascinated by Little Saigon ever since I first visited it in 1991. Little Saigon is the largest Vietnamese community outside of Vietnam and is the unofficial capital of the Vietnamese diaspora.[4] Located in the contiguous suburban towns of Westminster, Garden Grove, Midway City, and Santa Ana in Orange County, California, it is a site of intense mediation. By mediation, I refer to all the forms of technoculture found in everyday life, and I use the term to point to the ways technology is culture; culture is always mediated (i.e., dialogically shaped and filtered), but I am here concerned with the roles of technology in shaping Vietnamese American memory and political resistance. Driving along the broad suburban thoroughfares of Little Saigon, you can't miss these characteristics of the community: its strip malls are full of video and music stores devoted to Vietnamese American singers and songs. Every mall has at least one if not several stores stuffed full of cassettes, CDs, videotapes, and laser disks.[5] I have discussed the centrality of mediated music to Little Saigon elsewhere, in an article on the social importance of Vietnamese American karaoke (Wong 1994), and I consider this essay a companion piece—a look at a

Fig. 1. Little Saigon strip mall with karaoke store. (Photograph by Deborah Wong.)

rather intense site of production that complements my consideration of karaoke as both consumption and production.

Adelaida Reyes (1999a: 152) suggests that the Vietnamese American music industry has changed culture practices, writing that "for most Vietnamese, the love for music has been deflected from performing to listening. As one Vietnamese put it, their musical life is now lived largely through audiocassettes." She looks briefly at the Orange County industry and finds that most recording artists were their own producers, doing all their own marketing and distribution (1999a: 157). She describes the process as follows (ibid.):

> Once the recording and duplicating are finished, they contact the dealers to announce the availability of the new product. Some of the dealers are large audio stores but many are small establishments—bookstores or specialty shops. Artists distribute and sell locally, nationally and internationally. It is a tremendous amount of work and Pham Duy considers himself lucky because he has family members who share the work: he "creates"; his son [Duy Cuong] is the "fabricateur," running the publishing and recording enterprises; his daughters are singers; and his daughter-in-law travels abroad to manage the distribution which is worldwide.

Reyes also provides a fascinating glimpse of the crosscultural negotiations she witnessed between a Vietnamese singer and an American sound engineer during a recording session in a professional studio (1999a: 153–55). The singer was there to overdub a pre-prepared tape of American instrumental musicians playing the accompaniment and was thus locked into given tempos, unless the tape was played faster (thus raising the overall register). At one point, the sound engineer tried to dissuade her from the expected Vietnamese practices of sliding into pitches and ornamenting her line. Reyes suggests that their interaction mirrored Vietnamese/American cultural negotiation more broadly, noting that "There were no overt conflicts. There was considerable but not total loss of control on the part of the recording artist" (1999a: 154).

Little Saigon has a fascinating immigrant mass media industry in its own right, but it is also a terrific site through which to consider theories of mass communication. To oversimplify, one of the perennial questions in media studies is who drives whom? Does the mass media drive people, or vice-versa? Neo-Marxist theorists assert that media technologies create and sustain hegemonic holds over communities and populations, but cultural studies theorists look for fissures that suggest otherwise. I fall into the latter camp, not so much out of any romantic, totalizing belief that the lumpen proletariat ought to have a voice, but because I think both theoretical stances help us to see certain political realities: they help us to understand the consequences of certain technologies as well as to see how envoicing can happen (when it does), sometimes despite tremendously powerful counter-assertive forces. In *Big Sounds from Little Peoples,* Wallis and Malm argue that the multinational music industry has been very difficult to resist, noting that "Governments will . . . have to create systems for redistributing money to cover the expenses involved in keeping local music life alive and flourishing" (1984: 322). While issuing a stern warning about the might of multinational industry, Wallis and Malm maintain great optimistic belief in cultural response, taking pains to avoid demonizing technology: "Whichever way it goes, technology will play an important role. But technology alone will not determine the outcome. People and governments do that" (1984: 324). I will follow their lead and start by discarding the equation that technology equals hegemony: technology is culture, and culture is shaped by resistance as well as acceptance. We also need to look hard at what we call resistance, to be sure we see the messy internal politics that can grant resistance an even more interesting profile.

I consider a single technocultural artifact here as an example of these issues, and turn now to Pham Duy, one of Vietnam's preeminent composers; any Vietnamese or Vietnamese American would not only know his name but would probably be able to sing at least one of his songs. Adelaida Reyes (1999a: 68) writes, "I have not met a Vietnamese who did not know the name, Pham Duy, Vietnam's best known composer, and his use of folk and traditional materials in the music he created and performed—music subsequently heard all over Vietnam as a powerful rallying cry around Vietnamese nationalism during the war against the French." Pham Duy was born in North Vietnam in 1921 and emigrated to the United States in 1975; he is seventy-seven years old at the time of this writing and lives in Little Saigon (Midway City). He is a noted composer of songs as well as a folklorist: his book, *Musics of Vietnam* (1975), is one of the few extended English-language overviews of Vietnamese music. Pham Duy's life covers all of the momentous events of twentieth-century Vietnamese history. He fought against the French in the 1950s as a member of the Viet Minh (the Vietnamese resistance) and later fell out with the Communists over ideological matters;[6] his songs are still banned in Vietnam as a result. His name is inextricably linked to the creation of Vietnamese "new music" (*nhac cai cach*, "reformed music," or *tan nhac*, "modern music"), created from the 1920s to the 1940s by Vietnamese composers well versed in Western musical vocabularies and instruments but wanting to reinvigorate Vietnamese music. Ethnomusicologist Jason Gibbs (1997: 9–10) notes that "this reformed music was not clearly defined, but was generally used to denote the new western-style music composed by Vietnamese" and suggests that it led to the Vietnamese adoption of the artistic figure of the composer. Gibbs also notes that many of the Vietnamese composers associated with this movement made a point of studying traditional Vietnamese folk musics (Gibbs 1998) and that part of their motivation was to change the rather low regard in which many traditional musics were then held.[7]

Pham Duy is an outgoing, energetic man, fluent in French and English as well as in Vietnamese; he constantly goes on lecture tours to Vietnamese communities in the United States and in Western Europe to promote new projects. He is very comfortable making public presentations, and he clearly enjoys talking about his work; he is well known and admired for his promotional skills. He has lived in the Little Saigon area for almost twenty years: he is extremely resourceful and has continued to compose and to make sure that his music gets disseminated. A fiend for music technology, his home on a quiet residential street near the main drag of Little Saigon is also his studio and his business. He has state-of-the-art computers (a PC, a Macintosh, and a PC laptop), several MIDI keyboards, and any number of DAT recorders. At one point I asked him why he thought that the older generation generally doesn't explore new forms of technology, and he retorted, "Because they are fools! But I am a workaholic, a computerholic." A self-taught computerholic, he composes using a synthesizer and creates computer-generated sheet music that he publishes; he is well versed in HTML software and designed his own webpage, which is constantly growing (<http://www.kicon.com/Music.html>, then follow the links to Pham Duy's pages). With the help of his wife and their adult children, he records and packages cassettes and CDs of his works, which are distributed to countless shops in Orange County and far beyond.

Pham Duy's is one of many Vietnamese American home music businesses in Little Saigon, though it is certainly among the most technologically sophisticated. This is significant because Little Saigon is the center of the Vietnamese American mass media: if you buy a Vietnamese American cassette anywhere in the United States, it was almost certainly produced in Westminster, Garden Grove, or any of the townships that constitute Little Saigon. Vietnamese American newspapers, videos, and television news programs are also disseminated nationwide out of Little Saigon. The community is marked by a stunningly varied local mass media that permeates the area.

In fact, a major Internet project in Garden Grove has a central role in the Vietnamese diaspora. Some work in cultural studies has questioned whether virtual space is "real" social space, but this simply is not an issue for many Vietnamese Americans (and indeed for other immigrant communities, especially those created by forced migration). The Vietspace website (<http://www.kicon.com>) was created in 1996 and was logging over 10,000 hits a day in 1998; it has become "one of the premier meeting places—a 21st-century town square—for a worldwide community of refugees spread across thousands of miles," according to an article in the *Los Angeles Times* (Tran 1998). The site was created and is maintained by Kicon, a multimedia software company in Garden Grove that is owned and run by Vietnamese Americans. Although there are dozens of other web-

Fig. 2. Pham Duy in his kitchen, with CD packaging in progress. (Photograph by Deborah Wong.)

sites devoted to the Vietnamese community in diaspora, Vietspace is certainly the most extensive and up-to-date, as it is revised daily. It contains postings of articles from Vietnamese newspapers, wire reports from Vietnam, and radio broadcasts from the Vietnamese station in Westminster, all downloadable for free via audio and video hookups; samples of music videos, film clips, and songs by Vietnamese artists (including Pham Duy) are available, as well as a virtual art gallery of paintings by well-known Vietnamese artists. Most importantly, a missing-persons page allows Vietnamese to search for parents, children, friends, and colleagues who were separated after 1975: some people post photographs along with their personal information, and a number of reunions have resulted. In short, the site creates real connectedness, not simply imagined community, and is a vital link between farflung members of the Vietnamese community in the United States and beyond.

Still, questions of access and its relationship to socio-economic class must be asked for any form of technology, including the web. In some ways, the web is remarkably open and democratic; in other ways, it is linked to hardware that is not widely available in public institutions and that still requires a significant outlay of money. The Vietnamese American community is markedly differentiated and does not have unilateral access to mass communication forms. The major line of demarcation in Vietnamese American identity is relative generation and, after that, arrival date in the United States: the refugees who arrived between 1975 and about 1979 were largely middle- to upper-class Vietnamese whose connections to U.S. government officials facilitated their emigration; some were able to bring financial reserves with them. These immigrants founded Little Saigon and continue to play a leading role in (especially) the business life of the community. The second wave of emigration was characterized by many so-called boat people, who left Vietnam under dire circumstances, often enduring years of relocation camps (in the Philippines, Thailand, and Hong Kong) before resettlement in the United States or elsewhere; this group has found it harder to find a foothold in relocation, as many left Vietnam with few personal resources. Other differences include religion; Vietnamese can be Catholic, Buddhist, or Methodist, and the churches have become community centers for distinct groups of Vietnamese. Finally, respective generation creates fundamental lines of difference within the Vietnamese community, as the second generation and members of the 1.5 generation (those who were born in Vietnam but emigrated to the United States as children or young adults) are often more acculturated to American society and will ideally have different opportunities available to them. All these factors have effects on access to new forms of mass communication technologies. The newspaper article cited above contains a telling anecdote about an elderly Vietnamese man in Toronto who "made a ritual" (Tran 1998) of going to his son's home each morning and waiting for the son to log on to the computer so that he could listen to a radio program from Little Saigon posted to Vietspace; his son was his link to Internet access. Missing from the picture of a mediated Vietnamese America are the Vietnamese who can't yet afford a computer or the monthly fee for a browser (like several Vietnamese American undergraduates I know in nearby Riverside, California, where I teach). The community is thus heavily mediated (and mediating), but unilateral contact and participation in new forms of technology cannot be assumed.

Voyage through the Motherland *on CD-ROM*

Pham Duy has consistently experimented with different technological forms. Indeed, he has participated in virtually all of the major forms of twentieth-century mass communication. As a young man, he left his home in North Vietnam to travel as a *cai luong*[8] singer in a troupe called the Duc Huy Group, touring Vietnam from north to south from 1943 to 1945. He was the first to sing the so-called "new music" (*nhac cai cach*) for Radio Indochine in Saigon in 1944 on a twice-weekly program. Indochine was the first radio station in Vietnam, established by the French but featuring programs in French and Vietnamese from the beginning. Pham Duy was thus in on the ground floor of this fundamentally important form of mass media, and he quickly became well known through it. He joined the Vietnamese resistance in 1945, and broadcasts in which he sang his songs, many about a free Vietnam, were a regular feature of the Viet Minh clandestine radio, transmitted from a cave just outside Hanoi in the north. He writes, "a gun in one hand and a guitar in the other, I went to war with songs as my weapon" (1995: 25).

He is very forthright about the role of the political in his life and work, and he doesn't see its centrality as unusual. He said to me, "In Vietnam, everything—music, poetry—has to do with politics. You cannot avoid it. If you didn't have this situation in Vietnam, you wouldn't have *me*."[9] The "modern music" style in which he writes has an inherently political base, as its genesis was in Vietnamese intellectuals' rejection of French cultural hegemony. Although its Western influence is obvious to a Western ear, the fact that it was written by Vietnamese for Vietnamese listeners means everything for its followers. Its anticolonial ideological foundation had a powerful effect on Vietnamese audiences in the 1930s, as Pham Duy has written in his autobiography (1995: 18): "Modern music . . . had [a] strong psychological

Fig. 3. Pham Duy in the 1940s. (Photograph reproduced with the permission of Pham Duy.)

impact upon the mentality of Vietnamese youth. Gone were the patriotic songs written to ancient Chinese or French tunes, poorly made up, too simple and very much unpolished. The new musical language provided musicians with better means to express emotions and sophisticated feelings. Many realised how powerful music can be and used their songs to stir patriotic sentiment especially among the youth, who would play the key role in the fight against the French colonialists."

Since his arrival in the United States, he has formed several production companies, including "Pham Duy Enterprises" and "PDC Productions," named for himself and his son (Pham + Duy + Cuong). Duy Cuong came to the United States at the age of twenty-one in 1975; he is very adept with new forms of music technology (even more than his father, as Pham Duy readily admits), and the two have collaborated extensively. I have long been fascinated by musicians' home studios as actual sites of cultural production. Duy Cuong's skill with music software is the other half of Pham Duy Productions, their joint business. He works in a Macintosh platform, using Sample Cell II as his main program for mapping and numerous other programs for sound editing, including Sound Designer II, Alchemy, HyperPrism, Infinity Looping, Wave Convert, TimeBandit (for changing a tempo without changing pitch), HyperEngine, Medicine (which allows him to see sound waves and to visually edit them), and Sound Edit. His rack-mounted hardware, set up in his study at Pham Duy's home, lines one wall of the room and includes lots of MIDI sampling processors, e.g. Proteus I and II, Proteus World (for samples of non-Western instruments), two versions of Rack Mount with numerous piano samples, Vintage Keys (with samples of other kinds of keyboards), Music Workstation, a CD writer, and his latest purchase, a 24-bit mixer with a fiberoptic cable connecting it to his computer's CPU. He creates the finished albums, working closely with his father; no mere sound engineer, he is a collaborator of the closest kind.

When Pham Duy first began releasing his music in California (shortly after 1975), he used cassettes, and he continues to do so, but he was also the first Vietnamese American to issue his music on CD; CDs are now

Fig. 4. Duy Cuong in his studio. (Photograph by Deborah Wong.)

ubiquitous in the community, as other production companies picked up the medium once it became clear that they sold. His first CD-ROM appeared in 1995 and is centered on a song cycle whose title is variously translated as either *The Mandarin Road, The National Road,* or *Voyage through the Motherland* (*Truong Ca Con Duong Cai Quan*) and that contains elaborate discussions of his life and work. *Voyage through the Motherland* is arguably Pham Duy's most famous composition. It was composed between 1954 (the year that the Geneva Accord divided Vietnam into two countries) and 1960 and has since come to symbolize a united Vietnam, especially for the Vietnamese now abroad. The work features three large sections (titled "North," "Central," and "South," for the regions of Vietnam) and a total of nineteen songs; the piece depicts a traveler journeying from north to south Vietnam along the Mandarin Road, a highway that runs the entire length of the country. Each movement draws on regional folk melodies, giving the work tremendously strong affective power for Vietnamese Americans, who thus hear it as a nostalgic affirmation of a single Vietnam and, simultaneously, as a strong statement against Communism. When I played Duy Cuong's symphonic arrangement of his father's piece for the undergraduates in my course on Southeast Asian musics in 1992, for instance, the three young Vietnamese American women in the class all asked

Fig. 5. Cover of the CD-ROM version of *Voyage through the Motherland* (*Truong Ca Con Duong Cai Quan*). (Reproduced with the permission of Pham Duy and Coloa, Inc.)

if they could copy the tape, saying that they wanted to listen to it again and again. The work is deeply symbolic, but it arose from Pham Duy's personal experiences. He writes (1995: 68) that he has walked the length of the route four times: "In my life, . . . I have walked on "the mandarin route" four times. The first time when I was a singer of a drama and music troupe [*cai luong*] touring from Hanoi to Saigon. The second time when I left the south to come home after the return of the Expeditionary French Corps in Vietnam in 1945. The third time when, after a few months of training in Hanoi, I went back to the South, joining the resistance. Then in the fall of 1946, when I was slightly wounded, I left . . . to return to Hanoi."

He notes that he began work on the piece in 1954, just after "the great nations of the two capitalist and communist forces agreed, through the Geneva Conference, to divide Vietnam in two parts" (1995: 68). He heard the news while on a ship, en route to Paris to study at the French Institut de Musicologie. He writes in his autobiography that he "decided to protest" this political outcome with his music, through the piece that eventually became *Voyage through the Motherland*.

This work is well known in the different Vietnamese American communities. Adelaida Reyes (1999a: 84–88) describes its performance during a New Year (Tet) celebration in the Vietnamese American community of Woodbridge, New Jersey, in 1984, recounting its arrangement by a local Vietnamese musician into a musical, complete with costumes, scenery, and slides of Vietnam projected onto a screen. Working from the melodies alone (the event took place before the creation and dissemination of Pham Duy's published score), this musician harmonized it and created a production that emulated Broadway musicals, particularly *Nicholas Nickleby*, which he had studied closely through tapes. This example—let alone Pham Duy and Duy Cuong's numerous rearrangements and reissues of the work—suggests that "the piece" has undergone repeated reimagining, and that this process often takes place through other forms of the media and technology.

The CD-ROM of this work is remarkable in a number of ways: it surrounds a single musical work with a number of other histories and leaves it (like all CD-ROMs) up to the user to choose his or her own route through it. It is a profoundly historical document, and it is also intensely multimedia. Among other things, it represents a web of mass-mediated histories and exemplifies Little Saigon's status as the mass media center of Vietnamese America.

The CD-ROM album was a group effort, the first of a partnership that may result in five to ten more CD-ROMs of Pham Duy's work. The music and the entire content of the CD-ROM are Pham Duy's, including the choice of images. The actual programming and design was done by Bui

Minh Cuong, a staff member at Coloa, Inc., the company that produced the album.[10] The music was arranged for MIDI by Duy Cuong, who has arranged most of his father's albums after Pham Duy has written the lyrics, melody, and harmonic progressions.

The web site for Coloa, Inc., includes the following description of the contents of the CD-ROM:

- A multimedia presentation of the 11 pieces in the "Con Duong Cai Quan" cycle in which images of historic Vietnamese landmarks, beautiful scenery, bustling cities, quaint towns, and the Vietnamese people and their multifaceted character, will guide the user through the flowing lyrics of the songs.
- Included are multilingual musical narration and song lyrics presented in Karaoke form, in Vietnamese, English and French.
- The adaptation by Duy Cuong of this musical work "Con Duong Cai Quan," recorded in high fidelity stereo audio. Also included are other Pham Duy's songs in MIDI formats.
- Video clips of 1954 division of the country
- Several of Pham Duy's songs performed by Thai Thanh, Khanh Ly, Ngan Khoi Chorus, Bich Lien . . .
- Commentaries and critiques on Pham Duy's work by Tran Van Khe, Nguyen Ngoc Bich, Annie Cochet.
- Moments with Pham Duy
- An autobiography of Vietnam's prolific composer Pham Duy that will have rare insight into his life from the Khang Chien period, through his Tinh Ca works, up to his latest "Hat Cho Nam Hai Ngan" (Songs for the 21st Century). The presentation will be in Vietnamese, English and French.
- Press footage and reviews of "CDCQ" beginning from the '60's to present day. Some of the articles are translated to Vietnamese, English and French.
- Glimpses at many famous Vietnamese composers, poets and writers who were Pham Duy's contemporaries and how they influenced his music. Included in this section are photos of Van Cao, Vu Hoang Chuong, Dinh Hung, Duong Thieu Tuoc, Han Mac Tu, and many more along with their selected works.
- A short electronic book of "My Country Once Upon a Time," in which PD describes in depth his life journey from his involvement in Khang Chien to his travels throughout VN. Photos of historical landmarks and significant places and cities provide a colorful pictorial guide through VN's past.
- Complete index and references with hypertext jumps that allow quick and easy access to authors, cities, publications, etc. . . .
- Complete music notes of the whole *CDCQ* work that is printable.
- Bibliography and discography of other works by PD provided.

Though centered around the one long work, the CD-ROM obviously contains extensive related information. It is all in Vietnamese, but many sections are available in English and French translation (accessed by clicking). An extensive biography of Pham Duy, featuring lots of photographs, provides a colorful look at his life. The short biographies of his musical contemporaries (both singers and composers) offer an overview of the most famous Vietnamese musicians of the twentieth century. *Voyage through the Motherland* appears in two different mediums. First, the complete musical score in Western notation (done in music notation software) affords the possibility of playing the piece oneself (e.g., on piano). Second, a MIDI karaoke version of the work invites the viewer to participate by singing along while looking at photographs of Vietnamese landscapes and historical sites that fade in and out while song lyrics unfold with the music.

In yet another section, extensive cultural and historical information appears about each of the regions and cities referenced in *The Mandarin Road*—and, as the piece covers virtually the entire country, this section is essentially an introduction to the culture and history of Vietnam (all in Vietnamese). The bibliography and phonography of Pham Duy's work maps out a career that spans Southeast Asia, Europe, and the United States. Last but not least, the section offering numerous reviews and discussions of *Voyage through the Motherland* is staggeringly transnational in scope: the magazine and newspaper articles are in Vietnamese, English, and French; the video excerpt from a television interview with Pham Duy was originally broadcast on Little Saigon TV (showing Pham Duy and a commentator strolling through a park while chatting in Vietnamese); the videotaped excerpt of musicologist Tran Van Khe talking about Pham Duy was shot in Paris; and finally, the video clips of *Voyage through the Motherland* being performed by Vietnamese Americans were filmed in San Jose and Little Saigon in 1993–94.

This overview is only an indication of the kinds of items found on the CD-ROM album—there is much more, and it would take a viewer many hours to go through the entire document. I am particularly fascinated by the choice of a karaoke version rather than something more "complete," though vocabulary is a problem in trying to articulate the difference between this and any version *not* requiring participation. A karaoke version builds in the expectation of interactive involvement, much like CD-ROM technology in itself: both are "activated" by participation. As I argued elsewhere (Wong 1994), karaoke is central to many Vietnamese American social events, from weddings and other parties to informal socializing in restaurants; the Vietnamese American music industry produces hundreds of karaoke videotapes (and many, though fewer, laser disks) every year. One of my Vietnamese American students once invited me to his home for lunch, and the meal was followed by a karaoke session featuring a videotape from Little Saigon that alternated songs in English and Vietnamese, though all were American pop songs; we passed around a microphone that both amplified and added reverberation. Karaoke is central to Vietnamese American social life, so it is no surprise to find it on Pham Duy's CD-ROM, a strategy meant to encourage interaction beyond clicking the mouse.

Furthermore, the CD-ROM is replete with references to other forms of media—television, radio, etc.—and this is no surprise, as twentieth-century forms of mass communication tend to refer back to or even subsume the forms preceding them, thus creating links both metaphorical and actual. For instance, when CD-ROM drives are used to play audio CDs, the software is designed to look like a CD player on the monitor screen. In this case, the references link Pham Duy's work to the bigger mediated picture presented by Little Saigon: clips taken from the Little Saigon television station's programs are part of the connection to a community that exists partly in geographic and social space (in Orange County) and partly in virtual and mass-mediated space. However, I don't want to call this an imagined community, as much as it lends itself to Benedict Anderson's powerful construction, because Vietnamese Americans don't seem to see it as imagined but rather as part of a larger political stance of resistance to the current Vietnamese government.

On the CD-ROM, Pham Duy addresses the importance of computer technology to his work in the United States. *Voyage through the Motherland* was released on CD in a symphonic version on synthesizer, and an interview with Pham Duy about this earlier release is included on the CD-ROM. The interviewer asks him under what circumstances the work was composed, and after some explanation about the historical context, Pham Duy turns to its new, resourceful arrangement in an American context, saying:

My son (Duy Cuong) and I completed *Con Duong Cai Quan* in eighteen months using computer technologies. It would not have been a feasible task had it not been for the use of the Music Sequencer software. It is very costly to use a real symphony orchestra concert to compose. An orchestra can cost 260,000 dollars to perform. As a refugee, I don't have that kind of money, and therefore, had no choice but to use computer as a means to compose. The software help us to acquire all acoustic sounds, ethnic sounds and electronic sounds. One may deem these sounds are artificial, but it could not be done otherwise since I have only limited monetary resources. Had I been a musician in a prosperous country with strong cultures, I would have had an symphony orchestra concert at my disposal.

And yet, Pham Duy and Duy Cuong's commitment to exploring new forms of media technology is more than just financial resourcefulness. Pham Duy and Duy Cuong each said to me several times that they regard computer and sound technology as a means to cross over generational interests and concerns, to "bridge the gap" between the Vietnamese who grew up in Vietnam and those who grew up in the United States. Duy Cuong said that they rely on the computer and its associated technologies both "to preserve the musics of the past *and* to bring that music into the present." Clearly, Pham Duy and Duy Cuong don't use technology only as a means to an end; rather, they reflect on its cultural role in their community and make the most of its possibilities.

Pham Duy's CD-ROM in the Vietnamese American Community

Looking at a product alone provides a limited picture. By itself, a product is not a technoculture; it comes to life only when it is used and thus embedded in particular practices; any technological object—a cassette, a musical instrument, a computer, etc.—only has meaning in the context of praxis. Considering any of these things as autonomous objects is in itself a cultural practice, e.g., the turn-of-the-century Sachs-Hornbostel organological scheme for classifying instruments tells one more about the German scholars trying to organize a basement full of musical instruments from all over the world than about the people who actually played the instruments. Understanding Pham Duy's motivations and intentions in authoring the CD-ROM is certainly part of the picture but not all of it, and yet getting at reception is one of the most difficult and untheorized areas of ethnomusicology and performance studies (and it is completely overemphasized in marketing and advertising, both dominated by the late-twentieth-century aesthetic of overdetermination). John Mowitt is one of few critical theorists to focus on electronic reproduction and reception. He argues that focusing solely on production obscures the shaping force of reception (1987: 177):

> If recording organizes the experience of reception by conditioning its present scale and establishing its qualitative norms for musicians and listeners alike, then the conditions of reception actually precede the moment of production. It is not, therefore, sufficient merely to state that considerations of reception influence musical production and thus deserve attention in musical analysis. Rather, the social analysis of musical experience has to take account of the radical priority of reception, and thus it must shift its focus away from a notion of agency that, privileging the moment of production, preserves the autonomy of the subject.

In the case of Pham Duy's CD-ROM, the conditions and "radical priority" of reception are a diasporic Vietnamese community that believes it will return home and that is dedicated to reproducing the place itself and its metonymic relationship to cultural memory. Mowitt points out that in electronic reproduction, "the production and reception of all music is mediated by the same reproductive technologies" (1987: 194), i.e., electronic media carries with it political implications for control and channeling. The Kicon website and Pham Duy's CD-ROM are clearly dedicated to the cultural maintenance of memory and are predicated on the expectation of similar priorities on the part of its audience, but that audience produces, in turn, Pham Duy and the Vietnamese web specialists who create mediated community. Mowitt argues for an "emancipatory dialectic of contemporary music reception" (1987: 174) that acknowledges the institutions that generate social memory and experience, describing a "socio-technological basis of memory" that has certain political ramifications; I suspect that this is taken for granted by the Vietnamese American cultural producers who work so passionately to reestablish community through technoculture.

In June 1995, just after the CD-ROM was released, Pham Duy invited me to come watch him present the album to the members of a Vietnamese music club in Little Saigon, less than a mile from his home. Completely in his element, Pham Duy talked to some fifty Vietnamese American audience members ranging in age from their twenties to fifties and including an approximately equal number of men and women; he addressed the process of putting this well-known work onto CD-ROM and then the room was darkened and a computer projector threw the monitor image of the CD-ROM program onto a large screen at the front of the room. Pham Duy sat in the front row, watching the screen and commenting through a microphone on what went by; Bui Minh Cuong, the engineer from Coloa, Inc., who designed the CD-ROM, sat at the computer and silently ran through the contents of the album, clicking the mouse in response to Pham Duy's narration and commentary. The presentation took about forty-five minutes; when the lights were turned on, Pham Duy and Bui Minh Cuong went up to the front of the room and took questions from the audience.

The emcee first apologized for problems with the sound blaster on the computer used during the presentation (not the engineer's own) and then said, "Please encourage Pham Duy's efforts to help the Vietnamese community by buying the CD-ROM or encouraging others to buy it." The first question from an audience member was exactly what I would have asked: "Why do you think technology is important for the Vietnamese?" Pham Duy answered at length:

> I consider myself the first [Vietnamese] person to have made a CD. I did it for the community—I didn't mean to sell thousands or millions of them. Now the different music stores, Lang Van, Lang Vo, and others have come out with a million CDs and have squashed us! But at this point I can only count five people who work on CD-ROM—it's not too crowded.
>
> Walking on my way to exercise, I say to a friend, "Please come see my CD-ROM this afternoon!" [He answers,] "What is a CD-ROM?" Which means he doesn't understand what a CD-ROM is. I am really happy that I jumped in, and I guarantee that I will open up a new way for many other people to jump in, too. I think that we only need to have lived in America for two years to adopt this technology. If we lived in America for twenty years and didn't adopt this technology, we would do better to have stayed home. If we have technology going hand-in-hand with art

then we can do much better, for the sake of Vietnam. I really hope Vietnam will change quickly so I can give a hand to Vietnamese culture.

The audience member and Pham Duy each assumed that technology had a specific importance for Vietnamese Americans beyond its obvious uses and focused instead on *the* driving question for Vietnamese Americans of a certain generation, i.e., how can Vietnamese culture be "helped," maintained, preserved, defined in an American context? All other questions move out from this central concern. A doctor in the audience then stood up and offered his thoughts on the relationship between technology and the Vietnamese community:

I don't have any question, but I do have some suggestions I want to share with everyone today. About five days ago I read a San Francisco magazine article saying that we Americans should use CD-ROMs for educating the young, so I feel very happy and proud to see that Pham Duy has put his music on CD-ROM. . . . Pham Duy, you have filled a need of the Vietnamese people living in America with the new technology. This could be a great and effective benefit to the Vietnamese people living in America. The young will follow your CDs and thus understand Vietnamese culture and the history of our country. We will all understand it because he uses both the Vietnamese and English languages [in his CD-ROM]. In my family, I can see that the computer is not only for adults and students but also for my grandchildren, only six, seven, eight years old, who already play on the computer every day. I only use my grandchildren as an example, but I think that the thing this musician [Pham Duy] has given us will continue for a thousand generations. The musician [Pham Duy] deserves our congratulations and respect.

The doctor thus acknowledged that CD-ROM technology has the potential to bridge the perceived generation gap between those brought up in Vietnam and those not; he celebrated its new importance for its role in cultural maintenance. Pham Duy modestly acknowledged the doctor's praise by saying that half the congratulations should go to the engineer, Bui Minh Cuong, because he wouldn't have known what to do without him. "This is a very helpful marriage between technology and art," he said.

The emcee then said that they were out of time but noted that Pham Duy would be going on tour to Washington, D.C., and to Paris the next week to promote his CD-ROM, and he urged the audience to learn the new technology, to "jump in." He then announced contests for writing awards in the local newspapers and an upcoming meeting of the club to introduce the Internet. He said, "What is the Internet and how can we access its programs? What are the dangers and how can we avoid them—the good of the Internet and the bad of the Internet? A group of Vietnamese engineers will discuss this and will show us how to get on to the Internet. This is one of the fastest modern ways to make contact and to send information—it is super fast." He then thanked everyone for coming and the session ended.

Brief though it was, I found this exchange fascinating for its assumptions: technology was not celebrated simply for what it was or for its perceived modernity. Rather, everyone tacitly agreed that it was a new way to uphold and to preserve cultural information and memory. Whereas fear of cultural gray-out is a common middle-class American intellectual response to the expanded role of technology in everyday life, the music club audience was closely focused on the specific ways emergent technologies could help them to cohere as a community with ties to another culture. Also, they did not discuss CD-ROM technology in relation to matters of cultural assimilation: if anything, a marked confidence in maintained Vietnamese identity characterized the discussion.

Places and Non-Places in the Vietnamese Diaspora

I am fascinated on two counts by French anthropologist Marc Augé's little book (really an extended essay) *Non-Places: Introduction to an Anthropology of Supermodernity* (1995). Written in the style used by so many French intellectuals—intimate, poetic, yet assertive—the book outlines the growing presence of "non-places" in our lives, that is, supermarkets, airports, hotels, cash machines, and freeways, or sitting in front of a television set or computer, all described by Augé as having no "organic" social life. He sees these non-places as a consequence of "supermodernity," his term for the cultural logic behind late capitalist phenomena. Non-places are distinguished by Augé from "space" (1995: 77–78): "If a place can be defined as relational, historical and concerned with identity, then a space which cannot be defined as relational, or historical, or concerned with identity will be a non-place. The hypothesis advanced here is that supermodernity produces non-places, meaning spaces which are not themselves anthropological places and which . . . do not integrate the earlier places: instead these are listed, classified, promoted to the status of 'places of memory,' and are assigned to a circumscribed and specific position."

Augé writes that supermodernity is the "obverse" of postmodernity (1995: 30)—the other side of the same coin—and that excess of time and space is its defining characteristic, an "overabundance of events" brought about, among other things, by technology run amuck in a postindustrial world. Augé is particularly bothered by the way that other social spaces are brought into the home by television, creating a "spatial overabundance" (31) of overlapping places. He has a great nostalgia for what he calls "anthropological place" even while recognizing that it is an "invention" and a "fantasy" (75–76). He provides an intellectual history of place and space (referencing de Certeau, among others) and notes that he would "include in the notion of anthropological place the possibility of the journeys made in it, the discourses uttered in it, and the language characterizing it" (81). He argues that space, however, is defined by movement and travel, even (85–86) "a double movement: the traveller's movement, of course, but also a parallel movement of the landscapes which he catches only in partial glimpses, a series of snapshots piled hurriedly into his memory and, literally, recomposed in the account he gives of them, the sequencing of slides in the commentary he imposes on his entourage when he returns." Spaces are thus empty, solitary, ahistorical, shifting. Augé is most of all concerned with spaces that become non-places in the context of supermodernity: over and over again, he evokes "transport, transit, commerce, and leisure" (94) as the effects of supermodernity and the forces that create certain spaces/non-places and shape an individual's relationship to them. The user of a non-place or a traveler through it enters into a certain contractual agreement to allow his or her identity to be defined by the space. Augé sums up these conditions by asserting that "the space of non-place creates neither singular identity nor relations; only solitude, and similitude" (103).

I have gone into Augé's argument at some length because it offers a compelling model for misunderstanding Pham Duy's CD-ROM. I can't help but try to read Pham Duy's activities through Augé's concerns, as I am fairly certain that Augé would consider Pham Duy's CD-ROM a non-place, presumably read in solitude and driven by supermodernity. Augé says that in non-places "there is no room for history unless it has been transformed into an element of spectacle, usually in allusive texts" (103–104), and thus invites us to regard the CD-ROM as a collection of spectacular, nostalgic texts, inviting a certain gaze but offering little more than similitude. Augé does not address the politics of class or movement and travel: his travelers are unmarked, and his example *par excellence* (at the beginning of the book) is a French businessman going to his cash machine, driving along the freeway to the airport, and boarding a plane. Middle-class travel through supermodern non-places is thus presented as emblematic, but, at this point in my essay, its problematic relationship to the forced, traumatic migration of Vietnamese refugees is obvious and troubling. Pham Duy is both forthcoming and reflexive about the place of travel and movement in his life. I asked him why several of his major works are song cycles about roads and epic travel (*Voyage through the Motherland, Song of the Refugee's Road*, etc.), and he answered, "I am the old man wandering, the old man on the road. It is my destiny and the destiny of my people—always moving. The Jews and the Chinese went everywhere, but slowly, gradually. The Vietnamese went all at once—in one day, one hour! *Viet* originally meant to cross over—like an obstacle—to overcome. So this is the essence of the Vietnamese spirit. Now *Viet* just means 'people,' though its real meaning is 'the people who overcome, who cross over.'" Pham Duy thus theorizes the place of movement in his own work as well as in his culture, putting it in the context of historical comparison and the memory implicit in language. Augé renders difference invisible by flattening out travel and thus establishes universalized non-places by unthoughtfully making the businessman's plane flight and the refugee's journey equivalent. In a conversation with James Clifford, Stuart Hall refers to "the fashionable postmodern notion of nomadology" (Clifford 1997: 44), the theorized uprooting of peoples from fixed time and place. By inference, mass communications technologies are part of the apparatus of supermodernity that create non-places.

Music technocultures are considered "real" social spaces by many Vietnamese Americans. Arguing the matter in the abstract, or without an explicit consideration of history and cultural politics, has, it seems to me, little point; I have chosen here to work at a closely local level—some might feel overly local—but I think this is the only way to get at the social workings of any music technoculture. Pham Duy's work could be seen as a concerted attempt to create not only places but "places of memory" that fill a particular role in Vietnamese American culture. Clifford writes (1997: 250), "People whose sense of identity is centrally defined by collective histories of displacement and violent loss cannot be 'cured' by merging into a new national community." Pham Duy's places of memory stand between four different sites: a nation-state that he and many overseas Vietnamese refuse to recognize (the Socialist Republic of Vietnam), the united but colonized nation of pre-1954 Vietnam, the divided Vietnam of 1954–75, and the diasporic communities of overseas Vietnamese in the United States, Australia, and France. Vietnam does not stay put as a spatial or temporal location in any political terms, and landfall in the United States or any other overseas Vietnamese community hasn't resolved the problems of place for many Vietnamese, especially those of Pham Duy's generation. Little Saigon is almost certainly a different place for the Vietnamese American undergraduates now taking my classes than it is for Pham Duy. Clifford has eloquently argued for treating travel as a major condition of the twentieth century rather than an anomaly in relation to historicized places (1997: 3):

> During the course of [my] work, *travel* emerged as an increasingly complex range of experiences: practices of crossing and interaction that troubled the localism of many common assumptions about culture. In these assumptions authentic social existence is, or should be, centered in circumscribed places—like the gardens where the word "culture" derived its European meanings. Dwelling was understood to be the local ground of collective life, travel a supplement; roots always precede routes. But what would happen, I began to ask, if travel were untethered, seen as a complex

and pervasive spectrum of human experiences? Practices of displacement might emerge as *constitutive* of cultural meanings rather than as their simple transfer or extension. . . . Virtually everywhere one looks, the processes of human movement and encounter are long-established and complex. Cultural centers, discrete regions and territories, do not exist prior to contacts, but are sustained through them, appropriating and disciplining the restless movement of people and things.

But even as I feel that I understand something better about the Vietnamese predicament by reading Clifford, I must acknowledge that this explanation would be unacceptable to Pham Duy and many overseas Vietnamese. The route to the United States would be readily traded in if possible; the route of the Mandarin Road constituted Pham Duy's understanding of Vietnam as a place; the route chosen by a reader through the CD-ROM is understood to communicate that dedication to place. The technomusical artifact is thus metaphorical for so much, all at once, that it forces us to confront the supermodernity of the angry refugee.

Until recently, Vietnam was a non-place for overseas Vietnamese: it was not to be returned to, politically beyond reach, and morally repugnant. However, the resumption of diplomatic relations between Vietnam and the United States has changed everything. Vietnamese Americans can and do return to Vietnam, though most only for visits. Duy Cuong recently spent over two years in Vietnam (1995–97), sampling the sounds of Vietnamese instruments to bring back to Little Saigon for digital editing and reproduction;[11] these sounds are an integral part of his latest project with his father, *Minh Hoa Kieu* (The Tale of Kieu), a song cycle telling the quintessentially Vietnamese story of the young woman Kieu. Pham Duy could not return to Vietnam for twenty-five years; indeed, the present government persists in banning his songs despite their continued circulation through pirate cassettes, often brought by overseas Vietnamese, and his visit in 1999 was controversial. Duy Cuong painstakingly weaves the sampled sounds of Vietnamese musicians into the song cycle, over and under the voices of the Vietnamese American singers who sing the main parts. The resulting pastiche could be seen as a postmodern collapsing of past and present, Here and There, but Duy Cuong's purpose is to establish something authentically and unequivocally Vietnamese; as his father said, Duy Cuong is a "fabricateur," but this implies not artifice but rather the technowizard who maintains the past and its places by pulling them into the present.

The proliferation of technologies and their affordability and accessibility to consumers of any sort, including immigrant communities, poses a challenge to theories of mass mediation *and* of travel. Localized uses of technology in Orange County are tied into a diasporic community that has increasingly strong ties to its origin country as economic relations open up again.¹² We must be able to move theoretically from Pham Duy's study and kitchen table to Little Saigon's shops and TV stations, to Vietnamese American communities in other parts of the United States, to Paris, and indeed to Vietnam itself. This transnationalization of the local is far from coincidental—it is central to Pham Duy's sense of purpose. This producer and his product encourage us to theorize the movement of technology, power, and agency through different registers of place.¹³

Notes

1. Pronounced "Fum Tzuuee."

2. I use the term "Vietnamese American" throughout this essay in an attempt to stand outside the terminological identity politics of Vietnamese outside Vietnam. Some prefer to be called "overseas Vietnamese," or *Viet Kieu*, "Vietnamese citizens residing abroad."

3. The editors address these matters in great depth and detail in the introduction to this volume.

4. Though not incorporated as a township, the area is recognized by a highway sign reading "Little Saigon" near the exit off Route 22 in Orange County.

5. See Lull and Wallis 1992 for a discussion of the presence of popular music in the Vietnamese American community in San Jose, California.

6. As Pham Duy describes it, his songs about the personal sufferings and tragedies of the common Vietnamese people led to his parting ways with the Viet Minh. In his autobiography (1995: 40), he writes, "I had sung about the glory, now I sang about the tragedy—to the stern disapproval of the Viet Minh leaders who saw these songs as negative and potentially damaging to the spirit of the Resistance. I was let known of their disapproval in subtle but no uncertain terms, to which I simply did not agree, and did not even care. It was not my desire to be moulded [*sic*] into a war-glorifying propaganda machine.... My unyielding attitude had put me at odds with the Viet Minh leaders and would see me leaving the Resistance movement not long after."

7. Since the 1950s, a further development along similar lines has emerged from the Vietnamese conservatory system, called *nhac dan toc cai bien*, "modernized traditional music." Its blending of Vietnamese musics with European art music and Western pop is discussed in Arana 1994.

8. This is a form of music drama with historical relationships to Chinese opera and to other Southeast Asian forms of music drama.

9. All quotations from Pham Duy are from interviews I had with him at his home in 1995, 1996, and 1998.

10. The CD-ROM is available for $29.95 from Coloa, Inc., which can be contacted at PO Box 32313, San Jose, CA 95132; <email nxbcoloa@aol.com>; website <http://members.aol.com/nxbcoloa/page1.htm.>

11. Duy Cuong's wife, Phoebe Pham, was sent to Vietnam by the advertising company that employs her. They lived in Hanoi from June 1995 until December 1997, and Duy Cuong set up a studio using the computer he brought from home in the United States. He sampled over eighty different Vietnamese instruments and singing styles, asking musicians to play single pitches as well as entire pieces. He recorded them on a professional DAT recorder and then edited them on the computer, taking out the sounds of street noise. He told me that he kept in the additional sounds that make the samples sound "human," e.g., a flute player taking a breath; he even edited this back in so that the sample wouldn't sound "too elec-

tronic," as he put it. He has no plans to copyright or to package the samples, saying, "They're all here [on my computer], and that's good enough." He emphasized that many of the musicians were old and hard to find and thus feels that his archive of samples represents a record of traditional Vietnamese musics that would be hard to match.

12. U.S. trade sanctions against Vietnam were lifted in 1994 and had begun significantly to change the Vietnamese American music scene in Orange County by 1999–2000. Popular music imported from Vietnam raised contentious issues for Vietnamese Americans. Young people flocked to it, while the older generation regarded it as an arm of the Communist government; local radio stations largely refused to play it, even though it sold very well in stores. Imported CDs cost as little as $2, whereas those locally produced were $8–12. Local companies found they couldn't compete: whereas about thirty Little Saigon record companies dominated the diasporic Vietnamese market for many years (reaching a peak in 1995), eight companies had closed by 2000 and more were in trouble. This chapter thus depicts an ethnographic present and a historical moment that has already passed. See Marosi 2000 for more.

13. I am not a specialist in Vietnamese culture, so I am thankful for the help and advice I found along the way. Pham Duy was extraordinarily generous and patient, spending long hours over numerous interviews explaining his work to me; my thanks also to his son, Duy Cuong, for a fascinating tour of their home studios and to Duy Cuong's wife, Phoebe Pham, for help with translation. My student Duc Van Nguyen selflessly provided a detailed translation of the question-and-answer period described above. Jason Gibbs kindly shared his work on Vietnamese popular song history and provided detailed feedback on this essay. I have learned much from Adelaida Reyes's compassionate work on Vietnamese music in diaspora. As always, René T. A. Lysloff provided comments and feedback that made all the difference.

References

Arana, Miranda. 1994. "Modernized Vietnamese Music and Its Impact on Musical Sensibilities." *Nhac Viet: The Journal of Vietnamese Music* 3 (1 and 2): 91–110.

Augé, Marc. 1995. *Non-Places: Introduction to an Anthropology of Supermodernity.* Trans. John Howe. London and New York: Verso.

Castelo-Branco, Salwa el-Shawan. 1987. "Some Aspects of the Cassette Industry in Egypt." *The World of Music* 29 (2): 32–45.

Catlin, Amy. 1985. "Harmonizing the Generations of Hmong Musical Performance." *Selected Reports in Ethnomusicology* 7: 83–97.

———. 1992. "*Homo Cantens:* Why Hmong Sing during Interactive Courtship Rituals." In *Text, Context, and Performance in Cambodia, Laos, and Vietnam,* ed. Amy Catlin. *Selected Reports in Ethnomusicology* 9: 43–60. Los Angeles: Department of Ethnomusicology, UCLA.

Clifford, James. 1997. *Routes: Travel and Translation in the Late Twentieth Century.* Cambridge and London: Harvard University Press.

Gibbs, Jason. 1997. "Reform and Tradition in Early Vietnamese Popular Song." *Nhac Viet: The Journal of Vietnamese Music* 6: 5–33.

———. 1998. "Nhac Tien Chien: The Origins of Vietnamese Popular Song." *Destination Vietnam,* <http://www.destinationvietnam.com/dv/dv23/dv23e.htm>.

Greene, Paul. 1995. "Cassettes in Culture: Emotion, Politics, and Performance in Rural Tamil Nadu." Ph.D. dissertation, University of Pennsylvania.

———. 1999. "Sound Engineering in a Tamil Village: Playing Audio Cassettes as Devotional Performance." *Ethnomusicology* 43 (3): 459–89.

Hay, James, Lawrence Grossberg, and Ellen Wartella. 1996. "Introduction," in *The*

Audience and Its Landscape, ed. by Hay, Grossberg, and Wartella, 1–5. Boulder, Colorado: Westview Press.

Jones, Steve. 1990. "The Cassette Underground." *Popular Music and Society* 14 (1): 75–84.

Kottak, Conrad Phillip. 1990. *Prime-Time Society: An Anthropological Analysis of Television and Culture.* Belmont, California: Wadsworth, Inc.

Lull, James and Roger Wallis. 1992. "The Beat of West Vietnam." In *Popular Music and Communication,* 2nd ed., 207–236. Newbury Park, Calif.: Sage Publications.

Lum, Casey Man Kong. 1996. *In Search of a Voice: Karaoke and the Construction of Identity in Chinese America.* Mahwah, New Jersey: Lawrence Erlbaum Associates.

Manuel, Peter. 1993. *Cassette Culture: Popular Music and Technology in North India.* Chicago: University of Chicago Press.

Marosi, Richard. 2000. "Vietnam's Musical Invasion." *Los Angeles Times,* August 8, pp. A1 and A16.

Mowitt, John. 1987. "The Sound of Music in the Era of its Electronic Reproducibility." In *Music and Society: The Politics of Composition, Performance, and Reception,* ed. Richard Leppert and Susan McClary, 173–97. Cambridge: Cambridge University Press.

Pace, Richard. 1993. "First-time Televiewing in Amazonia: Television Acculturation in Gurupa, Brazil." *Ethnology* 32 (2): 187–205.

Penley, Constance, and Andrew Ross. 1991. "Introduction." In *Technoculture,* viii–xvii. Minneapolis: University of Minnesota Press.

Pham Duy. 1975. *Musics of Vietnam.* Edited by Dale R. Whiteside. Carbondale and Edwardsville: Southern Illinois University Press.

———. 1995. *History in My Heart/Lich Su' Trong Tim.* Edited by Nguyen Mong Thuong. Unpublished manuscript.

Reyes, Adelaida. 1999a. *Songs of the Caged, Songs of the Free: Music and the Vietnamese Refugee Experience.* Philadelphia: Temple University Press.

———. 1999b. "From Urban Area to Refugee Camp: How One Thing Leads to Another." *Ethnomusicology* 43 (2): 201–20.

Rodgers, Susan. 1986. "Batak Tape Cassette Kinship: Constructing Kinship through the Indonesian National Mass Media." *American Ethnologist* 13 (1): 23–42.

Slobin, Mark. 1982. *Tenement Songs: The Popular Music of the Jewish Immigrants.* Urbana: University of Illinois Press.

Sutton, R. Anderson. 1985. "Commercial Cassette Recordings of Traditional Music in Java: Implications for Performers and Scholars." *The World of Music* 27 (3): 23–45.

Tran, Tini. 1998. "Local Link Gives Scattered Vietnamese a Meeting Place." *Los Angeles Times,* June 1, pp. A1, A16.

Wallis, Roger and Krister Malm. 1984. *Big Sounds from Small Peoples: The Music Industry in Small Countries.* London: Constable.

Wong, Deborah. 1989/1990. "Thai Cassettes and Their Covers: Two Case Histories." *Asian Music* 21 (1): 78–104. (Reprinted in *Asian Popular Culture,* ed. John Lent. Boulder: Westview Press, 1995.)

———. 1994. "I Want the Microphone: Mass Mediation and Agency in Asian American Popular Music." *The Drama Review* [T143], 38 (2): 152–67.

CHAPTER SEVEN

Technology and Identity in Colombian Popular Music
Tecno-macondismo in Carlos Vives's Approach to *Vallenato*

Janet L. Sturman

Todo que habia soñado estaba aquí (Everything I had dreamed of was here).
(Carlos Vives, "Tengo Fe")

Reencountering a Forgotten World—The Music Video: La Tierra del Olvido

The peaks of the Sierra Nevada are barely visible. The "whoosh" of the high altitude winds as they rearrange billowy clouds against an azure sky intermittently breaks the silence. Like a great bird, the camera scans the North Atlantic coast of Colombia and then hovers by the window of a wood frame house, in which an indigenous couple stands. The pair faces the vast wilderness of the Tayrona Park and Land Preserve (near Santa Marta), gazing expectantly. The eye of the camera returns to take in the mountains on one side, rimming the ocean; on the other side, their bronze bases and emerald-colored tips majestically frame lush valleys, rich with dense vegetation.

    A blurred red haze streaks across the screen, interrupting this picturesque scenario. It is caused by the headlights of a jeep as it passes a trio of pointed thatched roof dwellings hidden in the hillsides. The jeep pulls into view. A tour bus follows it, bumping along a dirt road into the remote village. Carlos Vives emerges from the jeep; he hugs the indigenous residents and then turns to greet the bus riders, his friends. Eight casually attired musicians, whose dress and visages reveal a range of mixed ancestries (in contrast to Vives), emerge from the bus, which has cane sides and the sign "Mi Ranchito Tour" painted on it.

As the music, a *vallenato-son*, begins, the images on the screen shift between scenes of the group performing and scenes of the various steps involved in setting up a performance in this remote locale. While Vives sings "Tengo Fe," flashbacks show donkeys carrying musical instruments and sound equipment further inland, musicians building canvas party tents, and the sound crew arranging microphones, amplifiers, and loudspeakers. Refreshment has been prepared for the festivities, and Vives pauses to taste some fruit from one of the surrounding tables laden with food. These scenes are juxtaposed with others showing indigenous Aruacho men paddling canoes, and views of a young Aruacho woman standing alone in a clearing, waiting.

Scenes such as these comprise *La Tierra del Olvido,* the video version of Carlos Vives's 1997 CD *Tengo Fe* (I Have Faith).[1] The faith he expresses in the song "Tengo Fe" ostensibly concerns his faith in a lover and the love he shares with her, but as the video makes clear, he is also expressing faith in the potency of this land of overwhelming natural beauty, of the people who inhabit it, and of the customs they have built over time—including the mixing of racial and musical heritages. *Vallenato* was born of this mix, and this video, like the CD on which it is based, celebrates that genre above all, even as it includes numbers featuring *cumbia,* rock, and romantic pop ballad formats. The verse of "Tengo Fe" opens with Vives singing these lyrics:

Mi plegaria nace de adentro	My prayer is born from inside
Pido por los buenos tiempos	I ask for the good times
Recorrer juntos el cuento	Running with a story
De un dolor que no es eterno	Of a sorrow that's not forever

As he sings, the video images continue to juxtapose the high-tech concert scene with images of indigenous life in the province of Santa Marta. A mini-drama unfolds. Vives joins the lone Aruacho woman and holds her hands as he sings the chorus of the song:

De tu amor sueño liberado	Of your love, a free dream
Fue tu amor que me libertó	It was your love that freed me

In a shy, awkward gesture, the young woman removes the shell necklace from around her neck and hands it to Vives. Subsequent cuts show him staring at the necklace and sitting alone on the beach while daydreaming. Later the woman suddenly returns to stand at his side and silently offers him her hand.

The images on the screen make clear a running subtext of the song lyrics expanded throughout the video: Vives has returned to the land of his birth, because of his love for the land, the culture, and the people there, people with whom he identifies. Yet in no way does he reject the electronic and communications technology that has facilitated his cosmopolitan approach to music creation and allowed him international connections, resources, and stature.

Creating a Modern Macondo

This video represents a culmination of a perspective that the cultural theorist José Joaquín Brunner might label *macondismo*. The term refers to Macondo, the mythical coastal land based on the reality that Gabriel García Márquez portrays in his celebrated novel *Cien Años de Soledad* (One Hundred Years of Solitude). Reputedly, the name Macondo came from a plantation that García Márquez used to explore near his childhood home of Aracataca, which expanded in his hands to represent all of Cienaga (Vargas Llosa, 1971: 27, 104–105). Other critics have noted that the place name Macondo circulated in the popular imagination before *Cien Años de Soledad*, particularly in popular song, as a reference to a faraway place from which one never returns (Meija Duque 1996: 28, 39). However García Márquez came to the name, it is his literary vision that forms the point of departure for contemporary uses of the term. For Brunner, it is the syncretic perspective García Márquez used in creating his literary Macondo that is most interesting. The merger of past and present in a nostalgic, timeless vision of the simple as fantastic and the fantastic as both ordinary and familiar seems to offer a new way of facing the future: "Although García Márquez's Macondo may well have been interpreted by artists and scholars in ways that he would eschew, the impact of this real, yet created place and the narrative manner associated with it has had an undeniable impact on Colombian thought. Literary scholar Mejia Duque observes that with *Cien Años de Soledad* García Márquez gave Colombian writers the confidence to tell stories in their own way, restoring the time honored use of myth to frame modern concerns" (1996:34).

Macondo, Brunner claims, represents the means by which Latin Americans engage with modernity: via a paradoxical blend of nature and technology. *Macondismo*, he writes, "glorifies nature . . . it is a refuge where traditional culture [in this instance, European influenced] re-encounters primordial America" (Brunner 1991:21). Brunner sees this theoretical extension of Macondo as "the land of dreams and utopias; it offers an alternative vision of ourselves" (1991: 26). This is the kind of alternative vision of self that Vives explores in his music and videos. His cinematically inspired music creates a sonic narrative of these ideas put into practice, for while he may be using his own life as a point of departure, it is clear that he hopes to represent his country as well as other peoples who struggle to retain their connections to a local heritage in a global economy.

The goals and strategies of Carlos Vives directly resemble those of García Márquez in several ways. Songs by Vives include references to the famous Colombian author as well as to people, places, and perspectives from his writings. Both men admire *vallenato* singers, the troubadours of the laborers and rural residents of Colombian coasts and nearby valleys. García Márquez is frequently claimed to have said that *Cien Años de Soledad* is really a 350-page *vallenato* song (Samper 1995). Vives, like García Márquez, admires the way *vallenato* music captures a distinctive spirit of the Colombian people best felt in the provinces, and, indeed, both men grew up in the coastal provinces.

Yet despite their admiration for provincial customs, both men are advocates of modern technology. While García Márquez invites critique of the impact of industrialization in his novels, he has long been interested in the potential for mass media, particularly film and television, to reach new audiences, and he has worked extensively with film. He told one interviewer: "I came to realize that [despite its technical limitations] cinema had many more possibilities than literature" (García Márquez and Lemos 1988). It is in this domain that Carlos Vives's work stands out. Not for naught have people come to refer to his approach to *vallenato* as "techno-vallenato," for as a musician Vives has been able to use technology to transform and revitalize this working-class musical style and at the same time secure his position as a spokesperson for Colombian music (c.f. Luis Moya 1998). By harnessing the multifaceted nature of the entertainment industry and by using approaches ranging from the simple addition of electronic instruments to creating montages suitable for television broadcast, Vives has been able to reach audiences in ways that even his inspirational mentor might admire.

This merger of Carlos Vives's own technological savvy with García Márquez–inspired visions is what I shall call *tecno-macondismo*.[2] With this perspective Vives implicitly rejects the postmodern claim that geographical boundaries are incidental and replaces it with a mythic vision of real places, where the old and new coexist and people embrace their own heterogeneity. One of Vives's most expressive tools for conveying his views is his approach to *vallenato*.

Vallenato: *Initial Perspectives*

Vallenato takes its name from a kind of instrumental ensemble that features button accordion and percussion and developed in the valley (*valle*)

Fig. 1. Map of Colombia. Source: Kryzsztof Dydynski, *Colombia* (Oakland: Lonely Planet, 1995), p. 27.

at the base of the Sierra Nevada mountains (see fig. 1). As Peter Wade points out in his book *Music, Race, and Nation,* study of the development, distribution, and marketing of this music reveals that its promoters, from its very earliest stages to the present, have sought to balance a reverence for folkloric practice with the demands of a growing popular music industry.

Wade begins his study by reminding readers that *vallenato* is originally music associated with coastal, Caribbean populations. It shares this aspect of its origin with *cumbia*, the most widely recognized type of Colombian popular music. This regional, rural origin makes it somewhat surprising that people in the interior of the country, including those in the major urban centers, consider *cumbia* and *vallenato* to be national types that represent Colombian music for all the country and abroad. Wade argues that the wide embrace of coastal musics is additionally surprising because Bogotá, the nation's capital city, is dominated by people of white European descent who have typically celebrated traditions tied to their own ancestry. Certainly Bogotanos have a history of disdaining the traditions and manners of the coastal populations, which bear far more prominent evidence of African and Amerindian influence than does their own. Wade's view is that the homogenizing tendency inherent in claims to a national culture in Colombia is necessarily balanced by a competing concerning with maintaining the heterogeneity upon which so many aspects of class and social structures have been established (Wade 2000: 4). He also reminds us that Colombians developed their traditions of popular music in response not just to the hegemonic control of national elite culture but also in response to the foreign hegemonies of Europe and the United States. On the musical evidence of this national versus foreign tension, he remarks, "In Colombia this century, much of contemporary culture at various class levels was (and still is) seen as highly modern and also of 'foreign' origins. Much popular music, for example, came from Europe, the United States, Cuba, Argentina, and Mexico, to name only the most prominent [places]" (Wade 2000: 12). Examining the influx of foreign music and its impact invites consideration of the recording industry and its operations on local and international levels. This must be done if we are to fully appreciate the significance of Carlos Vives's approach to *vallenato*. He is certainly not the first to envision *vallenato* as something other than music of the rural or working class coastal peoples, but he has propelled the genre to new status at home and abroad.

New Views of Vallenato: *Control through Re-visioning*

Like Wade, I am interested in how contemporary manifestations of *vallenato* reflect new attitudes toward cultural identity, even as performers and promoters emphasize elements of older, regional practices. Furthermore, various approaches to *vallenato* performance illustrate ways that Latin American conceptions of modern identity contrast with those in North America or Europe, and even with those in other developing countries. My sense of these differences has been shaped by a growing acquaintance with theories regarding mass media, modernity, and identity presented by contemporary Latin American social scientists, among them José Joaquín Brunner, Nestor García Canclini, and Jesus Martín-Barbero. In particular, I am struck by how the performances and the consumer-oriented systems that support *vallenato* and related popular music in Colombia and abroad embody *macondismo*. Of particular interest is how these Macondo-like conceptions of identity share a nostalgia for closer bonds with nature mixed with a seemingly paradoxical eagerness to embrace whatever technology transcends the isolation of a regional position.

Technological Operations: Some Preliminary Perspectives

While a narrow interpretation of the term "technology" might refer primarily to electronic devices, computer-related operations, and the latest in telecommunications, in a broader sense technology refers to the tools people use to make work easier, enhance efficiency, or transform the nature or quality of the task. Most people feel that technology improves life in some essential way, even though the technology desirable for one purpose may produce unwanted, unanticipated, or unrecognized detrimental effects upon some other dimension of activity. In the case of the body of Colombian popular music examined here, technological advances in this bigger sense have affected (1) musical sound production (2) attitudes regarding music, and (3) the distribution of music. This essay will consider each of these effects.

In the realm of musical sound, the employment of electric guitars, synthesizers, computer-managed amplification, and sound mixing has certainly led fans to view the *vallenatos* of Carlos Vives as "techno-*vallenato*." However, non-electronic instrumental combinations should also be considered within the realm of technology in sound production. The history of the genre that follows notes the impact of performers' decisions to embrace musical instruments new to their region and new to the styles then in place.

The impact of instrumental choice invites a look at how technology affects how creators and listeners perceive music. To some degree, I have already introduced this perspective by linking Vives's technological development to Gabriel García Márquez's vision of Macondo. By using new tools, Colombian popular musicians have been able to construct new visions of who they are and who they intend to reach. As this historical overview will indicate, although Vives has a new approach to presenting local traditions, his attempt to reconceptualize *vallenato* is hardly unprecedented.

The impact of technology on ideas about music and representation makes it imperative to also consider distribution technology. Of particular interest are the technological developments that have shaped the growth of the recording and entertainment industry in Colombia. It is more than coincidence that the development of *vallenato* as a distinct form of Colombian popular music corresponds to the rise of Colombia's commercial recording industry. Indeed, the impact of *vallenato* upon Colombians and listeners abroad would be quite different if the style had not attracted industry interest. Carlos Vives's debt to this history and his knowledge of industry operations have allowed him to exert a measure of control over *vallenato* and related music that has greater cultural importance than his own stardom. Furthermore, Vives has been able to harness television and cinema technology to assist in his project.

The transformation of *vallenato* from a localized rural practice to its current international status illustrates how links between international economic networks and technological access have shaped the way people conceptualize themselves, their values, and their music. Thus I will present the history of *vallenato* first and the contradictory motives behind efforts to celebrate the genre in festival contexts second. The Festival de la Leyenda Vallenato (Festival of the Legend of Vallenato) relies on mass media and local customs to celebrate the tradition, but such efforts do not transcend the contradictions between a desire to maintain authenticity (understood as a connection to historic local and, above all, rural practice) and the need to secure commercial support.

Third, I will examine the somewhat unorthodox creative approach of Carlos Vives to *vallenato,* which reveals a personal vision of a collective Latin American imagination that uses international production values as a means of embracing and repositioning local practice. On recordings, on stage, and most certainly in his videos and television broadcasts, Vives presents a multimedia utopia that he generates to frame regional traditions and showcase his own talents. Vives's success with audiences at home and outside Colombia, particularly in the important media centers of Miami and New York City, illustrates the very serious role of so-called non-serious music as a tool for the construction of a national identity in a global economy. In many ways Vives's *tecno-macondismo* shares traits with the approaches of other artists, but his highly visible style amplifies the efforts of other makers of popular culture as they appropriate technology and redefine their products in light of a futuristic yet distinctively Colombian vision.

Recording Mestizaje: A Brief History of Vallenato

Most histories claim that the rural *vallenato* developed in the late nineteenth century, when the German accordion found its way to Colombia as a result of immigration, trade, and travel between Central Europe and the Americas.³ This statement, as Peter Wade, Jacques Gilard, and George List would argue, requires qualification. They insist that *vallenato* as a modern musical style emerged in the 1940s and is closely connected to the operations and interests of the emerging Colombian recording industry. Wade also argues that efforts by scholars and musicians to report a smooth, continuous history from a folkloric past centered on activity in Valledupar represent a willful reconstruction of history to accommodate cherished views of self and national identity (Wade 2000: 61–66).

This more recent view of the history of *vallenato* privileges the role of emerging technology in creating a popular music that represents regional flavor and a more inclusive view of national identity, and I find it convincing overall. On the other hand, the more generally circulated history of *vallenato*, when read not as a history of a specific style but instead as a history of musical activity, remains useful for indicating processes that helped shape the commercially crystallized *vallenato*.

Despite the mythic simplicity of the claim, Colombian music, like the people who create it, reflects a blending, or *mestizaje*, of European practices with indigenous and African ones. The accordion was first adopted as a solo instrument in the coastal regions and neighboring valleys in the 1880s and was initially played by musicians of European descent who migrated to the region (although the accordion used in these early years appears to have differed from the button *accordeon* used in *vallenato* since the 1940s [see Bermúdez 1996: 116]). The accordion may first have been played by immigrants in upper-class venues, but it was not long before *mestizo* (mixed race) musicians in the working-class neighborhoods borrowed the instrument to accompany their singing or to contribute to their dance bands. In some coastal ensembles the accordion was assigned the melody previously carried by flutelike instruments (most notably the *gaita*,⁴ borrowed from the indigenous people of the province) in traditional dances (including the *cumbia*) along Colombia's northeast Caribbean coast. Regional players developed their own ways of playing the accordion and used it to accompany a range of different dances. Two other instruments, the *guacharaca* (a kind of scraper) and the *caja* (a small, single-headed drum), completed what became the traditional *vallenato* ensemble, or *conjunto vallenato* (see fig. 2). Changing methods of playing *caja*, such as playing with the hands rather than with sticks, demonstrate the increasing African influences on the music as it became more associated with the region's working population than with the upper class.

The dance rhythms most popular among coastal populations were the *puya, paseo, son* (originally of Cuban origin), and *cumbia*. *Vallenato* ensembles quickly adopted all of these dances into their repertoire, and the designation for the ensemble and its characteristic instrumentation came to refer to a style of musical performance. It is important to emphasize that while *vallenato* is fundamentally dance music and its most popular, or signature, dance rhythm is the *paseo*, the style is not defined by the dance rhythm alone. Even in the 1990s, when musicians treat instrumentation with far greater flexibility, the term *vallenato* implies a core set of instruments and a performance style more than it does a particular rhythm. A *paseo* performed by a different kind of ensemble would not be labeled *vallenato*.

Vallenato's international currency is due to its dancibility, but locally it

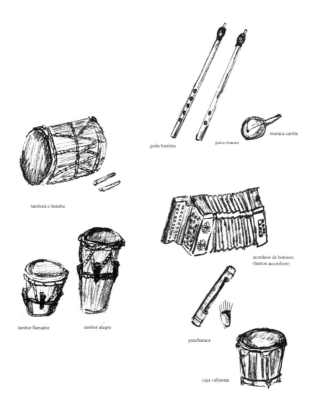

Fig. 2. *Gaita* flutes, folk percussion, and the trio of musical instruments used in classic *vallenato*. Source: Guillermo Abadía Morales, *ABC del folklore colombiano* (Santafé de Bogotá: Panamericana Editorial, 1995), p. 67.

was, and to some extent still is, as much a type of merrymaking and mode of narration as it is a musical genre, style, or set of characteristic rhythms. Three types of celebrations are associated with the early history of the *vallenato: colitas* (end-of-the-party dances), *piquerias* (song duels), and *parrandas* (song fests with virtually no dancing).

At the beginning of the century, reports Samper (1998), the *cola*, the final dance at well-to-do weddings, baptisms, birthdays, and religious festivals, was the opportunity for the servants to share their music with the hosts. The music that the servants had been enjoying during their breaks in the kitchen was brought out at the end of the party for all to enjoy. It was these situations, claims López Michelsen, that encouraged the combination of European rhythms with native ones, giving birth to *vallenato* songs (Weissbecker 1998: 1–2). *Las colitas* were popular primarily in the Valle de Upar and not elsewhere in the province.

More widespread was the practice of song battles between local artists. The lengthy *piqueria* with improvised verses is largely a thing of the past, except for its incorporation into the annual festival of *vallenato*. The famous song "La gota fria" (1938) describes such a face-off, and while the song reflects well-established practice, it also documents a waning tradition ill-suited to the demands of commercial recording. Contemporary performances sometimes feature a contest between *accordeoneros,* but usually only in their capacity as intepreters, not as creators of new songs pertinent to the specific occasion.

Such song duels were perfectly suited to the kind of performance event known as the *parranda,* which was especially important in the years prior to 1940, before the widespread circulation of audio recordings. A *parranda* is a kind of party for hearing music, with its own rules: one does not dance, and there is no schedule. The event can continue for days if desired, and there is lots of drinking (especially whiskey), which local listeners like to supplement with eating goat meat and cheese. Expectations focus on the act of performing or listening. In a *parranda* musicians play only the three traditional instruments of *accordeon, guaracha,* and *caja*. Anecdotes and stories can be part of the musical presentations. Reflecting on the nature of the *parranda,* Rafael Escalona, the legendary king of classic *vallenato,* recalls that hearing the history of a song was as enjoyable as, if not better than, hearing the song itself (Samper 1998: 2). Unlike *colitas, parrandas* still operate in many parts of the coastal provinces. According to Daniel Samper, they best represent the ideal social ambiance for *vallenato*—"its perfect microclimate" (Samper 1998: 3).

Despite accounts of the importance of the *parranda* in local and formal settings, its influence on the commercial promotion of *vallenato* was minimal. While locally *vallenato* developed as a vehicle for singer-songwriters with narratives texts telling stories or commenting on contemporary affairs, as opportunities arose for more widespread distribution singers began to emphasize romantic lyrics, especially favoring those suggesting their skills as lovers.[5] This new focus led to a new understanding of *vallenato*.

Peter Wade and Jacques Gilard argue that not only was *vallenato* affected by the commercial recording industry of the 1930s to the 1950s but that the recording industry actually prompted the creation of *vallenato* as a distinctive musical style in the 1940s. Again focusing on *vallenato* as a specific style, they locate its legitimate roots in the conscious efforts of songwriters from 1935–1960 who worked to attract the attention of sophisticated urban listeners. They hoped also to attract attention to the provincial region of Valledupar. Singers who came from this region, including Rafael Escalona and Freddy Molina, were tired of being considered a second class to elites from Santa Marta or Bogotá (Wade 2001: 63–64). This view does not negate the fact that musical practices leading to the development of what we now consider *vallenato* were underway long before recording technology helped establish it as musical style.

So-called dance academies, a euphemism for houses of prostitution, provided another important venue for *vallenato* performance, largely from 1900 to 1950 (although Marre notes modern examples into the 1980s). In the academies *vallenato* musicians were hired to help attract and entertain customers.[6] Shifts in labor and commercial activity influenced the operations of the academies and the musicians associated with them. In the 1920s and 1930s, the department of Magdalena along the north Caribbean coast experienced a boom in banana cultivation. Workers from the inland region of the *valle* (in the department of Cesar) moved to the coastal region for work, as did immigrants from elsewhere, including the Caribbean islands. During this period *vallenato* ensembles began to absorb more Afro-Caribbean influences, a process that Wade reminds us was augmented by the popularity of Caribbean (especially Cuban) music on commercial recordings that circulated throughout Latin America (Wade 2001: 101–103).

According to Samper, a second wave of musical development followed the migration of tailors from the interior of Valledupar to the coastal provinces (1998: 4). With the decline of the banana trade in the 1940s, cotton was cultivated in the *Zona Bañaera* (the banana-producing region), and people from the interior came to work the plantations. When they returned to their native regions for holidays, they brought with them the new musical customs they had learned along the coast. Hence the coast became connected to the previously isolated interior.

The elite segments of society also transported and transformed *vallenato* by opening the doors of the old salons in Bogotá to these regional dance rhythms. In the 1950s, *vallenato* musicians from wealthy families—Molina, Villazon, Castro, Murgas—traveled to study in the capital. There they became acquainted with elites who were fascinated with sensuous Costeño dancing and songs about the provinces and began to include dressed-up versions of *vallenato* in their celebrations in Bogotá and abroad. Wade confirms the growing fascination of Bogotanos with music from the provinces and links it to the circulation of recordings and to their new readiness to perceive themselves "as being progressive and modern while at the same time being the owners of a region with something particular about it" (Wade 2001: 105).

Festival de la Leyenda Vallenata

Today one of the principal mechanisms for supporting and disseminating *vallenato* is the annual Festival de la Leyenda Vallenata (Festival of the Legacy of Vallenato), held each April since 1968. Here legend and technology meet in subtle ways to establish a potent vision of regional Colombian music: the festival promotes a vision of *vallenato* as a music representing the people, while the recording industry turns to the festival as a means of validating and promoting new artists.

Wade reports that the festival had its roots in an informal gathering between the governor of the department of Cesar, Alfonso López Michelsen, and Gabriel García Márquez in 1963. The occasion celebrated García Márquez's return to Colombia after a seven-year absence; since the author yearned to hear again the music of the region, a festive series of presentations was arranged (Wade 2001: 178). The festival was formally established in 1968, when the governor agreed to transform the annual fiesta of Valledupar, held in honor of the patron saint of that capital city, by adding a competition for *vallenato* accordion players. His hope was to bring widespread attention to the isolated state of Cesar and at the same time make the state, and its representative music, internationally known (Monsalvo Riveira 1998: 1). That year Alejandro Durán was crowned the first king of *vallenato*. Subsequent festivals awarded first, second, and third place prizes for professional *accordeoneros*, as well as prizes for best amateur, best young musician, best unpublished song, and best author-composer.

The competitions at the festival generate attention from well beyond the local region. Judges have included the country's former president, Ernesto Samper, and Gabriel García Márquez.[7] The involvement and visibility of these figures only hints at the web of political, commercial, and artistic motives behind contemporary *vallenato,* which also relies upon its links to the nation's recording industry and commercial artist promotion. Charges have also been made that the festival has drawn support from the illicit drug trade (Marre 1985a). Jeremy Marre reports that the competition winners are frequently performers slated for new record releases, whose prizes have been pre-arranged in hopes of promoting forthcoming recordings (Marre 1985a: 124). While the validity of these charges is difficult to verify, it is clear that the division between the so-called authentic sounds of *vallenato* heard on the festival stage and the commercial variants of the music disseminated by recordings are not neatly separated.

Modern Times and Unsavory Associations

As we have seen, *vallenato* developed in connection with regional patterns connected to subsistence and commerce. Since 1970 the linkage to commerce has grown more complex. The valley of Upar was territory well suited to cattle ranching and certain kinds of farming, but in the 1970s and 1980s the area became known as the marijuana-growing region.[8] Activities associated with the region's controversial cash crop have affected the development and promotion of *vallenato.* Since the valley is not densely populated, it offers safe haven for drug dealers, many of whom took up residence in the area. These opportunists would support the local establishments, where they could gamble, launder money, and listen to *vallenato.* Soon local musicians found themselves with a group of wealthy supporters linked to the drug trade, who helped secure opportunities for recording contracts and performance opportunities outside of the region and especially in the central cities of Medellín and Bogotá, where the biggest domestic recording companies were located. Wade reminds us that while the influx of drug money into the industry undoubtedly affected the circulation of *vallenato* music and musicians, this was certainly not the only development that accounts for the boom of interest in the music in the 1970s and 1980s. Colombian recording companies had already mastered the complex art of turning governmental initiatives, violent politics, fluctuations of capital, and changing marketing procedures to their advantage.

Making *vallenato* widely available via commercial recordings did not automatically ensure widespread acceptance. To make it more urban and hip, this rural, working-class music was also modernized. Wade reminds us that this popularization was often initiated by the working-class musicians themselves, whose exposure to other forms of popular music, including rock, salsa, and merengue, informed their experimentation (2000: 11, 146).

Studio performers began using instruments and combinations typical of popular Latin dance music such as *salsa*. The core instruments of *accordeon, guacharaca,* and *caja* were supplemented by additional percussion, such as a trap drum set augmented by *bongos, timbales,* and *congas,* as well as synthesizers and an occasional saxophone.

Examples of this popularized *vallenato* abound among the recordings of Diomedes Díaz, a recording superstar whose fame began in the 1970s. He quickly developed a devoted following of supporters who viewed his music as both modern and authentic. His many nicknames include *el cacique de la junta,* head of the band (his band is known as "La Junta") and the "bad boy" or "Mick Jagger" of *vallenato* (see García Orjuela 1997, and *El Tiempo*). He records on Sonolux, the nation's largest label, and even when he performs classic *vallenato* songs of the 1950s (such as on his 1997 album *Mi Biografía*), his arrangements favor salsa-influenced enrichments to the accordion-led core instrumental ensemble.

Such transformations, and Díaz's high record sales, still did not remove the associations of the music with working-class patrons, late-night dance clubs, and mafia support and illicit drug traffic. Indeed, Díaz's reputation suffered in the nationwide debate regarding his involvement in the mysterious death of a young female admirer of his (c.f. García Orjuela 1995). Bogotanos dismayed at the saturation of *vallenato* on radio, TV, and blaring tape players cited the murder charge as evidence of the base nature of the music in general. Although Díaz was jailed in October of 1997, many originally thought he would escape prosecution because of his business affiliations and his huge following of admirers.[9]

The connections between *vallenato* and the underworld may be brushed aside by working-class fans, but the associations bolster the disdain of many educated middle- and upper-class urban professionals, who feel free to dismiss *vallenato* as vulgar, unsavory music for hoodlums as well as impoverished music for country folk. Transforming the music with studio production and enriched instrumentation did not remove the stain of corruption or its rural origins for some listeners. The sensual bravura in new songs by Díaz may be typical of a recent wave of *vallenato,* but such songs did little to reinforce the nobler aspects of the tradition (despite Díaz's reverence for the masters of the genre, and his fans' view that he represents a traditional *vallenato* sound).

There is a long tradition of the more cosmopolitan residents in the urban interior of Colombia dismissing the people and customs of the coastal regions as being provincial and unrefined. As one college student from Bogotá noted, "People at the higher economic levels felt embarrassed to listen to such country music, for fear of being categorized with the working poor—synonymous with being considered ignorant" (Betancourt 1997: 3). It took some other changes for people with this view to eventually purchase and play *vallenato* on their home stereos.

Carlos Vives—Beloved Star of Televison and High Society

> "Few countries have the capacity for digesting soap operas as does Colombia . . . it is a national vice." (Gabriel García Márquez, in García Márquez 1989)

In 1991, Carlos Vives, a dreamy-eyed star of *telenovelas* (soap operas) and a relatively unrecognized pop ballad singer, played the role of *vallenato* composer Rafael Escalona on a television drama about his life. The series prompted renewed general interest in the music of Escalona and his contemporaries from the 1940s and 1950s. Sony released a recording of the music heard on the docudrama that featured *vallenato*'s traditional instrumentation augmented by trumpet, flute, and trombone, as well as piano, strings, and synthesizer, under the title *Un canto a la vida* (1991). The success of this album and its followup *Escalona II* (1992) inspired Vives to record his own versions of classic *vallenato*. Wade reports that Sony rejected the artist's novel ideas to frame and back up the traditional accordion-led ensemble with a montage of new options including electric bass guitar, synthesizer keyboards, and *gaitas* and regional percussion, but Colombia's own recording company, Sonolux, agreed to let Vives try (Wade 2000: 217). Communications mogul Ardille Lulle, owner of Sonolux, backed the costly project, and Vives was able to contract with the best arrangers and performers to create a new vision of *vallenato*. The result was the highly successful album *Clasicos de la Provincia* released by Sonolux in 1993 (and licensed to EMI).

The album broke all sales records in Colombia and also did well in other Latin American nations. The experiment transformed Vives's career and propelled the singer to international stardom. His fascination with the country sound of *vallenato* brought a rough rhythmic energy to the overprocessed polish he had used on his earliest recordings of love songs. Vives's *vallenato nuevo* also brought a completely new audience to the genre. Many new listeners considered Vives' s approach to the genre more urbane: "He took music for truck drivers and made it acceptable for intelligent listeners," claims one fan (Morales, interview with author, 1997).

North Americans are likely to find striking parallels between this history of *vallenato* and the history of selected styles of popular music in the United States. The history of country music, for example, especially its transformation from music associated with hillbillies living in the hinterlands to music associated with the rugged western outdoors, shares similarities with the history of *vallenato*. Many contemporary *vallenato* performers (Diomedes Díaz, for example) continue to project a cowboy-like image. The media celebration of the rugged cowboy in the United States, first as frontier hero, then as urban hero, owes a great deal to the way performers presented themselves on radio, film, and perhaps most importantly, television. While Carlos Vives the TV star may bear some similarity to the American television cowboy singers who made country music acceptable for city listeners (such as Roy Acuff or Roy Rogers), he also bears a similarity to country-rock superstar Garth Brooks, whose rock-influenced country music performances blurred boundaries between pop genres. That approach helped attract waves of new young and urban listeners to music largely associated with middle-aged, rural, or working-class fans.

Some fans argue that Vives's approach to *vallenato* is ultimately less important than Vives himself (Betancourt, 1997: 4). Despite being born in Santa Marta, Vives bears no obvious physical evidence of a mixed heritage or the *mestizaje*[10] so characteristic of the region (and indeed most of Colombia). Just as important as his fashion model looks and white European heritage is his social status. Vives is from an upper-class family, and although he was born on the coast, he spent much of his youth living in an upscale district in Bogotá, the city where he now lives. His work in television is the kind of work to which poor and rich alike are drawn, and his long hair and casual outdoor clothing (he frequently performs in short pants and hiking boots) give him the look of a college-age rock star, though he was born in 1961. Vives bestows a privileged, youthful status on *vallenato*, made evident via high-tech media, particularly television. In no way could this image be matched by the ostensibly academic and civic support of the festival *vallenato* or even the high-tech opportunities provided by recording industry alone.

Vives's experience with television cannot be overemphasized. In examining the impact of television on music in Oman, Dieter Christensen wrote, "Among the effects of the mass media is a validation of traditional values, which in turn supports the maintenance of traditional practices" (Christensen 1992: 197). While the conditions surrounding media operations differ in Colombia, the role of television in validating Vives's new vision for *vallenato* and his status as a spokesperson for the genre and the people he claims to represent is equally significant. Television not only provided a mechanism for presenting or even defining tradition, it also offered Vives a means for reaching a much broader local audience than he would have reached with audio recordings alone.

Work in television trained Vives for how to tap into the collective imagination of the Colombian people. Vives began his television career in 1982 acting in the *telenovela Tiempo*. In 1986, he filmed *Gallito Ramirez,* one of the most popular of Colombia's *telenovelas* (Martín-Barbero 1992: 61). The Colombian communications scholar José Martín-Barbero argues that in Colombia, as elsewhere on the continent, the special Latin American brand of soap opera known as *telenovela* offers viewers a universe parallel to their own, a mythic world that allows them to imagine alternative endings for their own dilemmas. While this is a claim that North American viewers might make regarding their soap operas as well, the status that the *telenovela* enjoys in Latin America is much higher than that of the soap opera in the United States. In Latin America the comparatively compact *telenovela* tends to run from eight to ten weeks and is more of a purposeful (though melodramatic) mini-drama than is the North American soap opera, with a much wider collective audience.

The Colombian *telenovela* courts first and foremost a national audience and constructs a national identity by reference to regional customs and habits (Martín Barbero 1992: 75). Though it might appear to be a contradiction in a country with an urban population as large as Colombia, the vast majority of soap operas center around rural scenes and customs, and Martín Barbero argues that this is no accident (ibid.). The fascination with the countryside and its intersection with modern life, portrayed via the adoption of new technologies such as electricity, theater, radio, and motorbikes, is not simply a matter of nostalgia. Rather, it reflects visions of the ever-modernizing urbanite with the heart of a *campesino* (Martín Barbero 1992: 77). Surrounding this contradiction is what Martín Barbero identifies as political reality in Colombia—a powerfully centralized but disintegrated country. He writes, "Colombia will only integrate itself from the bottom levels up, and for this reason, must construct a national identity from regional perspectives. This task demands a constant re-acquaintance with the nation's own plurality" (1992: 77). The makers of the *telenovela*, concludes Martín Barbero, know this secret. And so, apparently, does Carlos Vives. A new fan of Carlos Vives and *vallenato* explains, "Carlos Vives has a special way of helping us through his music to come to terms with our characteristic identity as Colombians, as members of a race with its own roots that should feel proud of itself. He also teaches us that to succeed, it is necessary to know who one is and what one feels, forgetting the fear of what others may think, to be able to express oneself" (Betancourt 1997).[11]

Martín Barbero explains the popularity of the *telenovela Gallito Ramirez* by noting that in it the people of the Caribbean coast saw their own images portrayed positively. At the same time the entire nation overlooked the traditional marginalization of this region by sharing the feelings of *costeños* and seeing their customs as belonging to a general Colombian identity (1999: 75). Similarly, Vives juxtaposes *vallenato*, music of the Santa Marta school yards, and the melodies and instruments of the tribal peoples along the Sierra Nevada, with the tropical rhythms of salsa, the smooth sounds of Latin pop ballads, and modified rock beats, and he implicitly asks Colombians to accept this mix as theirs. While the *telenovela*, along with the visions of Gabriel García Márquez, helped shape Vives's integrative vision of culture, the popularity of these TV dramas contributed to Vives's televisual power, to use Ann Kaplan's term (c.f. Kaplan 1987). His ability to reach a wide general audience that included poor and wealthy alike was already tested, and this helped promote his later influence on the music industry.

As Vives has moved on to new projects (in 2000, he completed *Clasicos de la Provincia II*), he continues to bolster his recording efforts with film and video and to favor angles that document the people and legends representing regional Colombian music as well as those that blur the distinction between past and present. In April of 2000, Vives announced that he was planning a film about the life of the *vallenato* composer José Barros in which he hopes to play the part of Barros (*Reforma* 2000: 24). The linkage is more than vanity on Vives's part; many parallels exist between his life and that of Barros. Born 1915 in Mompox, a city close to the Caribbean coast, Barros was one of the first musicians from the provinces to take advantage of recording opportunities in the interior in the 1940s. During that era he recorded *cumbias* and other coastal genres with dance orchestras in Medellín and Bogotá and helped give coastal genres a new sound (Wade 2000: 121–22). This project promises to build on approaches that Vives explored in the 1997 video *La Tierra de Olvido*, which was created for special television broadcast. In this work he stars in a sequence of narrative dramatic segments merging the techniques of *telenovela* with those of documentary filming. The music, as well as the visual component of *La Tierra de Olvido*, conveys Vives's macondo-like visions as options for an entire nation.

Sounds from La Provincia *and Techno*-Vallenato

In an interview published in *El Espectador*, Vives stated, "What unites us is that we are looking for new ways to create music that constantly bring us back to ourselves." He continues, "It is clear that *vallenato* never left. *Porro* and *cumbia* embodied *vallenato*, and these rhythms can't be separated, even today" (*Reforma*, 2000: 24). These views are evident in the distinctive sound Vives brings to *vallenato*, as a closer look at the video *La Tierra del Olvido* makes clear.

Most striking is the shifting instrumentation used by Vives. La Provincia, his band, is a group of eight versatile musicians. *Provincia* translates as "province" and has customarily been used to refer to the outlying regions of the Caribbean coast and the adjoining territory, thus distinguishing them from inland centers of governmental and social power (Bogotá and, to a lesser extent, Medellín). Ignoring the pejorative interpretations of provincial, Vives defiantly positions the region, like his band, as a site of power. Like García Márquez, Vives turns to the provinces as the source for the most representative voices of Colombia.

The video *La Tierra del Olvido* includes twelve selections, the majority of which could be classified as *vallenato* due to characteristic rhythm patterns (three are *cumbias*). Only three truncated numbers, presented in a scenic section where Vives and friends recall the operations of a *parranda,* feature the traditional instruments of accordion, *guacharaca,* and *caja* alone. All the others feature a shifting array of instruments added to this core trio, including guitars (electric and acoustic), string bass, violincello, viola, violin, piano, keyboards, *tambora* (a type of local bass drum), trap drum set, congas, *gaita,* and sea shells. The instrumental blend cuts across class, ethnic, regional, and stylistic divisions. The bowed strings in "Tengo Fe" add a classical component to the harmonies of the accordion, while the percussion team contributes rock and salsa references. "Cumbia Americana" opens with a wailing electric guitar riff, adds congas to the basic *vallenato* mix, and prominently features the indigenous *gaita* flute in dialogue with the lead guitar, played with a whammy bar.

The blending doesn't stop with instrumentation; Vives also carefully blends old and new. "Tierra del Olvido" is based on an Ika (Aruacho tribal) melody. In the video this song opens as a solo sung by a young Aruacho woman (the same woman who appeared in the previously discussed beach scene). After she sings, her melody is played on *gaita,* with *caja* accompaniment. As the scenes with the trio of indigenous musicians fade, the video cuts to Vives and his paraphrase of the song. His arrangement includes lyrics recounting his desire to return to the land of these forgotten people and to the nearby land where he was raised. The relationship between the melody sung by Vives and those performed by the indigenous musicians is unmistakable, although they are not identical. While the identity of the woman is not clearly identified, Vives tours with a female *gaita* player named Mayte Montero, for whom he composed the song "Pa' Mayte."

The video and many of the songs, especially those written by Vives himself, concern the recapturing of fading memories. The middle of the video includes an homage to *vallenato* artists of the past (the *parranda*-like sequence mentioned above). Here Vives and his friends recall "Monte de la Rosa" by Emiliano Zuleta, "Tierras del Sinu" by Carlos Huerta, and "La Parrandita" by Leandro Díaz. The references to specific locales and to the artist's own activities serve as models for similar approaches in Vives's original compositions. In an interview (*Revista Semana* 1997), Vives says "I like the new people [performers] but there is something I like more about the old ones, and that is they are elementary, not in technical quality, but in that earlier way of thinking. [Today] they lack humility. Our music is not elegant. Making *vallenato* elegant is not the way. It is necessary to return to our primary values. In saying that I am not saying that the music should be poor. Poor is different from humble"[12] (Rincon Vanegas 1997).

The song that follows the homage set, "Malas Lenguas," composed by Vives, recounts him singing *vallenato* for people in Bogotá with his accent from the *Valle*, but only after they know him well enough to excuse his rural speech patterns. The lyrics continue in a traditionally self-reflexive mode: "Emiliano heard me sing 'La Gota Fria' with Egidio's accordion." The Emiliano he refers to is Emiliano Zuleta Díaz, the winner in 1985 and 1987 for author of best unpublished song, and "La Gota Fria" is the most famous *vallenato* song of all time. Egidio is Egidio Cuadrado, now *accordeonero* for La Provincia and the 1985 king of the *vallenato* festival. Vives's musical reflection recalls the mix of shame, honor, and pride felt by those who perform *vallenato*. Urban sophisticates may distain Vives's rural accent, but they admire his sophisticated image and his command of new technology. However, it is his detailed knowledge of regional style, repertoire, and practice that most pleased the older masters of tradition. Asked to defined success, Vives replies, "to lose our shame, to lose the shame of being who we are"[13] (Rincon Vanegas 1997).

Comparing Stances: Other Contemporary Performers

I am not looking to be a star here on earth, as much as a star in the universe.
(Totó la Momposina, singer)

Carlos Vives may be the one of the most ambitious of the contemporary Colombian musicians, but his expressed goal of changing how the world views Colombians through their music is shared by other performers, including those who have not so ostensibly embraced technology as an ally. Totó la Momposina, celebrated in the 1980s for her efforts to promote Colombian folk music, especially Afro-Colombian genres, began in the 1990s to blend folkloric instruments with those associated with urban popular music. With her 1996 CD *Carmelina*, Totó la Momposina performs *cumbia* with the *gaitas* and regional drums but also with saxophones and in salsa-inspired arrangements. Her 1999 release *Pacanto* also includes among the roots-style *cumbias* complex modern arrangements drawing on a range of Latin music sources. The examples with Congolese guitar, West African singers, and Afro-Cuban drumming remind listeners of how tropical music from Latin America circulated to African and back again. Momposina has long championed the musical treasures of Colombia's coastal provinces, but her latest approaches suggest her recognition of new options that take advantage of the latest studio recording technology. Her recordings, though less sentimental and commercially manipulative than those of Carlos Vives, are like his in that they suggest a Colombian model for accepting modernity, a model that keeps the old while embracing the new.

Curiously, the sentimentality of Vives's Macondo-like vision is more optimistic than the unblinking views of the living past that García Márquez presented in his novel. Furthermore, Vives's implied claims of shared destiny on the local and global level seem too sunny a vision even for the members of his band. In 1997, the eight musicians of La Provincia began to record independently from Vives, appearing under the name Bloque. Bloque's more clearly rock-based sound continues to include local rhythms and instruments, but the group's songs tend to attack modern experience rather than wrap it in a film-inspired mix of sound. Bloque's singer and songwriter, Ivan Benavides, explains the difference: "Carlos was very clear about what kind of songs he wanted me to write for him—catchy songs with very optimistic lyrics. It was meant to be commercial music and it was very successful. But I don't think in these kinds of messages: 'We are brothers, bring me your hand and walk together for a new world.' I don't think this way" (Kot 1999: 47). Benavides's vision of modernity differs from that of Vives, but it still shares his respect for local practice.

The Industry at Home and Abroad

The *tecno-macondismo* of Carlos Vives is not just a dream of influence and connection expressed musically. Vives understands the integrative nature of the entertainment industry, and it may be that his influence on industry operations themselves has been as important as his recordings and videos. Although he recorded for Colombia's Sonolux and now the multinational major label EMI, in the early 1990s he also established his own recording label, Gaira Música Local. The company gives him publishing and editorial control and allows him to manage production. Using the operating slogan *"¡Únete con los locales!"* (Unite with local people!), he dedicated the label to promoting the work of rising Colombian stars. Such artists include the female rock sensation Shakira, who now records for Sony and won a Grammy Award for Latin Rock in 2000, and Aterciopelados, a new wave Colombian rock band that, like Vives himself, was nominated for a Grammy award but did not win.

Significantly, these artists have all succeeded first at home and then moved into a global arena. Gabriel García Márquez explained in a 1989 interview that when he began writing there was a movement to create a universal Latin American literature: "Authors wrote with the objective of having their work translated" (García Márquez and Lemos 1989; see also Mejia Duque 1996: 35–36). He goes on to explain that what he instead sensed, and soon discovered to be true, was the importance of first reaching local readers. "When Latin Americans read our works, universalization is automatic" (García Márquez and Lemos 1989). Ultimately his goal has been to change the global market, not to change for it: "We [Latin American writers] are changing North America" (ibid.). That change can be seen in the flow of musical influence as well. Colombian artists have developed the tools to do more than render their experience credible—they have also made it alluring.

Recording sales statistics reveal some of the changes. Of all of the world recording markets, Latin America had the most accelerated growth since the middle of the 1990s, and the Colombian recording industry has become the fourth largest in Latin America (Yúdice 1999: 186–87, 205). The development of Sonolux parallels the career of Vives. The company moved its operations to Bogotá in 1992, the same year that Vives released *Escalona, vol. 2,* and then backed the artist in his groundbreaking release of *Classicos de la Provincia*. In 1995, Sonolux established the most technologically sophisticated recording studio in South America, the same year that Vives released the CD *La Tierra de Olvido*. Sonolux, along with Codiscos (Compañia Colombiana de Discos) has helped make Bogotá second in importance only to Miami in the Latin American recording industry.

Growing hand in hand with increased industry share is the Colombian listener's preference for local voices. Colombians are increasingly supporting domestic music over imported releases; in Colombia, domestic and regional recordings now account for three quarters of the market, with *vallenato* being the most popular in sales. Carlos Vives has been the topselling artist in this market (Yúdice 1999: 203). Despite the dominance of mainstream English-language rock on Colombian radio stations, in the past decade sales for English language rock have dropped in Latin America from 65 percent of the market share to only 32 percent (Yúdice 1999: 190).

This allure has made Vives and his techno-*vallenato* a player in Miami's recording industry. George Yúdice identifies Miami as the center of all Latin American operations. Artists who aspire to international stature must succeed here, where major recording offices, studios, producers, and television channels are concentrated and geared to the Spanish-Latin market. While the market for Latin recordings is fragmented elsewhere in the

United States, all styles, from romantic *boleros* to Texas-Mexican to *música tropical* find support in Miami (ibid., 218, 232). In this domain, Carlos Vives has successfully plied his polished sound and camera-pleasing visions of Colombian musical identity. Working with influential producer Emilio Estefan, he again broke sales records at home and abroad with his own brand of *vallenato* and *cumbia* in the 1999 album *La Tierra de Mi Amor*. The album garnered six Latin Grammy nominations, and although Vives did not win an award for best traditional or tropical pop singer, Estefan did capture the award for best producer of the year. As part of the promotion for this album, Vives toured the United States and Europe with La Provincia and Gloria and Emilio Estefan, increasing visibility for his work and for a vision of Latin music as a means of asserting local values.

Conclusion: Making Experience Both Visible and Credible

The crux of our solitude has been our lack of resources to render our lives credible.
(Gabriel García Márquez, interview, *Tales beyond Solitude*, 1989)

Through his contacts in the entertainment industry, Carlos Vives has mastered the resources to make at least a partial view of Colombian life visible to both a regional and a global audience. In this regard Vives has succeeded beyond the simple measure of personal visibility. He has opened doors for other Colombian artists to share their perspectives with larger audiences. The danger, however, is that the almost visceral appeal of Vives's *tecno-macondismo,* again rather like the reception of García Márquez's literary vision, prompts some to reduce his vision to a slogan. The technological support and delivery modes that Vives and his industry associates apply to *vallenato* have contributed to the genre's international currency, but they have also contributed to a perception of the music as a simplified symbol for an idealized (albeit shareable) vision of regional identity. Some might argue that Vives's polished production has contributed to an overly simplified vision of *vallenato* as one vast category of music, and that the details of rhythm separating a *puya* from a *paseo* have been lost in his new arrangements. Furthermore, his arrangements, with their connections to other tropical musics, especially salsa, as well as rock, jazz, and even classical music, have been criticized for detaching *vallenato* from its regional framework and making it subject to generic institutionalization. These complaints resonate with García Canclini's criticisms of the entertainment industry's tendency to institutionalize local difference as the latest commercial product.

Building on his concept of the McDonaldization of the international entertainment industry, García Canclini offers the term McOndo as a rubric under which to group recent examples of media efforts to institutionalize aspects of local difference emerging from Latin America (García Canclini 1999: 162). The casualty of the new multinational entertainment industry that most disturbs him is not its potential homogenizing effect (*vallenato* that sounds like international salsa, for example) but the fact that local difference becomes just one more product in an industry controlled from outside the localities represented (1999: 163).

To date, Vives has been able to escape this kind of commodification of his vision, in part because he remains involved with production even when he is working with major producers. He has been able to emphasize his appealing message, that respecting the past and honoring local genius does not require eschewing modern technology. Quite the opposite: for Vives, film and video production have provided models that made it possible to return to La Provincia and celebrate its resources (including its myths), simultaneously honoring the old and the new. He has used this model in live and recorded audio performance, essentially splicing together old and new approaches and working with musicians who are comfortable with both worlds. Vives's understanding of popular audience demands, media production options, and performance technology may well prevent what some would feel is an inevitable slide from macondo toward McOndo. For all the romantic dimensions of his vision, he manages to avoid reducing local tradition to some special sauce that distinguishes an otherwise generic product. Ever since his *Clasicos de la Provincia*, local practice has formed the base of Vives's work, not its decoration, and his goal of reaching Colombians first has succeeded. His concerts abroad are packed with Colombian-born immigrants, and his song "La Tierra del Olvido" has become an anthem for Colombians living outside their country (García 2000: 10). High-tech production and performance techniques have made it possible for him to respect the diversity of Colombian experience and to musically render it believable at home and abroad.

Notes

1. Vives released an album titled *La Tierra del Olvido* in 1995, but only the title song from that recording is included on the video.

2. I have chosen to use the Spanish-language spelling of techno (i.e., tecno) when combining it with a Spanish word of any kind.

3. This story has parallels in countless new world traditions of social dance music (for example, the appearance of the accordion in Mexico paved the way for the development of the contemporary *música norteña*).

4. According to Abadia Morales (1995: 66), these "beaked" flutes are adaptations

of the original *kuisis* of the Kogi people of the Sierra Nevada in the province of Santa Marta. Traditionally played as a set of one male flute (*gaita macho* or *kuisi sigi*, with two fingerholes) and one female flute (*gaita hembra* or *kuisi bunzi*, with five fingerholes), they are made from a hollowed cactus stalk that is capped at the extreme top by carbonated beeswax in which a turkey quill is placed. This quill is the mouthpiece into which the air is blown.

5. Wade comments on the widespread view held by Colombian elites living in the interior that Costeño people are characterized by their seductive sexual manners and extroverted nature (Wade 2001: 21–23, 114).

6. There is an obvious parallel to the flourishing of ragtime music as a result of opportunities for performers in New Orleans brothels in the early 1900s.

7. Even though García Márquez now resides in Mexico City, he was born in northern Colombia and considers himself *mestizo* and Caribbean (see García Márquez 1984: 177).

8. The name of Jeremy Marre's provocative film *Shotguns and Accordions: Music of the Marijuana Regions of Colombia* highlights this heritage, as it explores the role of the drug trade in fostering the commerical popularity of *vallenato*. Although he may sensationalize the issue, he is nevertheless one of the few scholars to directly address it.

9. An outlaw image is not all bad for a *vallenato* musician; it may enhance his status as a force to be taken seriously.

10. Wade uses the term *mestizaje* to describe the favoring of light skin color resulting in the so-called "whitening" of racial mixture. I am using the term here in its more widely accepted and general sense, meaning mixed race. Vives shows little physical evidence of mixed racial heritage, and his celebrated status does correspond with the tendency of the upper class to favor and emphasize white European (peninsular) ancestry.

11. Carlos Vives, a través de su música nos ayudo en cierta medida (por lo menos a mi) a encontrar nuestra propia identidad como colombianos, como miembros de una raza con raíces propias, que debe sentirse orgullosa de las mismas y además nos enseño que para triunfar hay que estar seguros de lo que se es y de lo que se siente, olvidando el temor a lo que piensen los demás y por lo tanto a expresarlo.

12. Me gusta la gente nueva, pero hay algo que me gusta más de los viejos y es que son elementales, no en la calidad técnica sino en el concepto de antes. Hace falta humildad en todo. Nuestra música no es de smoking ni de gala. Elegantizar el vallenato no es el camino. Hay que regresar a nuestros valores primarios. Con esto no quiero decir que la música tiene que ser pobre. Pobre es diferente a humilde.

13. Mi exito? Perder la verguenza por lo nuestro, perder la verguenza de ser lo que somos.

References

Abadía Morales, Guillermo. 1995. *ABC del folklore Colombiano* (An ABC of Colombian folklore). Santafé de Bogotá: Panamericana Editorial.

Anonymous. 2000. "Sigue con su vallenato" (Continuing with his vallenato). *Reforma*, April 2: 24.

Bermúdez, Egberto. 1985. *Los instrumentos musicales de Colombia* (Colombian musical instruments). Bogotá: Universidad Nacional de Colombia.

———. 1996. "La música campesina y popular en Colombia: 1880–1930" (Rural and popular music in Colombia: 1880–1930). *Gaceta* 32–33: 11–120.

Betancourt, Tania J. 1997. "Commentarios de la influencia de Carlos Vives en la música colombiana" (Comments on the influence of Carlos Vives on Colombian music). Selections from an unpublished student essay. Bogotá: Universidad de los Andes.

Bravo 57 Online. 1996. Carlos Vives (biografía). <http://bravo57.com/biografias/latinos/carlosvives.html>.

Brunner, José Joaquín. 1992. *América Latina: Cultura y Modernidad* (Latin America: Culture and Modernity). Mexico City: Editorial Grijalbo, S.A. de C.V.

Brunner, José Joaquín with Enrique Gomariz. 1991. *Modernidad y Cultura en América Latina* (Modernity and Culture in Latin America). San José, Costa Rica: FLASCO—Facultad Latinoamericana de Ciencias Sociales.

Christensen, Dieter. 1992. "Music Worlds and Music of the World: The Case of Oman." In *World Music—Musics of the World: Aspect of Documentation, Mass Media and Acculturation* (Intercultural Music Studies 3). Max Peter Bauman, editor. Wilhelmshaven: Florian Noetzel Verlag, 107–122.

Darling, Juanita. 1998. "New, nostalgic kind of music brings hope to Colombia." *The Detroit News,* September 26 <http://detroitnews.com/1998/nation/9809/26/09260076.htm>.

Del Castillo, Luis Alberto. 1997–98. "Reyes del Festival de la Leyenda Vallenata." <http://www.geocities.com/Paris/9478/festival.htm>.

Diomedes Díaz y Ivan Zuleta. 1997. *Mi Biografía* (My Biography). Cassette tape. Bogotá: Sonalux.

Dydynski, Krzysztof. 1995. *Colombia.* Oakland: Lonely Planet.

Estefan, Gloria. 1995. *Abriendo Puertas* (Opening Doors). CD #Ek 67284. New York: Sony/Epic.

Fine, Elizabeth C., and Jean Haskell Speer. 1992. *Performance, Culture, and Identity.* Westport, Conn.: Praeger.

García Canclini, Néstor. 1995. *Consumidores y Cuidanos: Conflictos multiculturales de la globalización* (Consumers and Citizens: Multicultural conflicts in globalization). Mexico City: Editorial Grijalbo, S.A. de C.V.

———. 1999. *La Globalización imagindada* (Imagined Globalization). Buenos Aires: Paises Estado y Sociedad.

García Canclini, Néstor, and Carlos Juan Moneta, editors. 1999. *Las Industrias Culturales en la Integración Latinoamericana* (The Culture Industy in the Integration of Latin America). Mexico City: Editorial Grijalbo, S.A. de C.V.

García, Carla. 2000. "Conquistan NY: Carlos los pone a bailar" (Conquering New York: Carlos makes them dance). *El Norte,* August 15: 10.

García Orjuela, Gabriel. 1995. Diomedes Díaz webpage, including "El Polémico Caso de Diomedes Díaz" (The Controversial Case of Diomedes Díaz), a collection of articles published in various Colombian newspapers concerning Díaz's involvement in the death of a 27-year-old woman. <http:www.geocities.com/Athens/Acropolis/5012/diomedes.htm>.

García Márquez, Gabriel. 1970. *Cien Años de Soledad* (One Hundred Years of Solitude). New York: Harper and Row.

———. 1984. "Fantasía y Creación Artística en América Latina y el Caribe" (Fantasy and Artistic Creation in Latin America). In *Cultura y Creacíon Intelectual en América Latina* (Culture and Intellectual Creation in Latin America). Edited by Pablo González Casanova. Mexico City: Siglo Veintiuno Editores, S.A., 174–78.

García Márquez, Gabriel, with Sylvia Lemos. 1988.*Conversations with Latin American Writers.* Princeton Films for the Humanities—8289.

García Márquez, Gabriel, with Sylvia Stevens, producer, and Holly Aylett, editor. 1989. *Tales Beyond Solitude.* Video recording. Chicago: Luna Films/Home Video.

Kaplan, Ann. 1987. *Rockin' Around the Clock.* New York: Routledge, 1987.

Kot, Greg. 1998. "Not Fusion but Friction: Colombia's Bloque Forges Volatile Rock from Ancient Rhythms." *Chicago Tribune,* April 16: 47.

Levitt, Morton P. 1986. "From Realism to Magical Realism: The Meticulous Modernist Fictions of García Márquez." In *Critical Perspectives on Gabriel García Márquez*. Bradley A. Shaw and Nora Vera-Godwin, editors. Lincoln: Society of Spanish and Spanish-American Studies, 73–90.

Marre, Jeremy, and Hannah Charlton. 1985a. *Beats of the Heart: Popular Music of the World*. London: Pluto Press in association with Channel Four Television Company, Ltd.

———. 1985b. *Shotguns and Accordions: Music of the Marijuana Growing Regions of Colombia*. Beats of the Heart Video Series. New York: Shanachie—SH-1205.

Martín-Barbero, José. 1981. "Transformaciones del Género; de la telenovela en Colombia a la telenovela Colombiana" (Generic Transformations; from Soap Opera in Colombia to Colombian Soap Opera). In José Martín-Barbero and Sonia Muñoz, *Televisión y Melodrama: Géneros y lecturas de la telenovela en Colombia* (Television and Melodrama: Genres and readings of the soap opera in Colombia). Bogotá: Tercer Mundo Editores, 61–106.

Mejia Duque, Jaime. 1986. *Gabriel García Márquez: Mito y Realidad de América* (Gabriel García Márquez: Myth and Reality in America). Buenos Aires: Editorial Almagesto.

Monsalvo Riveira, Cecilia. 1997. "El Festival de la Leyenda Vallenata" (The Festival of the Legend of Vallenato). <http://www.geocities.com/Paris/9478/festival.htm>.

Moya, Luis L. 1997–98. "Vallenato: An Overview." <http://www.uic.edu/~lmoya/vallenato.htm>.

———. 2000. "Vallesounds." <http:/www.vallesounds.com>.

Revista Semana. 1997. "Carlos Vives—Más que un cantante, Carlos Vives se ha convertido en el símbolo de una nueva generación en Colombia" (Carlos Vives— More than a singer; a symbol of a new generation). Bogotá: Revista Semana, May 26. <http://semana.com.co/users/semana/may26/personajes/vives/html>.

Rincón Vanegas, Juan. 1997."El Vallenato se meterá en el corazón del mundo" (*Vallenato* will capture hearts around the world). In *Revista del XXIX Festival Vallenato*. <http://www.geocities.com/Paris/9478/cvives.html>.

Samper Pizano, Daniel. 1995. "Historia del Vallenato." *El Tiempo*, reprinted on <http:www.geocities.com?Paris/9478/samper.html>.

Sonolux, S.A. 1999. Sonolux Website V. 4.0—La historia de Sonolux. <http:://www.sonolux.com>.

Totó la Momposina. 1996. *Carmelina*. CD #YHCD1. Suffolk: Yard High Records (WOMAD/Virgin/EMI) with license from MTM Ltda. Colombia.

———. 1999. *Pacanto*. MTM Ltd. Colombia. Co-produced with Nuevos Medios (Madrid) and Yard High (Suffolk). Now available on Harmonia Mundi/World Village CD #470005 (2001).

Vargas, Héctor Gómez. 1995. "La configuración de la mirada cultural: Medios de comunicación, transformaciones culturales y progresiones orgánicas" (The Configuration of a Cultural View: Communications Media, Cultural Transformations, and Organic. Progressions). *Razon y Palabra*. Generación McLuhan— Primera Edición Especial. <http://www.cem.itsm.mx/. . . logos/mcluhan/confi.htm>.

Vargas Llosa, Mario. 1971. *Gabriel García Márquez: Historia de un Deicidio* (Gabriel García Márquez: History of a Deification). Barcelona: Barral Editores.

Vives, Carlos, and La Provincia. 1991. *Escalona—Un Canto a la Vida* (Escalona—A Song to Life). Sony: CDC464696.

———. 1992. *Escalona, vol. 2*. Sony Latin: CD 81315.

———. 1993. *Classicos de la Provincia* (Classics from the Province). EMI-Latin: CD 27398.

———. 1995. *La Tierra del Olvido* (The Forgotten Land). Gaira Música Local, Sonalux: CD #01–0139–02038.

———. 1997a. *Tengo Fe* (I Have Faith). Bogotá: Gaira Música Local, Sonalux: CD #01–0139–02266.

———. 1997b. *La Tierra del Olvido.* Music video. Rafael Noguera and Carlos Vives. writers. Manuel Rivera, producer. Bogotá: Gaira Música Local, Sonalux, for RCN TV.

———. 1999. *El Amor de Mi Tierra* (Love of My Land). EMI-Latin: CD 22854.

Wade, Peter. 2000. *Music, Race, and Nation: Música Tropical in Colombia.* Chicago; University of Chicago Press.

Weissbeckerhaus, Tommy. 1995. "El vallenato: una leyenda hecho realidad" (*Vallenato:* a legend made real). *Colombia Popular—Servicio Informativo del Ejercito de Liberación Nacional.* <http://www.berlinet.de/eln/vallen-e.htm>.

Yúdice, George. 1999. "La industria de la música en la integración América Latina-Estados Unidos" (The music industry in the integration of Latin America and the United States). In *Las Industrias Culturales en La Integración Latinoamericana,* Néstor García Canclini and Carlos Juan Moneta, editors. Mexico City: Editorial Grijalbo, S.A. de C.V., 181–244.

CHAPTER EIGHT

The Nature/Technology Binary Opposition Dismantled in the Music of Madonna and Björk

Charity Marsh and Melissa West

Within feminist theory there has been an ongoing body of work that speaks to the subject of gender and technology as well as to nature and technology. This latter topic has increased relevance in the music world due to the rising interest in electronica, techno, and hip-hop in the last decade. What is considered "natural" is often thought of as opposed to what is considered "technological." Nevertheless, there are musicians interested in working with technologies without compromising stereotypically "natural" issues such as self-awareness of one's origins.[1] Björk and Madonna are two female artists who have chosen different methods to narrow the culturally constructed division between nature and technology. Through an investigation into the issues surrounding electronica, nature, and culture, as well as appropriate theories regarding them, we will illustrate how Madonna and Björk fuse the two seemingly opposing forces. Analyzing the efforts and successes of Madonna and Björk further establishes a crucial stepping stone in the process of "de-gendering" nature and technology in popular music.

The musical connections between Madonna and Björk date back to 1994. The title song of Madonna's album *Bedtime Stories* was written by Björk, and Madonna's vocal delivery in it unmistakably reveals Björk's influence.[2] In several interviews, each has stated her respect for the other's music. However, despite these nexus points within the popular music community, Madonna and Björk are viewed differently. Madonna is labeled a mega pop star, charting in the Top 40, whereas Björk established herself as a new wave and alternative artist and has moved into the realm of electronica in her solo career.[3] Regardless, however, they share an ability to produce music that includes elements traditionally viewed by many in Western societies as conflicting.

Defining Nature and Technology

The nature/technology dichotomy is in a continual state of flux, and nature and technology are also included within other categories of difference, such as feminine/masculine and subjective/objective. Clearly, within the realm of popular music there has been a general acceptance that some aspects of technology are viewed as more natural than others. Because cultural definitions of nature and technology are fluid, we must define these terms carefully and contextualize them.

In Western society, the categories of nature and technology are socially constructed respectively as feminine and masculine.[4] In the field of electronic music, likewise, there is much discussion of "warm" and "cold" sounds—warm is aligned with the feminine or nature and cold is aligned with the masculine or technology.[5] Another factor particular to Madonna and Björk's music is the dichotomy existing between acoustic (coded as natural) and electronic (coded as technological) sounds. Part of the problem with binary oppositions is the appropriation of the elements that are in alignment with nature by those that fall under the heading of technology. Because women's biological capacity to bear children is considered natural, Madonna problematizes the distinction between nature and technology by bringing motherhood into a technological realm. Björk's connection to nature extends to her homeland, Iceland, and arises from Björk's descriptions of Iceland's geographical and social characteristics.[6]

Although Madonna and Björk have different reasons for, and varying methods of, synthesizing nature and technology, both contribute to the dismantling of the nature/technology dichotomy promoted within the realm of popular music by using electronic technology in a non-traditional manner. This paper will focus on Madonna and Björk's views of nature and technology through an analysis of two of their later albums, Madonna's *Ray of Light* (1998) and Björk's *Homogenic* (1997). After a discussion of Madonna and Björk's views on electronic music, we will analyze Madonna's track "Nothing Really Matters" and her embracing of motherhood within a technological setting. We will then move to *Homogenic*, examining Björk's mingling of the concept of Iceland with electronica through her music, album art, and stage presentation. Donna Haraway's concept of the cyborg will be used to theorize the manner in which Björk deconstructs the nature/technology dichotomy as well as to call into question Madonna's relationship with technology.

Binary Oppositions

Binary oppositions comprise the culturally defined value system used predominantly in Western society to categorize difference. This system is comprised of a list of components classified as opposing elements; however, in many instances these elements can be understood through a less value laden approach as actually related to one another. The nature/culture dichotomy relies upon the definition of nature as feminine, subjective, and of the earth, whereas culture is defined as masculine, objective, and controlling the earth.[7] At the center of binary oppositions is gender.[8]

There are a number of factors in popular music that perpetuate the masculine/feminine dichotomy.[9] One of the most influential of these is the division between rock (masculine) and pop (feminine). In her article "Out on the Margins: Feminism and the Study of Popular Music," Mavis Bayton suggests that "far greater value [is] placed on rock as 'serious' music, in contrast to the 'light' and seemingly 'feminine' frivolity of pop" (Bayton 1992: 52). Because both rock and pop are "gendered" using these prominent characteristics, pop musicians are more likely to be women while rock musicians are more likely to be men.[10] This promotes the idea that pop music is less important than rock and also perpetuates the myth that women cannot be "serious" musicians.

Culture's appropriation of nature is manifested through technology, a concept that moves beyond physical objects or machines to a system of relationships and exchanges between the machine, its designer(s), and its users (Terry 1997: 3). Paul Théberge discusses the idea of control over sound and, subsequently, control over technology in his book *Any Sound You Can Imagine: Making Music/Controlling Technology* (1997). Théberge considers an advertisement where a woman is portrayed as half human, half machine (Théberge: 1997, 123)—the presence of the woman is intended to furnish the "naturalness" considered absent in technology alone.[11] In the male-dominated field of technology, the image of woman as machine not only implies male control of technology but also male control of woman, nature, and sound.[12]

Although feminist theorists have tried "to undo [the binary] oppositions, to revalue their terms, to cast them as contrasts rather than as strict dichotomies, or to negotiate a path that avoids too close an alignment with either side" (Code 1995: 191), many of these dichotomies continue to exist as the popular conception of electronica demonstrates.

Electronica

Electronica reached new heights within the culture of rave and techno music in the 1990s.[13] Because of the innovative uses of technology in electronic music, it is often deemed a "masculine" art form, and this leaves little room for women (Marsh 1999: 17).[14] Nevertheless, Björk has managed to challenge the persistence of these dichotomies in electronica. *Homogenic* is a synthesis of electronic sounds, techno noises, string orchestration, percussion, and her piercing voice. Often on a quest for new sounds, arrangements, and approaches in her music. Björk opposes the common belief that technology is cold and soulless, instead believing it to be warm and sentimental. She equates the way most people see technology with their fear of change. In an interview, Björk stated, "People saying 'techno is cold' is rubbish. Since when do you expect the instruments you work with to deliver soul? You do music with computers and get a cold tune, that's because nobody put soul into it. You don't look at a guitar and say, 'Go on then and do a soulful tune.' You have to put soul into it yourself" (Micallef 1997). The analogy she uses to explain the misconceptions of technology contains first a computer, an example of technology, culture, and the masculine, and second a guitar, which represents nature and the feminine.[15] Adding "soul" to her music is another criteria for Björk's compositional process. Björk is a pop musician who grew up listening to and playing electronic synthesizers in the 1980s. Goodwin suggests that this relationship with the synthesizer for Björk's generation was crucial to a shift in thinking about electronic instruments. He suggests that "the very technology (the synth) that was presumed in the 1970s to remove human intervention and bypass the emotive aspect of music (through its 'coldness') became the source of one of the major aural signs that signifies the 'feel'!" (Goodwin 1990: 265). Through her refusal to hear electronic music as cold and soulless, Björk has upset another of the characteristics that help define electronica as a "masculine" form of music. Björk stresses the tools are not responsible for making the music, rather it is the responsibility of the producer and the performer.

With the album *Ray of Light,* Madonna released her version of electronica, combining traditional pop melodies with an ambient groove. During the promotion of the album, it became clear that Madonna intended to bring something very new to the electronic music genre. "She [Madonna] was quoted on VH1 as saying she'd always been interested in electro/

techno music, but felt that it generally lacked emotion. Working with Orbit, she hoped to 'prove that it could be emotional.'" (Rule 1998: 34).[16] Madonna also goes on to say, "My intention was to marry that scene with something personal and intimate. If I have any complaint about so-called electronica, it's the lack of warmth. I like the textures, but sometimes it sounds alienated and cold" (Gunderson 1998). Clearly Madonna felt limited by the current state of electronic music.

Do you remember saying in an interview that techno equals death?

"Yes."

Do you still believe that?

To a certain extent. There was a type of techno I was listening to that had a real emotional void. But I think it's developed into something else and now there's feeling and warmth to it. You can attach it to humanity and before I couldn't. I couldn't feel anything. (Walters 1998: 74)

The dichotomies of warmth and cold emerge in her words here, alluding to the nature/technology binary opposition. Through this statement, Madonna shows an interest in reducing the distance between the two polarities. As is often the case in her career, it is assumed that the men working with Madonna are responsible for the creative output. Clearly, Madonna does not view the production of her albums in this manner:

How do people like William Orbit or Marius DeVries bring warmth to a synthesizer or a machine?

They don't; I do. They bring the cold, I bring the warmth [laughs]. (ibid.)

By assuming that Madonna does not have a role in the compositional process, this journalist in effect robs her of any agency in the production of *Ray of Light*.[17] Madonna's assertion that she brings warmth to the cold electronic sounds actually embraces the feminine of the feminine/masculine dichotomy, reflecting ecofeminism's celebrations of women's connection with nature. She also disrupts the assumption that technology is purely masculine in popular music.

The Power Relations between Producer and Performer

William Orbit produced Madonna's album, while Björk worked with LFO's Mark Bell to produce *Homogenic*. In order to analyze how Madonna and Björk disrupt major dichotomies through their work, it is necessary to explore the consequences of a musical relationship between a male producer and a female performer.[18] Do the general imbalances of power in male/female relationships of everyday life transfer to the recording studio? In a majority of cases male producers have the advantage in the studio due to the technological aspects of the music-making process.[19] Although the power struggle between producer and performer continues to prevail in the popular music industry, Björk and Madonna challenge the gender hierarchy by maintaining control on all levels and by participating in every aspect of the creation of their albums.

Madonna's Relationship with Producer William Orbit

William Orbit was the main producer with whom Madonna worked on *Ray of Light*. "[Orbit's] been breaking boundaries in the ambient underground for years. His *Strange Cargo* series laid the groundwork for countless modern electronic artists, and his remix list reads like a who's who of rock and pop" (Rule 1998: 30). Madonna's relationship with Orbit is not a new one. Orbit mixed several of Madonna's singles in the past, including "Justify My Love" (1990), "Erotica" (1992), and "I'll Remember" (1995). With Orbit as producer, one might even doubt Madonna's agency in the production of *Ray of Light*. Keith Grint has an interesting perspective on this issue: "If Foucault is right that truth and power are intimately intertwined, those seeking to change the world might try strategies to *recruit powerful allies* rather than assuming that the quest for revealing 'the truth' will, in and through itself, lead to dramatic changes in levels and forms of social inequality" (Grint 1994: 71).[20] To these ends Madonna recruited William Orbit, a powerful ally in the technological realm. Madonna's choice of William Orbit in itself represents her creative and active role in achieving an ambient sound. Madonna is often dismissed as having little or no role in the compositional process, yet the choices she makes, from her producers to the sounds she will use, are an integral part of the creative process.[21] On the other hand, Madanna's control over her career is often celebrated. Through Orbit, Madonna is able to change the perception of music and technology, creating her version of electronic music.

Many studies focus on the composition of a piece without looking beyond that to other important aspects of the music. Grint, for example, criticizes feminist studies of technology that only focus on the design and development phases of an artifact's life, as these constructivist studies of technology fail to see women at all (Grint 1994: 18). This is an important point to keep in mind while examining Madonna's album. If we were only to look at William Orbit's role as the producer, we would fail to see the influence Madonna had on the production of the album. William Orbit may produce Madonna's album, but the final product bears Madonna's name and ownership. Grint goes on to propose that studies in the realm of technology focus on the impact technology has on women, rather than the ways women impact the uses of technology (Grint 1994: 18). *Keyboard* magazine had this to say about Madonna's role: "Star artists often keep tight control over their producers. When asked how much creative latitude Madonna allowed Orbit in the studio, she told us that 'William had a very long leash, but I was firmly holding on to the end of it.' Her analogy of the process: 'I was the anchor, he was the waves and the ship was our record'" (Rule 1998: 34).[22] When asked exactly what he did for Madonna in the production of the album, Orbit replied, "It ran all the way from complete tracks, really, to just bare bones backing tracks on which she subsequently put her lyrics and melodies" (Rule 1998: 33). Also, Orbit was not the only person that was brought in to work on the album. Marius DeVries helped with the production,[23] Craig Armstrong orchestrated the violin arrangements,[24] and longtime collaborator Patrick Leonard was brought in to assist with both writing and production. The credits for each and every song on the album name Madonna as a producer along with William Orbit, Marius DeVries, and Patrick Leonard, depending on the song.

Madonna is rarely asked compositional questions in interview situations. Contrast this with an article in *Keyboard* magazine that spends several pages outlining Orbit's compositional techniques on (her!) album. This type of media coverage continues to perpetuate the myth that men are in charge of the technology and compositional work.[25]

Björk's Relationship with Her Co-Producers

Over the years Björk has collaborated with a number of producers. In an interview she explains how she only wants to work with musicians who are as strong as her or stronger.[26] Björk already has a solid understanding of electronica and how the technologies work.[27] The fact that Björk is particular about her co-producers illustrates her desire to maintain a high level of control over her albums. She acknowledges her own strength as a musician and will not relinquish control unless she is matched in musical ability.[28] In order to challenge the gender hierarchy, it is important for artists to completely understand the technological processes used to create music. For example, if Björk is unsure of how a piece of equipment works, she searches for the best person to teach her.[29] Björk's musicianship is well respected by many of the prominent producers and musicians in the field of electronica and techno.[30]

Björk's first two albums, *Debut* (1993) and *Post* (1995), were almost entirely collaborative projects. On *Debut* she worked exclusively with Nellee Hooper. He produced or co-produced ten out of the eleven songs. *Post* was more multicollaborative; Björk co-produced two tracks with Graham Massey, two with Tricky, one with Howie Bernstein, one that uses all three of them, one was produced by Hanslang and Reinsfeld, and four were produced by herself. Björk's first two albums "weren't as much solo projects as duets with the producers who had inspired her: Nellee Hooper, Graham Massey, Tricky, Howie B" (Van Meter 1998: 96). Although Björk's first two albums can be considered both solo efforts and "duets," the male producer's involvement lies just below the surface of each. Björk's musicianship always enables her to maintain her identity in a technological realm.[31] In an interview, Evelyn Glennie, a percussionist who plays with Björk, proclaims her awe at Björk's ability to "hang onto her own identity no matter who she collaborates with."[32] When Björk talks about technology she also acknowledges those people from whom she learns. For example, in a documentary segment she explains how Mark Bell taught her how to use a QY20, a type of sampler for capturing noise and changing the pitch. It has over one hundred sounds, and she is able to create many of her tunes on this portable piece of electronic equipment. In an interview, she expressed her comfort with "taking from both masculine and feminine teachers, and [does not] see any problem with blending tales of conquest and nurture, fort-building and daydreaming" (Powers 1997: 339).

Electronic/Acoustic: Madonna's "Nothing Really Matters"

Madonna's recording of "Nothing Really Matters" reveals a merging of nature and technology. Her vocal delivery and other musical features bring a natural element to its electronic underpinnings. It opens with a synthesized instrumental introduction, where the use of the electronic is foregrounded through timbre (*Ray of Light*, track 6: 0:00–0:19). The first noise the listener hears is the percussive grinding sound of what reminds one of a computer processor. Added to this is a sound that exploits the overtones of the harmonic series.[33] This sound is electronically produced on the album, but it has its origins in a "natural" place—the harmonic series and the abilities of acoustic instruments to create overtones.[34] Accompanying the fluctuating overtones in the harmonic series is a single low string line that begins in an acoustic state and then is electronically distorted. Above all of this, there is a melodic riff, played in the upper range of the keyboard, that continues throughout the song. This sound is never altered or sequenced but always occurs in its original timbre with some melodic exceptions. The fusion of nature and technology clearly takes place on many levels in the instrumental introduction alone: there is the sound of the electronic processor against the natural harmonic overtones, as well as a melodic riff that becomes naturalized through its repetition against the sound of the distorted single low string tone.[35]

With the entrance of Madonna's voice, another natural element is added to the electronic sounds (*Ray of Light,* track 6: 0:19–0:45). Madonna's voice is not electronically manipulated here, the way a lot of sampled voices are on other albums.[36] Her lyrics are also in direct contrast to the typical lyrics in dance music. Most dance music lyrics are made up of short repeated phrases that are electronically manipulated throughout the song. Madonna, on the other hand, sings long, narrative phrases. Her voice tells a story, arguably asserting her authorship. The short lyrical ideas presented by women in dance music of the mid 1990s on the other hand, clearly expose the "control" of the producer/mixer over the woman's voice. Throughout this piece Madonna does not allow electronic sound to interfere with her voice. In the vocal introduction Madonna's phrases rarely coincide with the phrasing of the melodic riff. Also in this section the processing sound of the instrumental introduction fades into the background, subsumed by more discrete pitched material. In the final phrase of the vocal introduction, when Madonna sings, "I'll never be the same, because of you," the natural sound of the voice is played against the technological sounds (*Ray of Light* track 6: 0:44–0:53). The voice is juxtaposed against an unpitched repeated bass sound that is then played for an extended time while the pitch is electronically changed. In this section the "natural" of the voice is played against the "technological" of the electronic sounds.

The influence of dance music is undeniable in the first statement of the chorus. The use of a rhythmic bass backbeat most clearly makes this connection (*Ray of Light,* track 6: 0:53–1:13). Compared to traditional dance music styles with repeated lyrics and additive phrases, Madonna uses a more narrative approach.[37] The bass riff in Madonna's chorus alone, although electronic, is a very warm, round, fat sound. Although Madonna is influenced by the dance music genre, her approach is clearly very different. In the chorus it is once again important to note that, although there are a lot of electronic sounds, they occur at the end of Madonna's singing lines. The "wonky echo" (*Ray of Light,* track 6: 1:09–1:10) and ascending button sound (*Ray of Light,* track 6: 1:12–1:13), for example, are heard at the end of the chorus, avoiding contact with Madonna's voice.

The final example to be discussed in "Nothing Really Matters" is the instrumental interlude (*Ray of Light,* track 6: 2:55–3:12). The synthesized solo instrument in it sounds first like a xylophone and then a piano, signifying the natural through its associations with acoustic instruments. This solo has an improvised quality about it, like a jazz solo. The associations with jazz and acoustic piano and xylophone conjure up an earlier form of music, one less controlled by technology in the moment of performance.[38] The improvised instrumental section is played out against the electronic rhythmic section.[39] The beat is maintained electronically and thus juxtaposes the instrumental solo in this section. Also against this electronic beat and instrumental solo is Madonna's voice, repeated over and over again, making a humming sound.

Madonna as Mother in the Technological Realm

Madonna also brings nature to technology through the theme of motherhood on *Ray of Light.* The biological capacity to have children has constructed motherhood as a "natural" role for women. Therefore, electronic music has not traditionally celebrated motherhood as one of its themes; as Edna Gunderson states, "Electronic music is the unlikely vehicle carrying *Ray of Light*'s somber freight" (Gunderson 1998). Keith Negus illustrates through his example of Sinead O'Connor as a musical mother that the singer-songwriter has traditionally celebrated themes of motherhood with acoustic guitar (Negus 1997: 179, 180). Negus goes on to cite O'Connor's song "three babies," from the album *I do not want what I haven't got* (1989), as an example of her personal confessional style (Negus 1997: 179, 180).

> The confessional tone is very apparent in her song lyrics and arrangements, production techniques and musical textures. The confessional characteristics are signified musically in the use of a restrained, intimate voice, recorded softly and close to the microphone and with little echo and by the repeated use of the first person. It is also signified in the sparsity of many song arrangements—the sense of emptiness and silence which suggests that only the singer (rather than an ensemble) is present. (Negus 1997: 180)

This confessional tone is very different in Madonna's electronic music. She does not use a restrained, intimate voice, recorded softly and close to the microphone. Instead Madonna opts for a strong voice with plenty of reverb and echo. Madonna also turns away from the sparsity of the typical confessional song arrangement by employing many electronic sounds in combination with her themes of motherhood.

Madonna has continually credited her current spiritual plateau to her relationship with her daughter, Lourdes, often suggesting that the album probably would have been very different if she had not become a mother. When asked what influenced *Ray of Light,* Madonna replied, "The birth of my daughter has been a huge influence. It's different to look at life through the eyes of your child, and suddenly you have a whole new respect for life and you kind of get your innocence back" (Morse 1998). There are several songs on *Ray of Light* that reflect Madonna's new role as a mother, including "Nothing Really Matters," "Little Star," and "Mer Girl." Each of these songs makes heavy use of sounds associated with electronica.[40] Madonna states, "There's a song on the album called "Nothing Really Matters," and it is very much inspired by my daughter. It's just realizing that when the day is done the most important thing is loving people and sharing love . . ." (Gardner 1998). The introductory lyrics to "Nothing Really Matters," however, are problematic in terms of recent feminist queries on the issues of motherhood, which suggest that motherhood may not be a completely fulfilling role for all women.[41] Also, by referring to her selfish nature before she had her baby, Madonna perpetuates the unrealistic notion that mothers must be selfless. The lyrics of the chorus reflect Madonna's new role as mother. As listeners we can assume that the "you" she is referring to is Lourdes, as she has stated that "Nothing Really Matters" was inspired by her daughter (Morse 1998).

In the chorus, Madonna extends the role of female nurturer to all people by stating "Love is all we need," making it an all-inclusive emotion. The final verse of the song makes reference to her daughter when she says, "Everything I give you all comes back to me." In this statement she is articulating the rewarding role motherhood has brought to her life.

Clearly, there are both advantages and disadvantages to Madonna's celebration of motherhood. Women's biological capacity for motherhood is seen by ecofeminists as connecting to an innate selflessness born of their responsibility for ensuring continuity of life (Wajcman 1991: 6). Ecofeminists believe that nurturing and caring are essential to the fulfillment of this responsibility (ibid.: 7). Madonna often plays into these themes of nurturing and selflessness in her representation of motherhood. Since the birth of her daughter, Madonna has privileged motherhood over her career, turning down movie roles and publicity opportunities and even canceling a yearlong world tour. From her privileged position as a wealthy and famous white woman, she has the resources to continue a rewarding career while at the same time focusing on her relationship with Lourdes. By combining themes of nature with electronic music, Madonna's music brings nature to technology, blurring the dividing line.

Implementing Iceland with Electronica: Björk's Homogenic

Throughout her solo career Björk has established herself successfully within the realm of electronica. This, in itself, illustrates her powerful presence and "serious" musicianship, according to the assumptions of the genre.[42] Categorizing Björk's music using the binary system is a real challenge. By combining elements of herself and her Icelandic heritage with the technology of electronic music, Björk has created a unique space that blurs the line between nature/culture, feminine/masculine, body/mind, and self/

other. She has, in one sense, blurred the dichotomies and developed a new place in popular music.

The music on Björk's album *Homogenic* cannot be analyzed in a vacuum. An analysis of the traditional division between nature and technology must move beyond the music to other sources of evidence. Björk uses everything, from her music, videos, performances, and album art, to her appearances and representations in and by the media, to produce this enigmatic union.[43]

Björk's identity manifests itself through her Icelandic heritage and her music. Two of the main images associated with the feminine aesthetic are isolation and exoticism. Iceland is isolated geographically as well as culturally. Because of this isolation, what appears to be an exotic aura surrounds Björk. When Björk speaks about Iceland, she talks about her roots, the history, the elements, and her family. For her, Iceland is rejuvenating and inspiring. In one interview she claims, "I function in Iceland perfectly, it's got nature, mountains and winds, and I can at any moment have a walk and sing at the top of my voice without anyone finding me weird. But it's still a really modern place. It's a nice *combination of nature and techno*" ("Björk future lover" 1999: 22).[44]

Björk describes *Homogenic* as minimalist because it is composed with only beat, strings, and the voice. One of the distinguishing features of Björk's music is her vocalization. In many instances the beat is informed by, and the tracks are produced around, her vocals. Not only does her voice represent Western society's understanding of the natural, but she also incorporates various native Icelandic influences, such as the vocal technique, a combination of speech and singing, used to narrate the sagas from the twelfth and thirteenth centuries.

The history of Iceland is vital to Björk's juxtaposition of nature and technology in her music. Because Iceland only gained independence in 1944, its development as a nation was quick. "Out of this sped-up modernization sprang both an almost mythological relationship to nature and a brand-new fixation on technology" (Van Meter: 96). For over seven hundred years Iceland had been a colony of Denmark. The Icelanders were not allowed to sing or dance or play music because of its association with the devil; thus, they became obsessed with storytelling. "The core of Iceland's national culture was its literary heritage, whose main components were the sagas from the 12th and 13th centuries and the romantic and often nationalistic poems of the 19th century, and which included a nationalistic interpretation of Icelandic history" (Gudmundssohn 1993: 2). Njall Sigurason, a folklore specialist, explains how Björk uses her voice in a specific way, like the Old Icelandic choir men. These men used a reciting voice that was a combination of singing and speaking. Björk's adoption of this technique can be heard throughout the album; an excellent example is in "Unravel" (*Homogenic*, track 3: 1:32–2:28).

Analyzing Björk's vocal technique as natural sounding is not difficult. Various distinctive vocal characteristics occur throughout the album, including Björk's primitive-sounding screams, emphasized by a sampled and digitized beat (*Homogenic*, track 9: 1:38–2:28).[45] In this example there is a basic perception of her voice as "natural" and being manipulated by something completely technological. The distortion of the beats and her voice add qualities of hard techno music, yet the methodical rhythm of the voice and the beat also evoke chantlike characteristics.

The *Homogenic* concert stage was designed in a manner that specifically addresses the distinctions between Western perceptions of natural and technological. The stage is divided in half, with the Icelandic string orchestra (representing "nature" or "traditional" music) on one side and live mixer Mark Bell (representing "technology" or "non-traditional" music) on the other. Throughout the concert, Björk moves freely between the two realms, embodying their crucial link. Björk also establishes an Icelandic context and presence by opening the concert with a traditional Icelandic ballad. She proceeds to synthesize the two components throughout the concert and ends on a purely techno level with a remix of "Pluto".[46] Although Björk began the concert with an Icelandic ballad and ended with her most "techno"-sounding composition from the album, the concert does not project a theme of technology consuming nature, rather, the fluidity of her movement between the two realms disrupts the distinction of both, rejecting the idea that one has power over the other.

Some of the techniques that Björk adopts to incorporate Iceland into her electronic soundscape derive from folk music practice and Icelandic stories. In the first track, "Hunter," Björk uses the interval of a fifth continually throughout the work. For example, the cellos play the repeated two-bar motif a fifth apart (*Homogenic*, track 1: 0:00–0:30). Fifths were common in traditional Icelandic folksongs and their use was particularly relevant to performance. Björk explains, "'Hunter' is based on what my grandma told me at Christmas; about two different types of birds. One bird always had the same nest and partner all their lives. The other was always travelling and taking on different partners. At some point there was a conscious decision made to remain a hunter" (Walker 1997). That decision is most important to Björk and her music.[47]

There are also elements of Ravel's *Bolero* in "Hunter." One of the three main sections of the whole song is the Bolero ostinato (*Homogenic*, track 1: 0:00–1:36). The sounds Björk uses to cover the rhythmic pattern from

Bolero are tightly interwoven. The same beat is repeated continually throughout the piece, with electronic sounds and the strings adding multiple layering. The electronic melody flows with the beat and takes on an "organic" feel—the perceived artificial sound changing to a perceived natural sound (*Homogenic,* track 1: 0:11–0:30). She exaggerates the strings by using sliding notes that are sluggish and slurring, drawing out specific notes similarly to how she draws out syllables with her voice. There is a sequence to her electronic sounds, with each sound taking its turn to weave in and out of the ostinato. The beats throughout the album are simplistic; Björk made "a conscious decision that the beats would be almost naïve, very natural but explosive, like still in the making." She suggests, "This force is Iceland" (Walker 1997).

Multiple Fusions of Nature and Technology: Homogenic *Album Art*

Another significant medium Björk uses to engage with the nature/technology dichotomy is her album art. The *Homogenic* album cover features a shocking computerized image of Björk on the front in which she appears half-human. She is a hybrid, a human/machine—a cyborg. Her costume is a kimono, made of shiny silver and crimson red. The necklace elongates her neck, making her head seem as though it is not quite attached to her body. Her fingernails are longer than claws but perfectly manicured and polished. The hair is divided on her head into what looks like two large satellite dishes. Her appearance subsumes representations of various cultures. Her dress associates her with the Orient, her necklace with Africa, and her hair with Asia. There is not one line on her face, and her eyes are black and silver. At first glance one may be tempted to suggest that Björk is merely appropriating other cultures. However, she is not; rather, she has created an image of cultural synthesis. The background is silver with little blue flowers. The flowers are representations of nature, whereas the silver alludes to the metallic and to images of technology.[48] The image seems to have caught her in a state of metamorphosis. Perhaps it is here where the fusion of the "nature goddess" with "technology's cyborg" is most evident.[49]

Inside the jacket there appears to be a microscopic vision of a living organism or plant. The shades are deep reds, like blood, and it appears fleshy. This image is fluid and organic, scientific and natural at the same time. But the idea of the microscope changes the vision to a scientific one. The blood is now controlled and gazed upon by the "masculine." When you turn to the backside of the cover, the scene depicted inside is now viewed from a distance. The image is still organic and scientific but, from this distance, the "masculine science" is less oppressive. The background appears similar to that displayed on the front cover. There is a larger image of a flower surrounded by the fluid, which contains a glowing light resembling an entity made of energy. The shape of this design is also reminiscent of the uterus and the womb. Across this light is Björk's trademark initial. The letter "b" is a graphic design that reminds one of hip-hop, techno, and the realm of electronica. This "masculinized" genre is set in an organic image of fluidity. The glowing light represents the creation of something new or perhaps the combination of the "feminine" and the "masculine." Björk's synthesis of nature and electronica is evident even here.

Conclusion

One of the most interesting concepts used to attack the nature/culture dichotomy is Donna Haraway's cyborg theory. In "A Cyborg Manifesto: Science, Technology, and Socialist-Feminism in the Late Twentieth Century," Haraway describes what she calls "an ironic dream of a common language for women in the integrated circuit" (Haraway 1991: 149). The cyborg theory is a theoretical ideology that contains no gender and is thus a better space for women and men to inhabit. Haraway defines a cyborg as a "cybernetic organism, a hybrid of machine and organism, a creature of social reality as well as a creature of fiction. The cyborg is a matter of fiction and lived experience that changes what counts as women's experience in the late twentieth century" (ibid.). It is imperative to remember that "the machine is not an *it* to be animated, worshipped, and dominated. The machine is us, our processes, an aspect of our embodiment" (ibid.: 180). Consequently, technology can be the machine elements that we use to adjust or alter our physical form or appearance.[50]

Haraway's cyborg theory relates to Björk's accomplishments in the music composition, production, and publicity of *Homogenic*.[51] Through a synthesis of natural and technological elements in her music, Björk is able to compose without necessarily adhering to characteristic boundaries. Creating her music in an electronic industry does not preclude her Icelandic heritage from emerging, nor does her electronic music feel void of emotion.[52] Björk, like Haraway, suggests an alternative escape from the oppressive forces of gendered binary oppositions. Björk manages to accomplish this by juxtaposing the elements traditionally considered by Western societies as opposing. Because "the cyborg [or Björk's *Homogenic*] is no longer structured by the polarity of public and private ... Nature and culture are reworked; the one can no longer be the resource for appropriation or incorporation by the other" (Haraway 1991: 151). Haraway describes the cyborg manifesto as "an argument for *pleasure* in the confusion of boundaries and for *responsibility* in their construction" (ibid.: 150).[53] As the cyborg theory develops, women's pleasure will be developed in the dismantling of the socially constructed categories of gender. The illumination of rigid gender categories initiates the process of freeing women from their subordination. By obscuring the boundaries between nature and technology, Björk disrupts the traditional role of women in popular music. She exposes gender in electronica by both using and moving beyond the stereotypes, challenging the assumptions that are attached to it.[54] As Haraway suggests, "It means both building and destroying [musics], identities, categories, relationships, [and the way we listen]" (ibid.: 181).

By embracing both "organic and technological components" (Haraway 1991: i) in her music, can Madonna be described as inhabiting the cyborg world? Madonna combines motherhood and acoustic sounds (coded as natural) with technological music. The cyborg lives in a community that breaks down the binary of public and private: it "defines a technological polis based partly on a revolution of social relations in the oikos, the household" (Haraway 1991: 151). Blurring the boundaries between public and private through motherhood lessens the strict distinctions for creating a new place; however, Haraway's theory moves beyond this concept. Haraway envisions people occupying a new space, in a new way: as a hybrid "of machine and organism" (Haraway 1991: 150). Despite Madonna combining elements of nature and technology in her music, does she exist in the new space that Haraway alludes to? Through the techniques used to compose the music on *Ray of Light,* Madonna appears to resists the hybridization: Madonna's voice remains relatively free from any radical electronic alteration. Because of Madonna's reputation for maintaining power over all aspects of her work, we believe that Madonna resists being controlled by the machine.

Haraway's utopian world envisions a society removed of all power imbalances: "I do not know of any other time in history when there was a greater need for political unity to confront effectively dominations of race, gender, sexuality and class" (Haraway 1991: 157). For these reasons it is important to problematize Madonna and Björk's cultural positions. Both Madonna and Björk benefit from their dominant position in most of these categories. Presently Madonna suffers from few issues of inequality in these power relations: she is a wealthy white woman, and although she has expressed bisexuality she is primarily playing out the heterosexual role. Björk's privilege also extends to class, sexuality, and race, however she is often regarded as exotic Other because of her Icelandic heritage. Despite their privilege Madonna and Björk still continue to be marginalized as women in the male dominated field of electronica.

Women's experimentation with technology is not entirely new. There are a number of other women who incorporate technological aspects in their music making. Performance artist Laurie Anderson is well known for her relationship with technology. "Only through [abusing and playing with technology] could she have invented her famous tape bow, where a tape loop is bowed across a violin with cassette heads instead of strings" (O'Brien 1996: 150). Recently in R&B and hip hop women have pushed beyond the boundaries of gender stereotypes. Missy Elliot utilizes various technological means to produce her albums. TLC's transition from R&B to hip hop also connects with technological components. In their music videos "She's a Bitch" and "No Scrubs," Missy Elliot and TLC embody futuristic personas through costume and movement. Toronto-based musician Esthero is another artist striving to move beyond gender barriers in electronica. Although these women also experiment with technology in their music, Madonna and Björk are unique because of their synthesis of nature *and* technology. Yet in spite of their success, women musicians are far from the majority in electronic music. Their positions are marginalized in a number of ways: Madonna is often dismissed as somebody flirting with the latest, hottest trend (in this case techno-culture) rather than as a "serious" musician employing technology, and, although Björk composes in the electronica realm, she is often described as a gypsy, child queen, pixie, or a sprite-like enigma,[55] which perpetuates the idea that women are incapable of harnessing technology to compose "legitimate" music.

Feminists have criticized the topics of analytical debate concerning gender and technology, suggesting that only those practices that reinforce or reproduce existing patterns of gender relations are noticed theoretically (Grint 1994: 17). "While gender and technology have been constructed as macro-actors and shut away into black boxes, we must insist on opening them up for investigation, where the meaning and significance of technology and gender identity are reconsidered in all their variations as they exist for the actors" (ibid.: 44). Through our analysis of Madonna and Björk we have opened the black box of electronic music, exploring the way two women musicians used technology in a nontraditional manner and contributed to the dismantling of the nature/technology dichotomy that exists in popular music. Although working toward similar goals, Madonna and Björk use various methods and inspirations for combining the two poles. Merging electronic production with themes stereotypically assumed to be feminine, Madonna and Björk have initiated the degendering of electronic music. Through a synthesis of nature and technology, these women have carved unique spaces for themselves in the realm of popular music.

Notes

1. These two terms, self-awareness and origins, here refer to Björk's Icelandic heritage and Madonna's role as mother.
2. Please refer to track 10, "Bedtime Stories," on Madonna's 1994 album *Bedtime Stories*, 0:00–0:40, and track 8, "Possibly Maybe," on Björk's 1995 album *Post*, 0:00–0:50. From these examples it is evident that Madonna imitates Björk's vocal technique, using an intimate whispering tone. Although neither artist uses a wide melodic range, their timbral inflections are diverse.
3. "Electronica" refers to numerous genres and styles of electronic dance music.
4. For example, the voice—a human, "natural" instrument—is accepted as natural despite the fact that it is recorded through a microphone and put through filters.
5. These particular alignments will be discussed in detail throughout the paper.
6. Björk often promotes Icelandic principles as "natural." Within this context she includes Iceland's history of colonization by Denmark, Icelandic literary tradition, such as its sagas and folklore, and the Icelandic people's connection to weather and survival of extreme climates. Refer to Aston, Micallef, Van Meter and Walker.
7. See Derrida 1976, Terry 1997, and Code 1995 for examples.
8. See the chapter "Gender for a Marxist Dictionary" in Haraway 1991.
9. For example, acoustic musicians are regarded as feminine, while electronic musicians are seen as masculine. The same is true for the singer/songwriter genre.
10. Traditional rock bands with their electric guitar, keyboard, bass, and drums usually have been made up of male musicians, whereas female pop stars are often featured as solo vocalists, rarely playing these instruments.
11. This also relates to Donna Haraway's cyborg theory, discussed later. Haraway too suggests once the cyborg world comes into existence, there will be a combination of the natural with the technological.
12. See Andra McCartney, "Gender Symbolism," in *Creating Worlds for My Music to Exist: How Women Composers of Electroacoustic Music Make Place for Their Voices*. Master's thesis, York University, 1994.
13. See Simon Reynold's *Generation Ecstasy* (1998) and Sarah Thorton's *Club Cultures* (1996).
14. See Grint 1995, pp. 11 and 54.
15. Although, in the recent past the electric guitar has been associated with men in popular music.
16. Emotion is another characteristic coded as natural and feminine.
17. The role of journalists in authenticating Madonna's work became apparent after reading Emma Mayhew's "Women in Popular Music and the Construction of 'Authenticity.'" These ideas will be discussed further in her forthcoming article in *The Journal for Interdisciplinary Gender Studies*.
18. This power imbalance can be true regardless of gender; however, it is more frequent in a relationship between a man and a woman.
19. Advantages including technological knowledge, male authority, studio experience, accesses to technology.
20. Emphasis is our own.
21. Madonna's creative input in her albums and music videos are underestimated by her critics, yet, at the same time, those same critics celebrate her control over the marketing of her image and career.
22. Although Madonna emphasizes her control, her role in the recording studio is still unclear.
23. DeVries has also produced with artists such as Björk and David Bowie.
24. Armstrong has worked with Massive Attack.

25. For further information, see Théberge 1997.

26. Walker 1997.

27. Björk began her recording career at the age of twelve and spent several years as a keyboardist and lead singer in the punk band The Sugarcubes. She is well versed in music theory and music production techniques.

28. What is problematic with this statement is Björk's continual use of only male co-producers. However, she has been known to work with women mixers. Of course, a major problem that continues to exist is the exclusion of women producers from the "boys club" at the top of the field, which is maintained through the assumption that men have a better understanding of technology and how it works.

29. Walker 1997. In this section of the documentary Björk explains how she uses a QY20 to write many of her songs. She speaks of its capabilities and how Mark Bell taught her how to use it.

30. Please refer to Aston 1996, Van Meter 1997, and Reynolds 1998.

31. Because electronic music is coded as masculine, Björk requires a high level of musical knowledge and ability to maintain her identity in this field.

32. Walker 1997. Besides evidence stated on the album cover, including composition, instrumentation, lyrics, and production, the roles of each person in the studio are unknown.

33. The sound referred to can be produced by playing over the main sound of a pitch to produce the harmonic series.

34. Acoustic is coded as natural in this context.

35. One of the ways that electronic music is developed is by sampling phrases or motives. This particular melodic riff is repeated without being sampled.

36. For example, the way Cher's voice is sampled in "Believe" exposes the technological interference. We thank Jennifer DeBoer for pointing this out to us.

37. The narrative approach is less controlled by technology than the short manipulated phrases of dance music, which are controlled by technology.

38. Although jazz musicians use instruments, which could be considered technology, the instruments we refer to here are acoustic. Because jazz music is an older form of music than electronica, its sounds have become naturalized. Compared to electronica, where technology is exposed in the electronic sounds, jazz has a much more natural quality to it. Also the idea of an improvised solo implies a sense of freedom. The improvised section in the instrumental interlude seems to avoid the control of technology, specifically the electronic bass backbeat that accompanies it, thus aligning it with nature.

39. The improvised solo has the quality of being free from the control of technology since the improvised solo was originally designed in jazz to give the performer a chance to improvise on the melody.

40. See the section on musical analysis for more details on the electronic music in "Nothing Really Matters."

41. See Singer 1996: 70–72.

42. See Frith and McRobbie; Bayton et al.

43. Three of these mediums will be referred to throughout the paper.

44. Emphasis is our own.

45. The word primitive is used specifically here despite all of the problematics that result from its use.

46. Claudio Dell-Aere is another scholar researching Björk who has come to similar conclusions after watching the *Homogenic* concert. We wish to express our gratitude to him for a number of thought-provoking conversations on this topic.

47. When creating her first two albums, Björk went hunting outside of Iceland to find her inspiration and search out innovative ideas. For *Homogenic*, she returned to her Icelandic roots.

48. Although silver is a mineral, which comes from the earth, it has become valued as capital and therefore appropriated by and for "culture."

49. For a more detailed analysis of this idea, please refer to Marsh 1999.

50. For example, a razor, which is used to shave or cut hair from specific parts of our bodies depending on a person's sex and/or culture. Make-up is another mechanical element used to mold faces.

51. Björk's first two solo albums, *Debut* (1993) and *Post* (1995), contain a wide variety of musical styles and instrumentation. They did not encompass themes of Iceland in the same way as *Homogenic*.

52. Both Iceland and emotion are coded as natural in this context.

53. Italics are the authors'.

54. Although Björk is competent in the male-dominated field of electronic music, she also plays into stereotypes traditionally associated with the feminine, such as "pixie, elfin, natural mother, goddess, shrieking singer, diva" (Micallef 1997).

55. Refer to Micallef and Reynolds.

References

PRINT SOURCES

Aston, Martin. *Björkgraphy*. London: Simon and Schuster, 1996.

Bayton, Mavis. "How Women Become Musicians." In *On Record: Rock, Pop and the Written Word*, eds. Simon Frith and Andrew Goodwin. New York: Pantheon Books, 1990, 238–257.

Berry, Venise T. "Feminine or Masculine: The Conflicting Nature of Female Images in Rap Music." In *Cecilia Reclaimed: Feminist Perspectives on Gender and Music*. Chicago: University of Illinois Press, 1994, 183–201.

Bradby, Barbara. "Sampling Technology: Gender, Technology and the Body in Dance Music." *Popular Music* 12/2 (1993): 155–176.

Burnell, Barbara S. *Technological Change and Women's Work Experience: Alternative Methodological Perspectives*. Westport, Conn.: Bergin and Garvey, 1993.

Cockburn, Cynthia. *Machinery of Dominance: Women, Men and Technical Know-How*. Boston: Northeastern University Press, 1985.

Code, Lorraine. "Must a Feminist Be a Relativist After All?" In *Rhetorical Spaces: Essays on Gendered Locations*. New York: Routledge, 1995, 185–207.

Derrida, Jacques. *Of Grammatology*, trans. Gayatri Chaleravorty Spivak. Baltimore and London: Johns Hopkins University Press, 1976.

Gardner, Elysa. "Ray of Light." *Los Angeles Times* (March 1, 1998).

Glenn, Evelyn Nakano, Grace Chang, and Linda Rennie Forcey. *Mothering: Ideology, Experience and Agency*. New York: Routledge, 1994.

Goodwin, Andrew. "Sample and Hold: Pop Music in the Digital Age of Reproduction." In *On Record: Rock, Pop and the Written Word*, eds. Simon Frith and Andrew Goodwin. New York: Pantheon Books, 1990, 258–273.

Grint, Keith, and Rosalind Gill, eds. *The Gender-Technology Relation: Contemporary Theory and Research*. London: Taylor and Francis, 1995.

Gunderson, Edna. "Her *Ray of Light* Shines Earnestly in New Direction." *USA Today* (March 3, 1998).

Haraway, Donna J. *Simians, Cyborgs, and Women: The Reinvention of Nature*. New York: Routledge, 1991.

Jones, Steve. *Rock Formation: Music Technology and Mass Communication*. London: Sage Publishers, 1992.

Khayatt, Didi. "Talking Equity: Classroom Experiences." The Florence Bird Lecture, Carleton University, November 6, 1998. Unpublished.

Marsh, Charity. "*Homogenic:* Björk's Fusion of Goddess and Cyborg." Unpublished essay, 1999.
McClary, Susan. *Feminine Endings: Music, Gender and Sexuality.* Minneapolis: University of Minneapolis Press, 1991.
Micallef, Ken. "Home Is Where the Heart Is." *Ray Gun* (September 1997).
Middleton, Richard. *Studying Popular Music.* Milton Keynes: Open University Press, 1990.
Mies, Maria, and Vandana Shiva. "Ecofeminism." In *Feminisms,* eds. Sandra Kemp and Judith Squires. New York: Oxford University Press, 1997, 497–503.
Moi, Toril. "Feminist, Female, Feminine." In *The Feminist Reader: Essays in Gender and the Politics of Literary Criticism,* eds. Catherine Belsey and Jane Moore. New York: Basil Blackwell, 1989, 117–231.
Morse, Steve. "Madonna in the lotus position." *Boston Globe* (March 20, 1998): C1.
Negus, Keith. "Sinead O'Connor—Musical Mother." In *Sexing the Groove: Popular Music and Gender,* ed. Sheila Whiteley. London: Routledge, 1997, 178–191.
Nehring, Neil. *Popular Music, Gender and Postmodernism: Anger is Energy.* Thousand Oaks, Calif.: Sage Publications, 1997.
O'Brien, Lucy. *She Bop.* New York: Penguin Books, 1995.
O'Dair, Barbara, ed. *Trouble Girls: The Rolling Stone Book of Women in Rock.* New York: Rolling Stone Press, 1997.
Ortner, Sherry B. "Is Female to Male as Nature Is to Culture?" In *Feminist Cultural Theory: Process and Production,* ed. Beverley Skeggs. Manchester and New York: Manchester University Press, 1995, 256–276.
Reynolds, Simon. *Generation Ecstasy: Into the World of Techno and Rave Culture.* London: Little, Brown and Company, 1998.
Rule, Greg. "William Orbit: The Methods and Machinery behind Madonna's *Ray of Light.*" *Keyboard* (July 1998): 30–34, 36, 38.
Singer, Linda. "Bodies—Pleasures—Powers." In *Feminism and Sexuality: A Reader.* Eds. Stevi Jackson and Sue Scott. New York: Columbia University Press, 1996, 263–275.
Small, Christopher. *Music of the Common Tongue.* Hanover, N.H.: University Press of New England, 1988.
Strathern, Marilyn. "Less Nature, More Technology." In *Feminisms,* ed. Sandra Kemp and Judith Squires. New York: Oxford University Press, 1997, 494–97.
Straw, Will. "Characterizing Rock Music Culture." In *On Record: Rock, Pop and the Written Word,* eds. Simon Frith and Andrew Goodwin. New York: Pantheon Books, 1990, 97–110.
Terry, Jennifer, and Melodie Calvert, eds. *Processed Lives: Gender and Technology in Everyday Life.* London: Routledge, 1997.
Théberge, Paul. *Any Sound You Can Imagine: Making Music/Consuming Technology.* Hanover, N.H.: University Press of New England, 1997.
Thornton, Sarah. *Club Cultures: Music, Media and Subcultural Capital.* Middletown, Conn: Wesleyan University Press, 1996.
Tong, Rosemarie. *Feminist Thought: A More Comprehensive Introduction,* 2d ed. Boulder, Colo.: Westview Press, 1998.
Van Meter, Jonathan. "Björk: Animal? Vegetable? Mineral?" *Spin.* 13/9 (December 1997): 93–98.
Wajcman, Judy. *Feminism Confronts Technology.* University Park, Pa.: Pennsylvania State University Press, 1991.
Walters, Barry. "Madonna Chooses to Dare." *Spin* 14/4 (April 1998): 70–76.
West, Melissa. "The Aura of Madonna's *Ray of Light* a 'Progressive' Concept." Unpublished essay, 1999.

SOUND RECORDINGS

Björk. *Debut*. CD 61468. Elektra Entertainment, 1993.
Björk. *Homogenic*. CD 62061. Elektra Entertainment, 1997.
Björk. *Post*. CDE2 61740. Elektra Entertainment, 1995.
Björk. *Telegram*. CDE2 61897. Elektra Entertainment, 1996.
Madonna. *Bedtime Stories*. CDW45767. Maverick/Sire, 1994.
Madonna. *Ray of Light*. CDW46847. Maverick/Warner, 1998.

VIDEO

Walker, Christopher, producer and director. 1997. *Bravo Profiles: Björk*. South Bank Show. Video recording.

CHAPTER NINE

Before the Deluge
The Technoculture of Song-Sheet Publishing Viewed from Late-Nineteenth-Century Galveston

Leslie C. Gay, Jr.

In 1884 the Galveston music publisher Thos. Goggan & Bro. published, as part of this company's routine activities, a rather ordinary song sheet, entitled "Longing," by a scarcely known German American composer, a Mr. A. Jungmann.[1] Sometime after this song's publication, one copy of this song sheet—a copy found today in the archives of Galveston's Rosenberg Library—became a point of social interaction, for one notable aspect of this song sheet is a hastily penned inscription to "Lou Bonnot comp[liment]s of Mr. Reitmeyer." Like the song's composer, not much is known about these persons. Starting with the 1888–89 Galveston City Directory and continuing in directories well into the twentieth century, Mr. William F. Reitmeyer is listed as a piano tuner working for the Thos. Goggan & Bro. music store, a necessary complement to Goggan's music-publishing business. And Miss Louise Bonnot appears in the city directory of 1891–92, identified as a music teacher. This inscription, this trace of human interaction, and these bits of information about the lives of a young piano tuner and a music teacher suggest relationships and motivations about which I can only guess—an exchange of friendship or courtship, or just business savvy—nonetheless, possible relationships to which I shall return.

This inscription draws these two people together around the song sheet and the piano. The scenario, and the communications it implies, anchors to specific technological adaptations and cultural institutions. Its communication emerges from the technoculture of this time and place, carried in part on the technologies of writing, printing, and publishing, like the more obvious high-tech symbol of the late nineteenth century—the railroad—rides on the technologies of "gears and girders."[2] My focus here is on how songsheet publishing, viewed from late-nineteenth-century Galveston, constitutes a technocultural community, loosely bound by specific technological adaptations, uses, and meanings. I am particularly concerned with how those within this community were supported, while those outside, namely newly arrived immigrants and long-established minority groups in Galveston, were excluded. In this sense, song sheets and the technologies they encompass helped to incorporate an emerging transregional public culture and point to tensions among ethnic and class-based social groups, working to accommodate some while resistant to others.

My research on this technoculture follows studies of technologies that encompass more than examination of an isolated technological artifact. Rather, my approach necessarily includes associated cultural practices—a shift "from the instrument to the drama in which existing [social] groups perpetually negotiate power, authority, representation, and knowledge," as Carolyn Marvin puts it.[3] Most studies concerned with the nexus of music, culture, and technology exist, however, within a frame defined almost exclusively by the historical span of electronic media technologies. This study looks across this conceptual barrier.

Late-nineteenth-century Galveston serves as an especially well-suited locale for research into these issues. Located in a transitional zone, Galveston existed as a liminal place—geographically and historically—where predominately English- and Spanish-speaking cultures came together, where the memory of immigration and settlement was strong for many of its citizens, even as the notion of "frontier" began to wane, and at a time when technologies, among other factors, were transforming the United States. Yet, although it existed before the deluge of the omnipresent electronic media that so define our existence today, this Galveston shows cultural practices and social organizations surprisingly familiar to many of us.

A Technocultural Institution

Throughout the nineteenth century, sheet-music publishing and availability expanded dramatically in the United States. In the first quarter of the century some 10,000 pieces of music were published by U.S. commercial publishers.[4] By 1866 the Boston publisher Oliver Ditson alone had a catalog of 33,000 pieces of music, and within ten years, Russell Sanjek estimates, there were some 200,000 published compositions for piano and piano and voice.[5] Furthermore, with the growth of transcontinental railroads after the U.S. Civil War, in addition to long-established shipping routes, publishers could deliver sheet music cheaply and efficiently to individuals and other music businesses throughout the country.[6] Equally important, and corresponding to the growth of sheet-music publications, are increases in the sheer number of music-publishing houses. Beginning with a handful of active businesses after the American Revolution, music publishing expanded to at least sixty-five firms by 1869.[7] Significantly, rather than becoming centralized among a few publishers in New York City—as in the twentieth century's Tin Pan Alley—music publishing in the previous century became important in cities small and large throughout the country, from Boston to Cincinnati and San Francisco, and from Chicago to St. Louis and New Orleans.[8] A house publication of Chicago's Lyon & Healy makes the extent of music publishing clear: "Here [in Illinois] . . . every town of 20,000 inhabitants has its music publisher, and some have catalogues of a hundred or two."[9] Music publishing, not surprisingly then, was a part of the everyday activities of some folks in Galveston, Texas.

During the latter half of the nineteenth century, the city of Galveston gained prominence as the most important trading and banking center in Texas. As a seaport on the Gulf of Mexico it rivaled New Orleans.[10] Throughout this period it was a major city of the state, consistently greater in population than the older Spanish-Mexican San Antonio and nineteenth-century urban rivals of Austin, Houston, Dallas, and Waco.[11] Moreover, Galveston's shipping lines to New Orleans, Mobile, Havana, New York, Boston, and Europe made it not only a center of trade but also one of immigration, especially German immigration.

As a component of this commercial activity and immigration, Thomas and John Goggan, a pair of immigrants from County Kerry, Ireland, established Galveston's first music store, Thos. Goggan & Bro. Thomas set up shop in Galveston in 1866, just after the U.S. Civil War, and his brother joined him in the business in 1868. Although Thomas's route to Galveston remains unknown, John immigrated first to New York in 1862 and then on to Cincinnati, where he learned music printing and retailing with the William C. Peters Piano Company. Emulating the model of Peters's business, Goggan, nearly since its inception as a music-instrument retailer, also published song sheets. The brothers' store, located first in the Pix Building on 22nd and Post Office Streets, moved in 1877 to its own three-story building at the corner of 22nd and Market Streets, both of which were near the center of Galveston's commercial activities.[13] In both locations they sold pianos, organs, and other musical instruments besides song sheets, instrumental music, and pedagogical piano pieces. The store acquired stock from U.S. instrument makers; notably, they also imported instruments directly from European manufacturers. The Galveston store sold its first Chickering piano in 1866 (also the first sold in Texas), and early on Thos. Goggan and Bro. became the sole selling agent for Steinway pianos in Texas.[14]

The Goggans expanded greatly in the area. They engaged at least twenty employees, who included salespersons but also technicians, such as three piano tuners, one of whom was Mr. Reitmeyer. At John Goggan's death in 1908 (Thomas died in 1903 while on a visit to Ireland), in addition to the Galveston main office, five branch offices were located throughout the state, in Dallas, San Antonio, Waco, Austin, and Houston. Other Galveston business persons partly credited the brothers' success to their aggressive advertising in newspapers and circulars, and to the use of traveling salespersons working throughout the state, with the arrival of the railroads in Texas. Even as Goggan's music publishing in Galveston ended in 1930, the Houston branch continued as a music jobber through the early 1940s, and a few of the branch stores continued their retail business through the mid-twentieth century.[15]

This synopsis of Goggan's history points to institutional affinities with other song-sheet publishers. Research on publisher histories in the United States portrays important similarities and connections among publishers and links America's publishers with London models.[16] Like Goggan, most publishers began through the combination of a small-scale engraver or printer and someone with music interests, such as piano manufacturing, musical instrument sales, music performance, or all three. Generally, from an urban location favoring trade, publishers established themselves as, or emerged from, a music shop offering music instruction and the sale of instruments along with song sheets, instrumental dance music for the piano, and instruction manuals. Frequently, music publishers began sheet-music sales by first offering music of distant publishers. Those publishers that flourished developed close and long-standing working relationships with numerous selling agents and publishers elsewhere in the geographic region and beyond to tender their materials.[17]

These similarities among song publishers illustrate elements of a recognizable transregional technoculture. Corresponding operation histories point to shared technological adaptations, an aspect that helps unify songsheet publications. Similar engraving and printing techniques inevitably dispersed a familiar "look." Moreover, business relationships among publishers strengthened these connections. Not only was it common for publishers to acquire one another's music plates outright and thus reproduce exact copies of song sheets, publishers also shared publications through selling-agent agreements among distant publishers across a wide geographic area.[18] Hence, many song sheets published by Galveston's Thos. Goggan & Bro. appear in other publishers' catalogs—including Boston's Oliver Ditson and Company and Peters in Cincinnati. The reverse also occurred.

I choose to highlight the technological basis of this cultural institution with the designation "technoculture." However, such predictable social organizations and operations of music publishers could also identify a unified "industry," what Howard Becker labels an "art world."[19] Such worlds, in Becker's sense, each with its own well-defined and often restrictive modes of working, have been associated with the large entertainment and "culture" industries of this century, not those of the past, and often have been assumed to be related to, if not wholly products of, twentieth-century capitalist models of production. Goggan's business concerns also compare to other aspects of late-twentieth-century market capitalism. The transectorial interdependence between song-sheet publishing and music instrument retailing remains today as a familiar alliance between music publishers, manufacturers, and recorded-music retailers. Yet, its very familiarity obscures the importance of such interdependencies to Goggan and other song-sheet publishers within this emerging technoculture in the United States. Paul Théberge argues that recent innovations in musical instrument and recording technologies depend upon a technical interdependence among often-disparate industry firms, multinational corporations, and manufacturing sectors.[20] Rather than a new phenomenon, however, broadly based transectorial interdependencies long ago linked printing and instrument-manufacturing technologies with marketing and retailing. Moreover, just as recent innovations in communications technologies, such as computer user groups, have helped shape communities of music consumers,[21] newly expanded transportation routes after the U.S. Civil War, especially the railroad, accelerated print communications and helped draw some nineteenth-century Galvestonians into the public culture of song sheets and song performance.

Publishers of this technoculture also disperse similar musical styles. Within the United States, song sheets published in the latter half of the nineteenth century are surprisingly homogeneous; with few exceptions, they are similar in form and content. Goggan's song publications illustrate familiar song types of this period, including waltz songs, Scotch and Irish songs, and minstrel or "coon" songs.[22] Thomas M. Bowers's "Trilby and Little Billie," published by Goggan in 1895 and subtitled "Waltz Song and Chorus," illustrates conventions that mark this song type: triple meter, solo strophic song with chorus, clearly framed tonality in a major

Fig. 1. "Trilby and Little Billie"

key with a few diatonic harmonies, and reserved and regular melodic phrases with formulaic nonharhomic tones. Moreover, its limited vocal range, straightforward text setting of a somewhat exotic, sentimental text, and predominant rhythmic pattern consisting mostly of strings of dotted-half notes suggest this song was meant for home consumption (see fig. 1, mm. 9–25).

While the use of duple meter and, at times, more complex rhythms distinguished the other common song types from waltz songs, the bulk of all Goggan's song publications show the simplified structural, harmonic, and melodic practices that, according to Charles Hamm, most U.S. song composers used from the middle of the century on to accommodate "an even wider range of listeners and performers."[23] The very straightforwardness of this song, its lack of overtly technical or extravagant passages, makes it a music for amateur performance. The shared stylistic similarities of these songs across a wide geography suggest links among communities of performers and listeners, consumers of this public culture. Thus, despite existing well outside the major music-publishing centers of Philadelphia, New York, and Boston, Goggan participated in this public culture, selling music that largely conforms to that published in other distant cities. Shared song-sheet styles, coupled with predictable organizations and operations of publishers across the country, identify a unified technocultural institution.

Before the Deluge / 209

Women and This Technoculture

The exchange between Mr. Reitmeyer and Miss Bonnot, the piano tuner and the music teacher, emerges from the technological space defined in part by the song sheet and the piano. In nineteenth-century Galveston, as elsewhere, the performance of song sheets achieved a level of activity and musicianship that sustained publishers, piano technicians, and music teachers. Moreover, with the sales of song sheets and the requisite pianos, music publishers and educators encouraged music performance in the home, especially among women. These technologies thus imply activities, socially and musically performative, that illuminate aspects of this space, from broad practices of publishers and teachers to notions of self and place for women.

In this context, Mr. Reitmeyer's gift to Miss Bonnot could have been intended as an act of courtship, an exchange that references the relationship among domestic life, women, and the piano. Mary Burgan argues that for women in nineteenth-century England the piano functioned as "an emblem of social status," providing "a gauge of a woman's training in the required accomplishments of genteel society."[24] An 1856 publication in New York suggests a similar association in the United States. The etiquette book's title literally tells it all: *The lady's guide to perfect gentility in manners, dress and conversation, in the family, in company, at the pianoforte, the table, in the street, and in gentlemen's society.* . . . [25] Skilled performance at the piano for women thus demonstrated the virtue of "gentility," a model of "cultivated" personal excellence and social display.[26] Moreover, along with other furnishings in the parlors of nineteenth-century homes, the piano and the song sheet helped define a space of "culture" and "comfort" for women and their families, says Katherine Grier,[27] one that further illustrates a cosmopolitanism, marking associations and knowledge beyond the local even while contained by this domestic space.

Perhaps, too, Mr. Reitmeyer's gift of the song sheet gently encouraged romance, a means to woo Miss Bonnot, one function of popular song familiar to us in this century. Some scholars, notably Donald Horton, have argued that popular song in the twentieth century offers "conventional conversational language" for dating and courtship, a ready-made script to support such social interactions.[28] This was also the case in the late nineteenth century. For example, Russell Sanjek explains how the composer and Harlem Renaissance writer James Weldon Johnson shifted his emphasis from musical theater to the larger popular song market, writing songs that "young men took along when courting, to be played by their girls, giving both sexes an opportunity to vent their sentiments decorously."[29]

The gift equally could have been simple business savvy by Mr. Reitmeyer, encouraging or maintaining a relationship between a local piano teacher and his firm. Music education, piano sales, and music publishing were increasingly important careers or business concerns at the time, and they were substantially linked. The period from the Civil War through the Great Depression was a golden age for the piano business in the United States. In 1866, the founding year for Thos. Goggan & Bro., piano sales reached $15 million, accounting for some 25,000 new instruments. And in the 1900 census, the first time such information becomes available, music careers employed 92,000 Americans, Moreover, piano instruction established itself as an economically rewarding career at this time. In 1887 the Music Teachers National Association (founded 1876) reports that its members taught a half-million piano pupils. Connections between music publishers and teachers were tight, with the most successful piano teachers as active salespersons of music sheets, which they purchased at a 50 percent discount from most firms. Somewhat controversial is the fact that teachers realized almost half as much income from their music sales as from teaching, even as publishers profited 500 to 600 percent on each music sheet.[30]

For Thos. Goggan & Bro. as well as other piano retailers and song-sheet publishers, women certainly were important customers, both as teachers and general consumers who helped shape the firm's activities. Théberge notes that the marketing of pianos to women points to the home as "the center of family life and locus of individual consumption."[31] This connection finds correlates in other aspects of women's lives: for instance, music periodicals of this period at times overlap music with other household technologies, as seen in one periodical from 1880, the *Musical and Sewing Machine Gazette*.[32] Thus, it is not surprising that as tastes included other household technologies, the offerings at Goggan's retail stores changed to match them. Eventually, the stores sold all sorts of home appliances besides music wares, everything from sewing machines and toasters to gas ranges and freezers. Equally significant, after establishing the first radio station on Galveston Island in the early 1920s, Goggan also sold radios, reflecting shifts in home music consumption away from song sheets and pianos.

Looking at Boundaries

The circumscribed spectrum of musical styles seen in song sheets, beyond illustrating close ties among publishers, also reflects relationships between publishers and audience communities. Consideration of these complex relationships requires questions about how audience boundaries are marked and maintained: how do the publications help define this public culture, in what ways are they directed toward and away from specific communities, and what communities are represented or not represented? Goggan's publications reveal the publisher and the local songsheet audience working to define itself, to draw some Galveston citizens in while restricting others. The gift from Mr. Reitmeyer to Miss Bonnot exists within a social area of shifting and blurred boundaries. Technological aspects of song sheets, from musical and lyrical content to literacy (both English and music literacy), and the socioeconomic requirements to obtain such technologies define and support this community while excluding others.

Minstrel songs, what Hamm calls "the first distinctly American genre,"[33] began to outline the boundaries of this technocultural community. Minstrelsy emerged in the United States from those public spaces such as markets, historic and geographic points where in the tangent of European and African cultures "there was an eagerness to combine, share, join, draw from opposites, play on opposition," W. T. Lhamon shows us in his book, *Raising Cain*.[34] As a theatrical performance, minstrelsy became a public expression of racial relationships between European and African Americans. "Minstrelsy," writes Eric Lott, "brought to public form racialized elements of thought and feeling, tone and impulse, residing at the very edge of semantic availability, which Americans only dimly realized they felt, let alone understood."[35] This relationship was largely an expression of white middle-class men toward African Americans and thus points to a particular junction of gender, ethnicity, and class. As published song sheets—that is, without their multidimensional theatrical context—minstrel songs, and the later form "coon" songs, are less clearly gendered male but remain focused at intersections of ethnicity and class.

Minstrel songs published in Galveston and elsewhere make explicit a boundary between black and white. Music and song performance within minstrel shows, especially in early minstrelsy, has been characterized by its connections to oral traditions—based loosely on melodic and rhythmic gestures derived from African American dance and song with heterophonic renditions with banjo, fiddle, tambourine, and bones accompaniment.[36] The difficult translation of this theatrical musical style to song sheets reflects these roots in oral traditions. In an analysis by Katherine Reed-Maxwell, early minstrel publications, often not attributed to any composer, illustrate African American characteristics of asymmetrical rhythmic and melodic groupings structured by call-and-response patterns. Later

Fig. 2. "Paint All de Little Black Sinners White"

minstrel-song publications, those clearly meant for parlor performance with named, if not well known, composers, show more conventional groupings of notated, printed European American music—more rationalized rhythms and uniform melodic repetition with "correct" harmonic progressions within a formalized song structure.[37]

Fred Lyons's "Paint All de Little Black Sinners White," published by Thos. Goggan & Bro. in 1887, exhibits these print-music characteristics (see fig. 2, mm. 9–16). The song features a simple accompaniment with rigid rhythmic patterns, each beat of the quadruple meter squarely reinforced. Harmonically, its skillful modulation from the key of D minor to F major on each melodic phrase allows for a unique harmonic progression. Its use of syncopation and dotted rhythms in the vocal melody gives a rhythmic bounce, distinct from most waltz and other parlor songs, which, however, could just as easily derive from European American oral traditions as African American ones. Strophic verses with chorus, framed by a piano introduction and piano "dance" ending section, organize this piece. In sum, Lyons's piece bears few musical similarities to earlier minstrel publications and the more oral-based forms common to the minstrel stage.

Moreover, through its cover iconography of racist caricatures and lyrics in minstrel dialect, Lyons's more-than-ironic song portrays a transformation of African American "sinners" into what I can only interpret as "good white citizens." The lyrics begin:

> Oh! My troubles gone and my heart is feeling gay,
> Dey gwine to paint all de little black sinners.
> We'll have a happy time, and we'll celebrate the day . . .

The first verse continues after some repetition:

> We'll all mix together no matter where we go,
> When dey paint all de colored people white.[38]

The cover-page illustration, also shared with another Lyons song, "Dem Chickens Roost Too High," reflects the scenario of the lyrics by showing a railyard in which African Americans are unloaded from one train, literally given a paint job, and placed on another train (identified as the "T. G. & Bro." line; see fig. 3).

Such textual and visual aspects surely reference class and ethnic divisions—especially the way in which white workers used the minstrel show to distinguish themselves from blacks.[39] These divisions eventually mutated the "Jim Crow" stage character into a gloss for the apartheid-like segregation codified by law in the late-nineteenth-century United States. But the "redemptive" aspects of Lyons's text, the ultimate resolution made by transforming these caricatured persons into white citizens, coupled with its lack of overtly derogatory lyrics, made it acceptable for performance in the home while maintaining social divisions. All these aspects distinguish minstrel song sheets from risqué theatrical performances and moved minstrel song sheets away from traditional African American expressive forms. This boundary, however, was neither rock solid nor invariably fixed.

Throughout the last decades of the nineteenth century, African Americans began to appear regularly and in significant numbers on the minstrel stage and even to rival in popularity their white counterparts. As Robert Toll has shown in his book, *Blacking Up,* minstrelsy was an avenue for mobility to African Americans not available elsewhere, a domain in which black men found their "first chance to become entertainers."[40] Indeed, Fred Lyons was such an entertainer, a noted, if not well known, banjo player and comedian, and a song-sheet composer.[41] Minstrelsy also offered to African Americans avenues of moral, political, and economic discourse, what Dale Cockrell views as an agitating social "noise." Such discourse from the stage played a role in the fight against slavery.[42] Yet by mid-century minstrelsy had largely lost this capacity, according to some researchers. "Dialect blackface," writes Cockrell, "had become more a form of gross mockery."[43]

Fig. 3. "Paint All de Little Black Sinners White," cover illustration. Courtesy of the Lester S. Levy Sheet Music Collection, Department of Special Collections, the Milton S. Eisenhower Library, The Johns Hopkins University.

The widely exposed and standardized form and content of the minstrel stage, especially the derogatory stock characters, imposed a racist archetype upon all who chose to perform there, black or white. Such an archetype effectively delineated minstrel performance as a product primarily of white America and for white America, regardless of its performers. As song-sheet publications, minstrel songs perpetuated this segregation and marked a boundary that, like the minstrel stage, African Americans did not cross easily or without some self-consciousness.

There were, of course, African Americans in Galveston presumably involved with a range of other musics. The city directories from mid-century on list a minimum of two "colored" churches, implying a significant population of African Americans. Yet Goggan apparently directed no publications toward this audience; their musical interests existed outside this publisher's purview. For instance, there are no examples of black religious music, what had become known as "Negro spirituals," in Goggan's catalog. Within the nation, too, few publishers dealt with such music, mostly connected with the abolitionist movement.[44] This disregard certainly was due in part to the currents of increasing segregation throughout the country. Goggan, however, was also responding to commercial interests based on demographics: this audience was largely illiterate.

Exact knowledge of how socioeconomic and technological restraints effected literacy rates of African Americans in this period of Galveston's history is difficult to estimate. Through 1870 general literacy was higher in the nonslave states than in the South. Within the slave states, literacy was greater for white families holding slaves, that is, for wealthier families. Thus, throughout much of the southern United States, good reading skills were uncommon for working-class people, black and white, especially in the more remote, rural areas. In urban areas with the highest concentrations of commercial and manufacturing activity, however, literacy was consistently higher. In this light, Soltow and Stevens show areas of the Texas Gulf coast near Galveston with some of the highest illiteracy rates in the country.[45] Literacy in Galveston was probably higher, especially among wealthier whites. Between the 1870 and 1900 censuses, white illiteracy in the country as a whole had declined from 11.5 to 6.2 percent. For African Americans, while illiteracy remained high, the change was dramatic, from 81.4 to 44.5 percent.[46] If we make the uneasy assumption that music literacy correlates with general literacy, it is probable that only a few African Americans in Galveston could claim such "higher order" literacy skills as music reading, restricting many from song-sheet performance.

The technologies of literacy and language demarcate this community for the publisher Thos. Goggan & Bro. in Galveston. Texas's heritage as a former Mexican state, Coahuila y Tejas through 1836, coupled with the number of Spanish surnames found in Galveston city directories between 1861 and 1881, imply that a segment of its population, probably a large one, spoke Spanish rather than English. That Goggan's publications center on an English-language audience, with little regard for Spanish speakers, ties to deep prejudices of this public culture in Texas. Texas-Mexican populations in the brief period of the Texas Republic endured discrimination from their white neighbors that included the atrocities of forced migration, land dispossession, and random violence.[47] Texas Mexicans fared little better than African Americans during the period of increasing segregation after the U.S. Civil War, as Arnoldo De León makes clear: "Mexicans were held in no higher esteem than blacks and Indians. . . . Jim Crow signs read 'for Mexicans' instead of 'for Negroes' in South Texas. Though the idea was no longer vocalized as often as before the Civil War, it was understood that only minor differences separated 'greasers' from 'redskins' and 'niggers.'"[48] Yet Goggan did publish a series of piano pieces called *Choicest Mexican Music*, and one song with Spanish lyrics, "La Golondrina."[49] And, while the publications are decidedly directed to an English-language audience, not a Spanish one, such publications suggest relations and accommodations between whites and Texas Mexicans more complex than the strict segregation that Jim Crow signs and racist terms denote.

"La Golondrina," Goggan's one Spanish-language song, was published for an English-speaking audience to advertise the Mexican National Railroad and its connections within Texas. All text on its cover and back-page advertisement is in English, with the Spanish lyrics given in English translation. While this advertisement primarily shows transportation ties between Mexico and Texas, it also emphasizes other boundaries and connections between the United States and Mexico: first in a map on the back page, then in its cover illustration of the Mexican and U.S. flags, and finally with a statement comparing the song's Mexican popularity with the popularity of "Home, Sweet Home" in the United States (see figs. 4 and 5).

Goggan's instrumental series, *Choicest Mexican Music*, is no less directed toward an English-language audience. Mexican songs of this series are virtually stripped of Spanish-language referents, arranged for piano only, with titles often given in English translation with the original Spanish displayed secondarily, if at all. Such song-sheet publications support the case that relations between whites and Texas Mexicans in the late nineteenth century were often inconsistent and contradictory, especially when they concerned "Castilian" and landed elites.[50] Song publications reveal complex ethnic prejudices that distanced working-class Texas Mexicans while acknowledging elites, even promoting their commercial concerns.

German immigrants in Galveston are no better represented in the music

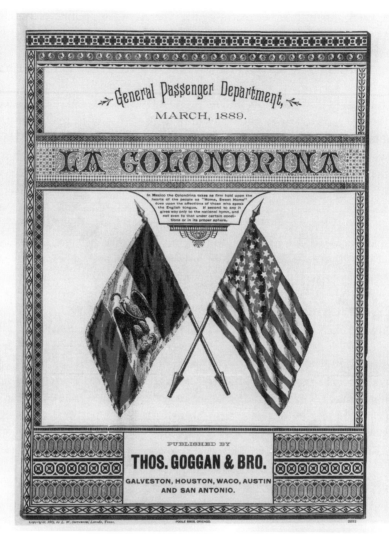

Fig. 4. "La Golondrina," cover illustration. Courtesy of the Lester S. Levy Sheet Music Collection, Department of Special Collections, the Milton S. Eisenhower Library, The Johns Hopkins University.

publications of Thos. Goggan & Bro. than Texas Mexicans, although German Americans crossed over as song composers—including our Mr. Jungmann. German immigrants (mostly farmers and craftspeople) and first- and second-generation German Americans were a significant segment of Galveston's population; by the mid-1850s, from one-third to one-half of

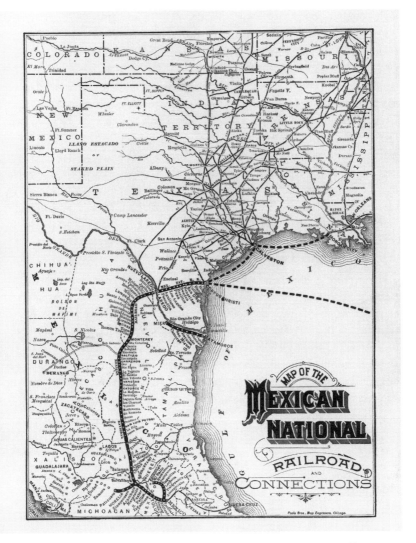

Fig. 5. "La Golondrina," back illustration. Courtesy of the Lester S. Levy Sheet Music Collection, Department of Special Collections, the Milton S. Eisenhower Library, The Johns Hopkins University.

Galveston's population was German. Local publications often cited these immigrants for their musicianship. The historian Earl Fornell notes that German immigrants' "proficiency in the use of musical instruments . . . inspired both admiration and chagrin among" non-German craftspeople in Galveston; German wind bands also were regular and important contributors to civic celebrations.[51] Moreover, German singing societies were active in the numerous German communities of the state, formally organized and concertizing statewide as the Texas *Sängerbund* by 1853.[52] Yet, although Goggan's city directory advertisement list German accordions for sale in the music shop, no song-sheet publications are in German or refer to Germans at all.

As stated already, many working-class Galvestonians likely found their lack of musical literacy a deterrent to participating in this technoculture. The portrayal of rural working-class Texans in Goggan's song sheets further suggests a border that contains a genteel urban class while also excluding farm- and ranch-hands. One published song about Texas cowboys, rather than embracing the experiences of rural life, presents a removed, romanticized picture of rural Texas. Comparisons between published "cowboy" songs and those collected at the turn of this century from working ranch-hands show few similarities.

"The Texas Cow-Boy," by Mrs. Robt. Thomson,[53] corresponds more to "cowboy" songs published elsewhere in the United States[54] than to songs sung by working cowboys such as those collected by folklorist John Lomax.[55] The awkward use of "cowboy" idioms, marked by Thomson in the text with quotations, show the song's composer, singers, and audiences as anything but ranch-hands. The song only gingerly associates with cowboys and cowboy life, not allowing them to get too close. Thomson's first and last verses with chorus follow:

> I am a Texas cowboy,
> Light hearted, gay and free,
> To roam the wide, wide prairie,
> Is always joy to me,
> My trusty little pony,
> Is my compaion true;
> O'er plain, thro' woods and river,
> He's sure to "pull me thro."
>
> *Chorus:*
> I am a jolly cowboy,
> From Texas now I hail,
> Give me my "quirt" and pony,
> I'm ready for the "trail";
> I love the rolling prairie,
> We're free from care and strife,
> Behind a herd of "long-horns"
> I'll journey all my life.
>
> And when in Kansas City,
> The "Boss" he pays us up,
> We loaf around a few days,

> Then have a parting cup.
> We bid farewell to city,
> From noisy marts we come
> Right back to dear old Texas,
> The cow-boy's native home.[56]

Such gentility is lacking in Jack Thorp's song, "Little Joe, The Wrangler," collected by John Lomax in his *Cowboy Songs*.[57] With its basis on an actual event, "Little Joe, The Wrangler" locates itself in a specific place—a camp on the Red River—with a real person, Little Joe.[58] Compare Thomson's text with Thorp's (verses 8 through 11).

> Little Joe the wrangler was called out with the rest
> And scarcely had the kid got to the herd
> When the cattle they stampeded; like a hail storm, long they flew
> And all of us were riding for the lead.
>
> 'Tween the streaks of lightnin' we could see a horse far out ahead
> 'Twas little Joe the wrangler in the lead;
> He was ridin' "old Blue Rocket" with his slicker 'bove his head
> Trying to check the leaders in their speed.
>
> At last we got them milling and kinder quieted down
> And the extra guard back to the camp did go
> But one of them was missin' and we all saw at a glance
> 'Twas our little Texas stray—poor wrangler Joe.
>
> Next morning just at sunup we found where Rocket fell
> Down in a washout forty feet below
> Beneath his horse mashed to a pulp his spurs had rung the knell
> For our little Texas stray—poor wrangler Joe.[59]

Its narrative structure and the deadly reality of the tale—within a stampede, horses are powerful and dangerous tools, not cherished companions—distinguish it from Thomson's published song.

Besides the contrast of the texts, there are other distinctions between the two cowboy songs. Thorp's song in its earliest publication appears as lyrics only, meant to be sung to a familiar tune, "Little Old Log Cabin in the Lane."[60] The tune, as published in Lomax's collection, also shows a musical distance between the two songs, suggesting a social distance as well. Thomson's song with its four-square meter and the tune's consistent rhythm on the verse—which contrasts ever so slightly in the chorus with the use of dotted rhythms—bears little in common with the looser, more varying rhythms, including some syncopation—in the Lomax transcription of "Little Joe" (see fig. 6, mm. 9–16, and fig. 7, mm. 1–8). Moreover, the wide range of an octave plus a fifth of Thomson's song, often outlining the chords of the accompaniment, and its use of a few chromatic tones, would do little to reinforce the narrative aspects of Thorp's song. The

Lomax version, with its narrow range of a sixth (an octave at one brief point), is constructed with the more limited pitch set of a hexatonic scale. Rural music of working-class life in Texas exists apart from the gentrified musical expression found within the public culture where song sheets are found.

Goggan's publications reveal, too, a segment of this population striving to place Galveston within an emerging transregional identity of which the technologies of music publishing were part, struggling to represent Galveston not as peripheral to but as engaged with this public culture. Thus, a performance of Thomson's "The Texas Cow-Boy," in part because of its remoteness from working cowboys, presents Galvestonians as genteel, urban Texans, both locally knowledgeable and distantly cosmopolitan.

Other song-sheet publications also emphasize the local while connecting to cosmopolitan themes, reinforcing or transforming a civic identity for Galveston and Texas. "The Pirate Isle No More," by W. A. Hogan and H. A. Lebermann (1889), a song written for the city's semicentennial celebration, sensationalizes its earliest history with Spanish and French buccaneers—"once was the home of the Pirate Lafitt"—while repositioning the city as a newly arrived mercantile center—now "Texas' proud queen of commerce." This theme continues on its cover, shared with Eduard

Fig. 6. "The Texas Cow-Boy"

Fig. 7. "Little Joe, the Wrangler"

Holst's "The Semi-Centennial Grand March" (1889). The illustration lauds the city's fifty-year history of expansion through "before and after" lithographs, comparing its early existence as a seaside hamlet with a later view of Galveston's business district. A waltz celebrating the new state capitol, the "Texas State Capitol Grand Waltz" by Leonora Rives (1888) and dedicated to Gov. L. Sullivan Ross, features a lithograph of the newly built, and here dramatically illustrated, capitol building in Austin. The architecture of the building, a rotunda flanked by opposite wings similar to the nation's Capitol, is shown in gargantuan scale, dwarfing the carriages and pedestrians on the surrounding streets. Another Goggan publication, "Roosevelt's Rough Riders: March—Two Step" (1898), written by L. F. Haaren for the regiment of U.S. cavalry volunteers who fought in the Spanish-American War, shows Galveston involved with an important national event along with the latest dance craze. Local interests surely influenced its publication—the regiment trained near San Antonio. However, the war signals the emergence of U.S. colonial expansion beyond its continental boundaries in the same decade as the notion of a geographic "frontier" in Texas had all but vanished, thus allowing for a more cosmopolitan Texas within an emerging public culture.

Summary and Conclusion

Song-sheet publishing viewed from the perspective of Galveston's Thos. Goggan & Bro. reveals a connected "industry" of publishers, both large and small, across the country, actively engaged in support of a technocultural community in which song sheets and pianos are important parts. This community is linked through institutions that create and support song sheets, aligning printing and instrument manufacturing technologies with transportation, marketing, and retailing.

This transregional community, moreover, accommodates those with the socioeconomic and corresponding technological wherewithal to participate while restricting those not so blessed. The song-sheet publications of Goggan, along with our piano technician Mr. Reitmeyer and our music teacher Miss Bonnot, thus exist within a social domain connecting Galveston homes and their household technologies with other music publishers and an emerging mostly white, English-speaking public culture across the United States. This technoculture deters the participation of portions of Galveston's population through its "mechanisms"—the economic requirements to access pianos and the song sheets themselves, along with the requisite performance skills and necessary literacy, both music and English-language literacy. My analysis of Goggan's publications suggests that for a less educated working class, and especially for immigrant and minority groups, full participation in this community was difficult. In the case of the African American population of Galveston, the explicitly racist topics and iconography of some song sheets further discouraged their participation.

In retrospect, the existence of such a technologically based public culture should surprise no one. Such social structures became more pronounced throughout the first half of the twentieth century and remain familiar, traceable still today. Song-sheet technologies, like more recent electronic ones that merge domestic spaces with a cosmopolitan world, may offer a prospect of crossing social boundaries, as the minstrel song composer Fred Lyons did, but they more often mediate among similar, like-minded folks near and far, connecting them and defining, in part, their social relationships. With their similar physical designs and unified musical styles, the technological basis of song sheets, too, seems well established,[61] if less obviously high-tech as vinyl LPs or digital CDs and MP3 recordings. Like these newer technologies, the importance of song sheets and their publication, however, does not lie with the artifacts alone but, rather, with their social use. What emerges here is the way in which these technologies and access to them delineate and maintain a public culture across the United States, even for small-scale music publishers such as Goggan, in cities such as Galveston, a point that bears emphasis, lest we let our familiarity with the "nuts and bolts" obscure the social significance of such old technologies.

I am skeptical, though, of views that take technological developments and adaptations—even such seemingly fundamental technologies as language, literacy, and print—as deterministic, as "thunderclaps of history" that predictably transform human societies.[62] Elsewhere I argue against such simplistic views by maintaining that relations between technologies and their cultural use are complex, with uses and meanings constructed and contested through the discourse of daily lives.[63] Yet communication technologies like song sheets are implicated within the myriad ways we build social relations, make exchanges, and create meaning. The gift from Mr. Reitmeyer to Miss Bonnot, whatever else it might denote as a point of exchange, exists in this context to maintain and conserve rather than transcend and radicalize, to buffer rather than challenge the social domain disclosed by the song sheets.

Finally, most discussions of music, culture, and technology focus on technologies of the electronic media. While there are profound cultural implications with the application of electricity to communications, including musical communications, a focus mostly on the technological extremes of the late twentieth century obscures important transformations of the more recent past before widespread use of electricity. We often take electronic-communications technologies—from the radio and audiocassette to cable television and computer networks—as radically altering how we interact. Within this wash of often noisy electronic "texts," it is easy to miss the ways in which technological adaptations of the past—notably writing and printing—have shaped social relations and remain critical to our lives. Not only in the old-fashioned newspapers or even stuffy, scholarly journals do these technologies remain significant. We find their influence in the film- and videotape-archive shows of VH1, which, no matter how hip or corny, rely upon written traditions of historiography. And we see them in the assumptions of literacy—I could add English literacy—modeled mostly on print specifications that, despite hypermedia claims, support the Internet's World Wide Web. The importance of even seemingly mundane and modest technologies as those concerned with song-sheet publishing should not be overlooked in the glare of electronic technologies today.

Notes

My research on song sheets began with the support of the NEH and its Summer Seminar "American Song and American Culture in the Nineteenth Century," support for which I remain grateful. I also wish to thank Anna Peebler and Shelly Henley of the Galveston & Texas History Center for their assistance with much of this research.

This publication comes from archival research at the Galveston & Texas History Center, Rosenberg Library, the Lester S. Levy Sheet Music Collection, Johns Hopkins University, and the Smithsonian Institution, Washington, D.C.

1. Scarcely known to us today, that is; according to the published opus number, this was Jungmann's ninety-sixth composition.
2. Cecelia Tichi, *Shifting Gears: Technology, Literature, Culture in Modernist America* (Chapel Hill: University of North Carolina Press, 1987).
3. Carolyn Marvin, *When Old Technologies Were New: Thinking about Electric*

Communication in the Late Nineteenth Century (Oxford: Oxford University Press, 1988), 5.

4. Russell Sanjek, *American Popular Music and Its Business: The First Four Hundred Years*, vol. 2, *From 1790 to 1909* (New York: Oxford University Press, 1988); Richard J. Wolfe, *Early American Music Engraving and Printing: A History of Music in America from 1787 to 1825 with Commentary on Earlier and Later Practices* (Urbana: University of Illinois Press, 1980).

5. Sanjek, *American Popular Music and Its Business*, 348.

6. Ibid., 290; John F. Stover, "Railroads," in *The Reader's Companion to American History*, ed. Eric Foner and John A. Garraty (Boston: Houghton Mifflin, 1991), 906–910.

7. Harry Dichter and Elliott Shapiro, *Early American Sheet Music: Its Lure and Its Lore, 1768–1889* (New York: R. R. Bowker, 1941); Dena J. Epstein, ed., *Complete Catalogue of Sheet Music and Musical Works, 1870* (New York: Da Capo Press, 1973); Wolfe, *Early American Music Engraving and Printing*.

8. Dichter and Shapiro, *Early American Sheet Music;* Sanjek, *American Popular Music and Its Business;* Wolfe, *Early American Music Engraving and Printing*.

9. Quotation from Epstein, ed., *Complete Catalogue of Sheet Music and Musical Works, 1870*, xi.

10. Earl Wesley Fornell, *The Galveston Era: The Texas Crescent on the Eve of Secession* (Austin: University of Texas Press, 1961); Howard Barnstone, *The Galveston That Was* (New York: Macmillan, 1966); David G. McComb, *Galveston: A History* (Austin: University of Texas Press, 1986).

11. E. Dana Durand, *Thirteenth Census of the United States: Statistics for Texas* (Washington, D.C.: Department of Commerce and Labor, Bureau of the Census, 1913).

12. Indicative of both trade and immigration, *Heller's Galveston City Directory* of 1874 lists consulates from Spain, Germany, England, the Netherlands, Sweden, Russia, and Denmark. See also Fornell, *Galveston Era*, 27, 89, 125–39.

13. The Pix Building, sometimes called the "old Tribune" building, was the home of the *Galveston Tribune*, a daily newspaper and a printer. It is not known, however, if any of Goggan's song sheets came off the *Tribune*'s press.

14. Information on Goggan's commercial history is compiled from *The Industries of Galveston* (Galveston, Tex.: Metropolitan Publishing Company, 1887); "Two Prominent Citizens Drowned: Messrs. John Goggan and John Moore Lost Their Lives at Redfish Reef," *Galveston Daily News*, Sept. 7, 1908; "Gengler's Fifteen Years Old When Goggan Opened First Texas Music Store," *Galveston Daily News*, June 16, 1926, Special Gengler Edition; "Goggan Observes 65th Anniversary," *Galveston Daily News*, Oct. 23, 1931, 14; Maury Darst, "Goggan Firm Wrote Songs about the Island," *Galveston Daily News*, March 6, 1967.

15. Darst, "Goggan Firm Wrote Songs about the Island"; Dichter and Shapiro, *Early American Sheet Music*.

16. Nancy F. Carter, "Early Music Publishing in Denver," *American Music Research Journal* 2 (1992): 53–67; Dichter and Shapiro, *Early American Sheet Music*; D. W. Krammel, "Music Publishing," in *Music Printing and Publishing*, ed. D. W. Krummel and Stanley Sadie (New York: Norton, 1990), 79–132; Ernst C. Krohn, *Music Publishing in the Middle Western States Before the Civil War*, Detroit Studies in Music Bibliography (Detroit: Information Coordinators, 1972); Ernst C. Krohn, *Music Publishing in St. Louis*, ed. James R. Heintze, Bibliographies in American Music (Warren, Mich.: Harmonie Park Press, 1988); Peter A. Munstedt, "Kansas City Music Publishing: The First Fifty Years," *American Music* 9, no. 4 (1991): 333–83; H. Edmund Poole, "*A Day at a Music Publishers:* A Description of the Establishment of D'Almaine & Co.," *Journal of the Printing Historical Society* 14 (1979/

80): 59–81; Sanjek, *American Popular Music and Its Business;* Wolfe, *Early American Music Engraving and Printing.*

17. This pattern began in the United States with the establishment in 1793 of J. C. Moller and Henri Capron's music shop in Philadelphia and their first subscription series of song sheets (*Moller and Compton's Monthly Numbers*). It continued through the establishment of Benjamin Carr in Philadelphia in 1794, and James Hewitt and George Gilfert (both the same year) in New York, and culminated with the development of the music-publishing giants of the nineteenth century in the partnerships of Firth, Hall & Pond (1814–75) in New York and the company of Oliver Ditson in Boston (1835–1931) (Krummel, "Music Publishing," 116).

18. Wolfe, *Early American Music Engraving and Printing.*

19. Howard Becker, *Art Worlds* (Berkeley: University of California Press, 1982).

20. Paul Théberge, *Any Sound You Can Imagine: Making Music/Consuming Technology* (Hanover, N.H.: Wesleyan University Press [University Press of New England], 1997), 59.

21. See ibid., 131–53.

22. See Charles Hamm, *Yesterdays: Popular Song in America* (New York: Norton, 1983), and Hamm, *Music in the New World* (New York: Norton, 1983). Representative examples from Goggan's catalog include the waltz song "Trilby and Little Billie" by Thomas M. Bowers, the Scotch song "Sweetest Lass in All the Land" by Ida Walker, and the coon song "Mr. Coon, You'se Too Black for Me" by S. H. Dudley.

23. Hamm, *Yesterdays,* 294.

24. Mary Burgan, "Heroines at the Piano: Women and Music in Nineteenth-Century Fiction," in *The Lost Chord: Essays on Victorian Music,* ed. Nicholas Temperly (Bloomington: Indiana University Press, 1989), 42–67, 42.

25. Emily Thornwell, *The lady's guide to perfect gentility in manners, dress and conversation, in the family, in company, at the piano forte, the table, in the street, and in gentlemen's society; also, a useful instructor in letter writing, toilet preparations, fancy needlework, millinery, dressmaking, care of wardrobe, the hair, teeth, hands, lips, complexion, etc.* (New York: Derby and Jackson, 1856). See also Ann Douglas, *The Feminization of American Culture* (New York: Knopf, 1977); Leslie C. Gay Jr., "Acting Up, Talking Tech: New York Rock Musicians and Their Metaphors of Technology," *Ethnomusicology* 42, no 1 (1998): 81–98; Arthur Loesser, *Men, Women, and Pianos: A Social History* (New York: Simon and Schuster, 1954); Judith Tick, "Passed Away Is the Piano Girl: Changes in American Musical Life, 1970–1900," In *Women Making Music: The Western Art Tradition, 1150–1950,* ed. Jane Bowers and Judith Tick (Urbana: University of Illinois Press, 1986), 325–48.

26. Katherine C. Grier, "The Decline of the Memory Palace: The Parlor after 1890," in *American Home Life, 1880–1930: A Social History of Spaces and Places,* ed. Jessica H. Foy and Thomas J. Schlereth (Knoxville: University of Tennessee Press, 1992), 49–74.

27. Ibid., 53–54.

28. Donald Horton, "The Dialogue of Courtship in Popular Song," in *On Record: Rock, Pop, and the Written Word,* ed. Simon Frith and Andrew Goodwin (New York: Pantheon Books, 1990), 14–26. See also Simon Frith, "Words and Music: Why Do Songs Have Words?" in *Lost in Music: Culture, Style, and the Musical Event,* ed. A. L. White (London: Routledge and Kegan Paul, 1987), 77–106.

29. Sanjek, *American Popular Music and Its Business,* 285.

30. Ibid., 347, 350, 370.

31. Théberge, *Any Sound You Can Imagine,* 99.

32. *Musical and Sewing Machine Gazette* (New York: Howard Rockwood, 1880).

33. Hamm, *Music in the New World,* 183.

34. W. T. Lhamon Jr., *Raising Cain: Blackface Performance from Jim Crow to Hip Hop* (Cambridge: Harvard University Press, 1998), 3.

35. Eric Lott, *Love and Theft: Blackface Minstrelsy and the American Working Class* (New York: Oxford University Press, 1993), 6.

36. Dale Cockrell, *Demons of Disorder: Early Blackface Minstrels and Their World* (Cambridge: Cambridge University Press, 1997); Lhamon, *Raising Cain*; Thomas L. Riis, *Just Before Jazz: Black Musical Theater in New York, 1980 to 1915* (Washington, D.C.: Smithsonian Institution Press, 1989).

37. Kathryn Reed-Maxfield, "Emmett, Foster and Their Anonymous Colleagues: The Creators of Early Minstrel Show Songs," paper presented at the Annual Meeting of the Sonneck Society, the University of Pittsburgh, April 4, 1987, quoted in Riis, *Just Before Jazz*, 4–5.

38. Fred Lyons, "Paint all de little Black Sinners White" (Galveston, Tex.: Thos. Goggan & Bro., 1887).

39. David R. Roediger, *The Wages of Whiteness: Race and the Making of the American Working Class* (London: Verso, 1991).

40. Robert C. Toll, *Blacking Up: the Minstrel Show in Nineteenth-Century America* (New York: Oxford University Press, 1974), 196.

41. Riis, *Just Before Jazz*, 9–10; Ike Simond, *Old Slack's Reminiscence and Pocket History of the Colored Profession from 1865 to 1891* (ca. 1891; Bowling Green, Ohio: Popular Press, Bowling Green University, 1974), 24–25.

42. Cockrell, *Demons of Disorder*. See also Lhamon, *Raising Cain*.

43. Cockrell, *Demons of Disorder*, 147. See also Riis, *Just Before Jazz*, 7. Other changes on the minstrel stage reflect the loss of subersive elements. In 1887, the publication year of Lyons's song, Primrose and West's Minstrels, one of the most "refined" white minstrel troupes, went on stage without blackface, signaling an important shift in popularity from blackface minstrelsy toward vaudeville-style performance (Simond, *Old Slack's Reminiscence*, 48).

44. Those publishers who did publish "Negro spirituals" did so without much commercial success. Antebellum Northerners, especially abolitionists, expressed interest in the music of slaves, and accounts of African American musical expressions began to appear in periodicals and newspapers. A December 1861 edition of the *New York Tribune*, according to Dena Epstein, carried the first published text of a spiritual taken from runaway slaves around Hampton, Virginia (Dena J. Epstein, *Sinful Tunes and Spirituals: Black Folk Music to the Civil War* [Urbana: University of Illinois Press, 1977], 45–46). Other publications followed, including a song sheet copublished by Horace Waters and Oliver Ditson entitled "The Song of the Contrabands—'O let my People go'" (Sanjek, *American Popular Music and Its Business*, 270). The most extensive early publication of African American spirituals, based on music collected in the early 1860s by Lucy McKim in Port Royal, South Carolina, was *Slave Songs of the United States* (William Francis Allen, *Slave Songs of the United States* [1867; New York: Peter Smith, 1933]). *Slave Songs*, published by A. Simpson & Co., contains words and music of 136 "shouts" or spiritual songs. Oliver Ditson, however, declined in 1871 an offer to bring out a second edition of *Slave Songs of the United States* because, Russell Sanjek concludes, none of the earlier publications of black spirituals "lighted the spark of public demand" (Sanjek, *American Popular Music and Its Business*, 270–71).

45. Lee Soltow and Edward Stevens, *The Rise of Literacy and the Common School in the United States: A Socioeconomic Analysis to 1870* (Chicago: University of Chicago Press, 1981), 148–92.

46. John K. Folger and Charles B. Nam, *Education of the American Population* (Washington, D.C.: Government Printing Office, 1967), 113–14.

47. David Montejano, *Anglos and Mexicans in the Making of Texas, 1936–1986* (Austin: University of Texas Press, 1987), 27.
48. Arnoldo De León, *They Called Them Greasers: Anglo Attitudes toward Mexicans in Texas, 1821–1900* (Austin: University of Texas Press, 1983), 221.
49. "La Golondrina," Thos. Goggan & Bro., Galveston, Tex., 1883.
50. Montejano, *Anglos and Mexicans in the Making of Texas*, 84.
51. Fornell, *Galveston Era*, 129, 133, 138.
52. Martha Fornell and Earl W. Fornell, "A Century of German Song in Texas," *The American-German Review* (1957): 23–31; Lota M. Spell, "The Early German Contribution to Music in Texas," *The American-German Review* 12 (1946): 8–10.
53. Mrs. Robt. Thomson, "The Texas Cow-Boy," Thos. Goggan & Bro., Galveston, Tex., 1886.
54. For example, George Cooper and Fred A. Rothstein, "Texas Charlie," Hitchcock's Music Store, New York, NY, 1885.
55. John Avery Lomax, *Cowboy Songs, and Other Frontier Ballads* (New York: Sturgis and Walton, 1910).
56. Thomson, "The Texas Cow-Boy."
57. Lomax, *Cowboy Songs, and Other Frontier Ballads*. "Little Joe" has its own literary/technological history. Jack Thorp self-published "Little Joe, the Wrangler" as part of a collection for fellow cowboys; later it was appropriated by Lomax for his collection. Thorp reports he wrote this song while on a trail ride in Higgins, Texas, 1898. See N. Howard ("Jack") Thorp, *Songs of the Cowboys*, ed. Austin E. Fife, Alta S. Fife, and Naunie Gardner (1908; New York: Clarkson N. Potter, 1966), 25, 29. A summary of N. Howard "Jack" Thorp's biography appears in this edition of the collection (Thorp, *Songs of the Cowboys*, 3–9).
58. Lomax, *Cowboy Songs, and Other Frontier Ballads*.
59. Ibid., 31–32.
60. Ibid., 25, 29.
61. A number of authors have argued or suggested this shared technological basis to differing degrees and within different scholarly contexts: see, for example, Richard Crawford, *The American Musical Landscape* (Berkeley: University of California Press, 1993); Jon W. Finson, *The Voices That Are Gone: Themes in Nineteenth-Century American Popular Song* (New York: Oxford University Press, 1994); Lester S. Levy, *Grace Notes in American History: Popular Sheet Music from 1820 to 1900* (Norman: University of Oklahoma Press, 1967); and, espcially, Hamm, *Yesterdays*.
62. Ruth Finnegan, *Literacy and Orality: Studies in the Technology of Communication* (New York: Basil Blackwell, 1988), 5–14.
63. Gay, "Acting Up, Talking Tech."

References

Allen, William Francis. "Preface." *Slave Songs of the United States*. New York: Peter Smith, 1933 [1857], i–xxxvi.
Barnstone, Howard. *The Galveston That Was*. New York: The Macmillan Company, 1966.
Becker, Howard. *Art Worlds*. Berkeley: University of California Press, 1982.
Burgan, Mary. "Heroines at the Piano: Women and Music in Nineteenth-Century Fiction." *The Lost Chord: Essays on Victorian Music*. Ed. Nicholas Temperly. Bloomington: Indiana University Press, 1989, 42–67.
Carter, Nancy F. "Early Music Publishing in Denver." *The American Music Research Journal* 2 (1992): 53–67.
Cockrell, Dale. *Demons of Disorder: Early Blackface Minstrels and Their World*. Cambridge: Cambridge University Press, 1997.

Crawford, Richard. *The American Musical Landscape*. Berkeley: University of California Press, 1993.
Darst, Maury. "Goggan Firm Wrote Songs About the Island." *Galveston Daily News* 6 March 1967.
De León, Arnoldo. *They Called Them Greasers: Anglo Attitudes toward Mexicans in Texas, 1821–1900*. Austin: University of Texas Press, 1983.
Dichter, Harry, and Elliott Shapiro. *Early American Sheet Music: Its Lure and Its Lore 1768–1889*. New York: R. R. Bowker Co., 1941.
Douglas, Ann. *The Feminization of American Culture*. New York: Knopf, 1977.
Durand, E. Dana. *Thirteenth Census of the United States: Statistics for Texas*. Washington: Department of Commerce and Labor, Bureau of the Census, 1913.
Epstein, Dena J. *Sinful Tunes and Spirituals: Black Folk Music to the Civil War*. Urbana: University of Illinois Press, 1977.
———, ed. *Complete Catalogue of Sheet Music and Musical Works, 1870*. New York: Da Capo Press, 1973.
Finnegan, Ruth. *Literacy and Orality: Studies in the Technology of Communication*. New York: Basil Blackwell, 1988.
Finson, Jon W. *The Voices That Are Gone: Themes in Nineteenth-Century American Popular Song*. New York: Oxford University Press, 1994.
Folger, John K., and Charles B. Nam. *Education of the American Population*. Washington, D.C.: Government Printing Office, 1967.
Fornell, Earl Wesley. *The Galveston Era: The Texas Crescent on the Eve of Secession*. Austin: University of Texas Press, 1961.
Fornell, Martha, and Earl W. Fornell. "A Century of German Song in Texas. *The American–German Review*. (1957): 24–31.
Frith, Simon. "Words and Music: Why Do Songs Have Words?" *Lost in Music: Culture, Style, and the Musical Event*. Ed. A. L. White. London: Routledge & Kegan Paul, 1987, 77–106.
Gay, Leslie C. Jr. "Acting Up, Talking Tech: New York Rock Musicians and Their Metaphors of Technology." *Ethnomusicology* 42.1 (1998): 81–98.
"Gengler's Fifteen Years Old When Goggans Opened First Texas Music Store." *Galveston Daily News*, Wednesday, 16 June 1926: Special Gengler Edition.
"Goggan Observes 65th Anniversary." *Galveston Daily News*, Friday, 23 October 1931: 14.
Grier, Katherine C. "The Decline of the Memory Palace: The Parlor after 1890." *American Home Life, 1880–1930: A Social History of Spaces and Places*. Eds. Jessica H. Foy and Thomas J. Schlereth. Knoxville: The University of Tennessee Press, 1992. 49–74.
Hamm, Charles. *Music in the New World*. New York: W. W. Norton & Company, 1983.
———. *Yesterdays: Popular Song in America*. New York: W. W. Norton & Company, 1983.
Horton, Donald. "The Dialogue of Courtship in Popular Song." *On Record: Rock, Pop, and the Written Word*. Eds. Simon Frith and Andrew Goodwin. New York: Pantheon Books, 1990, 14–26.
Krohn, Ernst C. *Music Publishing in St. Louis*. Bibliographies in American Music. Ed. James R. Heintze. Warren: Harmonie Park Press, 1988.
———. *Music Publishing in the Middle Western States before the Civil War*. Detroit Studies in Music Bibliography. Detroit: Information Coordinators, Inc, 1972.
Krummel, D. W. "Music Publishing." *Music Printing and Publishing*. Eds. D. W. Drummel and Stanley Sadie. New York: W. W. Norton & Co., 1990. 79–132.
Levy, Lester S. *Grace Notes in American History: Popular Sheet Music from 1820 to 1900*. Norman: University of Oklahoma Press, 1967.

Lhamon, W. T., Jr. *Raising Cain: Blackface Performance from Jim Crow to Hip Hop.* Cambridge, Mass: Harvard University Press, 1998.
Loesser, Arthur. *Men, Women, and Pianos: A Social History.* New York: Simon and Schuster, 1954.
Lomax, John Avery. *Cowboy Songs and Other Frontier Ballads.* New York: Sturgis & Walton Company, 1910.
Lott, Eric. *Love and Theft: Blackface Minstrelsy and the American Working Class.* New York: Oxford University Press, 1993.
Marvin, Carolyn. *When Old Technologies Were New: Thinking About Electric Communication in the Late Nineteenth Century.* Oxford: Oxford University Press, 1988.
McComb, David G. *Galveston: A History.* Austin: University of Texas Press, 1986.
Montejano, David. *Anglos and Mexicans in the Making of Texas, 1836–1986.* Austin: University of Texas Press, 1987.
Munstedt, Peter A. "Kansas City Music Publishing: The First Fifty Years." *American Music* 9.4 (1991): 333–83.
Poole, H. Edmund. "*A Day at a Music Publishers:* A Description of the Establishment of D'Almaine & Co." *Journal of the Printing Historical Society* 14 (1979/80): 59–81.
Riis, Thomas L. *Just before Jazz: Black Musical Theater in New York, 1890 to 1915.* Washington, D.C.: Smithsonian Institution Press, 1989.
Roediger, David R. *The Wages of Whiteness: Race and the Making of the American Working Class.* London: Verso, 1991.
Sanjek, Russell. *American Popular Music and Its Business: The First Four Hundred Years,* vol. 2, *From 1790 to 1909.* New York: Oxford University Press, 1988.
Simond, Ike. *Old Slack's Reminiscence and Pocket History of the Colored Profession from 1865 to 1891.* Reprint ed. Bowling Green, Ohio: Popular Press, Bowling Green University, 1974.
Soltow, Lee, and Edward Stevens. *The Rise of Literacy and the Common School in the United States: A Socioeconomic Analysis to 1870.* Chicago: The University of Chicago Press, 1981.
Spell, Lota M. "The Early German Contribution to Music in Texas." *The American-German Review* 12 (1946): 8–10.
Stover, John F. "Railroads." *The Reader's Companion to American History.* Eds. Eric Foner and John A. Garraty. Boston: Houghton Mifflin Co., 1991, 906–10.
Théberge, Paul. *Any Sound You Can Imagine: Making Music/Consuming Technology.* Hanover, N.H.: Wesleyan University Press/University Press of New England, 1997.
The Industries of Galveston. Galveston: Metropolitan Publishing Company, 1887.
Thornwell, Emily. *The lady's guide to perfect gentility in manners, dress and conversation, in the family, in company, at the piano forte, the table, in the street, and in gentlemen's society; also, a useful instructor in letter writing, toilet preparations, fancy needlework, millinery, dressmaking, care of wardrobe, the hair, teeth, hands, lips, complexion, etc.* New York: Derby & Jackson, 1856.
Thorp, N. Howard ("Jack"). *Songs of the Cowboys.* Eds. Austin E. Fife, Alta S. Fife and Naunie Gardner. New York: Clarkson N. Potter, 1966 [1908].
Tichi, Cecelia. *Shifting Gears: Technology, Literature, Culture in Modernist America.* Chapel Hill: The University of North Carolina Prres, 1987.
Tick, Judith. "Passed Away Is the Piano Girl: Changes in American Musical Life, 1970–1900." *Women Making Music: The Western Art Tradition 1150–1950.* Eds. Jane Bowers and Judith Tick. Urbana: University of Illinois Press, 1986, 325–48.
Toll, Robert C. *Blacking Up: The Minstrel Show in Nineteenth-Century America.* New York: Oxford University Press, 1974.

"Two Prominent Citizens Drowned: Messrs. John Goggan and John Moore Lost Their Lives at Redfish Reef." *Galveston Daily News*. Monday, 7 September 1908.

Wolfe, Richard J. *Early American Music Engraving and Printing: A History of Music Publishing in America from 1878 to 1825 with Commentary on Earlier and Later Practices*. Urbana: University of Illinois Press, 1980.

CHAPTER TEN

Stretched from Manhattan's Back Alley to MOMA
A Social History of Magnetic Tape and Recording

Matthew Malsky

> It is the music of fevered dreams, of sensations called back from a dim past.
> It is the sound of echo, the sound of tone heard through aural binoculars.
> It is vaporous, tantalizing, cushioned. It is in the room, yet not a part of it.
> It is something entirely new. And genesis cannot be described.
> (From Jay S. Harrison's *New York Herald Tribune* review of works for tape recorder
> by Vladimir Ussachevsky and Otto Luening; reprinted in Daniel 1952)

The Anecdotes[1]

It's small wonder, in retrospect, that by the age of ten Walter Murch was already fascinated with recorded sound. After all, he was later to become Hollywood's sound designer extraordinare and perhaps its most notably articulate pioneer in film sound.[2] By first borrowing a friend's new magnetic tape recorder, then by convincing his parents to get their own (ostensibly to "pirate" music off the radio), he indulged in a quintessentially postwar, middle-class activity: acquiring and then playing with a new gadget.

> I made a pest of myself at that [neighbor's] household, showing up with a variety of excuses just to be allowed to play with that miraculous machine: hanging the microphone out the window and capturing the back-alley reverberations of Manhattan, Scotchtaping it to the shaft of a swing-arm lamp and rapping the bell-shaped shade with pencils, inserting it into one end of a vacuum cleaner tube and shouting into the other, and so forth. (Murch 1994: xiii)

In the aftermath of World War II, consumer markets were burgeoning to almost mythic proportions with the goods resulting from wartime technology transfers, and in the home audio market the magnetic tape recorder was the new star. By 1948 it had already begun to revolutionize production and dissemination in broadcast television and radio, commercial music recording, and film soundtracks.[3] Replacing wire and transcription disc recording methods, magnetic tape recorders improved the quality of the sound, altered and expanded production and post-recording techniques, and even changed the repertory itself.[4] By 1950 magnetic tape machines designed for businesses and the home recording enthusiast were nationally advertised and widely available and affordable (through Sears & Roebuck, for example.) This machine is at the center of the postwar boom in home entertainment and has a rich sociological, musical, and technological history.

Yet Murch's personal account indicates an effect beyond its institutional one; it hints at a newly developing relationship that people had with their sonic surroundings. In contrast to the dominant media, the phonograph and the radio, magnetic tape and the recorder allowed people to record sounds themselves, rather than simply play back commercial recordings. While contemporary social critics such as David Riesman warned of the rise of the "other directed" personality type—and of its effects on listening—Murch's collecting and examining might be seen as proof of the tenacity of more traditional American character qualities.[5] It is hard to imagine his experiments as other than play, independence, and enjoyment, a sort of *jouissance* of direct engagement with the ethereal, ephemeral, and tactile qualities of sound in a newly tangible form. Not only does Murch seem engrossed with his immediate sonic environment, closely examining and considering the sounds produced within the confines of his own space, but he seems able to imagine, through technological extension, a whole new sonic landscape.[6]

Meanwhile, uptown we find another example of a young man who had an opportunity to play with a tape recorder, one that is notable if only because his mother didn't throw away the tapes. In 1951, as a young member of the composition faculty at Columbia University, Vladimir Ussachevsky recommended that the department's new music programs, the Composers Forum concerts, be recorded. Of course, the commercial equipment purchased for that purpose became his responsibility.[7] Being a self-described curious sort, he experimented with recording, editing, and two predominant tape manipulations. First, he made recordings through altered tape speeds, which he called transpositions.[8] He recorded single piano notes and then re-recorded them at slower tape speeds. Still recognizable as piano timbres, but now defamiliarized, these sounds had an unearthly, oceanic quality. He was introduced to the second technique, feedback, by an engineering student at the Columbia radio station. Peter Mauzey armed him with some homemade devices, and he used them to create the illusion of a swimming multitude of piano sounds.[9] Ussachevsky, despite his professional credentials, paralleled Murch's enthusiasm, techniques, and naive enjoyment.

At first, the results were tentatively presented on a May 5, 1952, program at McMillin Theater at Columbia as "experiments" (Ussachevsky 1977: 4). In 1952, Otto Luening and Vladimir Ussachevsky spent the summer in collaboration at Bennington College. In the fall, they both participated in a recording session at the home of Henry Cowell. Then, on October 28, 1952, in an American Composers' Alliance concert at the Museum of Modern Art in New York City, they presented the works that were quickly granted this historic designation: the first American art music compositions written specifically for the tape recorder. On the program were Luening's *Low Speed, Invention,* and *Fantasy in Space,* and Ussachevsky's *Sonic Contours.* The pair were dubbed "tapesichordists" by critic Oliver Daniel in his *Time* magazine review of the event, coining a short lived and nearly forgotten appellation (Daniel 1952: 63).[10] But the descriptive term that stuck, and tellingly so, was simply "composers."[11]

*Evading the Cultural Logic of Recorded Sound:
The Real, Reality, and a Surplus of Fantasy Space*

If we begin by stating the obvious, that there is a great deal of difference in the examples of Ussachevsky and Murch and that their activities are easily separated by the lines of hierarchical discourse (high vs. popular, the new new vs. the old new, professional art vs. amateur play, truth vs. ideology), we might be led to discount any significant congruences, despite a shared medium. Instead, it is more interesting to treat these anecdotes as equivalent at first and consider both through the lens of the same pair of questions. Only then will the ratiocination for the all-too-common divisions become clear. First, with regard to the technology itself, what is the shared material history that brought the machine to that point and engendered a common but multivalent narrative? The story is told through personal accounts, popular hobby manuals, magazines and the popular press, fictive depictions, concert and convention reviews, and the notes accompanying sound recordings. In dialectical terms, these were the contemporary texts that were generated by the practice and that rendered those social applications of the technology into words. Given that access to the type of tapes represented by Murch is limited, the story must of necessity be read from written texts.[12] However, and not coincidentally, while Murch's original recordings are inaccessible, the works by Ussachevsky and Luening are still available (see note 56). These recordings, along with other documentation, are also sources. Second, we should begin with some common framing questions: Why would anyone want to indulge in this sort of sonic representation? What were the stakes? What did the tape recorder really record?

The magnetic tape recorder rended the real. It upset the cultural logic of recorded sound. For a brief historical moment, as a free, as-yet-unencumbered signifier, it opened up a hole in reality and allowed us to shift our listening gaze.[13] In a manner analogous to the emergence of language at the gateway to the symbolic, the widespread dissemination of the tape recorder required a variety of strategies for resealing that tear, in the domains of mass culture as well as high art. To develop this notion of the introduction of new technology as a symptom, or to write a "case study" of its visible manifestations, it is necessary for us first to apply to recording the Lacanian concept of the "real," what Zizek calls "a black hole in Reality."[14] In terms of magnetic tape recording, how does what we call "reality" imply a surplus of fantasy space? How does tape recorder practice and history display the modalities of the real, which returns and answers and can be directly rendered?[15] The logic of recording operates as "a barrier separating the real from reality."[16] It is through these fantasy spaces, the sum total of what folks did with their tape recorders, that the "black hole" of the real was held at bay.

It Returns and Answers

Let's begin by considering another anecdote, one which is only slightly more aberrant. In Samuel Beckett's *Krapp's Last Tape,* the exposure to the real opened by the introduction of a tape recorder is not sufficiently repaired through its use, and the results appear as madness.[17] The play is set in the sparsely furnished den of the ancient, disheveled, and dirty Krapp, who is surveying and auditioning his voluminous collection of reels of magnetic tape. A tape recorder occupies center stage, both literally and figuratively, and there is little to indicate that any other characters exist in Krapp's world, or even that there is a world outside his room and reproduced voice. He is, perhaps, an unsuccessful Murch later in life. Almost as a party game in an ongoing secret anniversary celebration, Krapp records and re-records himself as he struggles, unsuccessfully, to articulate his vision of a single moment in his life: a probably imagined encounter with a woman, the object of his desire.

. . . What I suddenly saw then was this, that the belief I had been going on all my life, namely—[KRAPP *switches off impatiently, winds tape forward, switches on again*] — great granite rocks the foam flying up in the light of the lighthouse and the wind-gauge spinning like a propeller, clear to me at last that the dark I have always struggled to keep under is in reality my most — [KRAPP *curses louder, switches off, winds tape forward, switches on again*] — my face in her breast and my hands on her. We lay there without moving. But under us all moved, and moved us, gently, up and down, and from side to side.

[Pause.]

Past midnight. Never knew such silence. The earth might be uninhabited. (Beckett 1958: 60)

That last image is the Lacanian real itself, a space that is experienced as uninhabited, fluid and in motion, insubstantial, and devoid of any social impositions. Here, for Krapp, "the dark [he] has always struggled to keep under" has traumatically returned, disrupting his reality and impairing his ability to function. Krapp has abandoned normalcy (or perhaps, it has abandoned him) for a "stage" of his own making, one that exists solely through its reproduction. He has withdrawn from all other semblance of human exchange and has advanced the tape recorder as his only possible link to human activity. His connection to reality through language has all but failed him, and what remains is an untethered desire, the Lacanian "*objet a.*" It takes the form of a simple recorded moment, where the tape recorder itself functions as a fantasy space, "as an empty surface for the projection of desires; the fascinating presence of its positive contents does nothing but fill out a certain emptiness" (Zizek 1991: 8). In Lacanian theory, fantasy designates the healthy subject's "impossible" relation to *a*, the object-cause of its desire.[18] Krapp's desire is structured by a lack, one that in its unattainable, unfulfillable quality becomes circular — almost as if the tapes he was searching through were looped, with their heads spliced to their tails. That is, his desire creates a surplus of fantasy space.[19]

But Krapp's fantasies fail to help him locate his desires — as expected. He fails to record or to find the correct place on the tape that contains that previously recorded nonsymbolized kernel, that certain unspeakable something that will explain everything. The Lacanian real is, in part, that nonsymbolized kernel, an imaginary element that makes a sudden appearance in the symbolic order in the form of traumatic "return" and is understood through its "answer" even though it can only be acknowledged as a signifier of pure contingency — it can seemingly only be understood in terms of the existing symbolic order. Confronted with the reproduced return of his voice, Krapp is unable to make the recordings of his own voice serve as an appropriate "answer"; to convincingly construct a vision of reality through sound. As Zizek notes, madness or psychosis sets in with the failure of our barriers to keep the real at bay — the real overflows or is itself included in reality (Zizek 1989: 55–84; Zizek 1991: 20). As the blank space on the Nixon secret tapes also showed, the real can be a pretty scary Thing when it unexpectedly pokes through our tenuously constructed reality.

Krapp's symptom is a historical condition, or at least it is historically conditioned.[20] Magnetic, like mechanical, systems for recording sound date from just before the start of the twentieth century. The principles of magnetically recorded sound were probably first proposed by an American, Oberlin Smith, in an article for the engineering magazine *The Electrical World* of September 8, 1888 (Tall 1958: 1; Angus 1984: 28; Morton 2000: 91). The first practical application began in 1893 with Danish inventor Valdemar Poulsen's device, which used wire as the recording medium.[21] Poulsen applied for a patent first in Denmark on December 1, 1898, and for worldwide protection shortly afterward. The U.S. patent was filed on July 8, 1899 (and is reproduced as the frontispiece of Tall 1958). In 1900, Poulsen's telegraphone, based on his design and made by his company in America, was the first commercially manufactured magnetic recording device (Holmes 1985). Marketed as an office dictation recorder and an automatic telephone answering machine, the telegraphone competed with Thomas Edison's mechanical dictaphone (Tall 1958: 2–3; Morton 2000: 91–95, 109–10). The company struggled, not producing any significant number of machines before 1912, and moved from its initial location outside Washington, D.C., to Springfield, Massachusetts—near the home of Thadeus Cahill's groundbreaking (and similarly telephonic) musical instrument, the teleharmonium (1906).[22]

Beginning with the outbreak of World War I in August 1914, the Germans, one of the largest customers of the American Telegraphone Company, used magnetic recording in an unexpected way. Broadcasting from a station at Sayville, Long Island, to submarines off the coast, German spies first recorded Morse-coded intelligence reports and then broadcast them at twice their recorded speed (Angus 1984: 28–30).[23] This process improved signal transmission efficiency but also filled the airwaves with patterned static, with noise where meaning was expected. The mystery was finally deciphered in June 1915 by a puzzled ham radio amateur who recorded the signals using a hand-cranked cylinder phonograph and then discovered what they were while replaying them slowly.[24] That ham radio amateur, like Krapp, displays a symptomatic understanding of recording. The already arbitrary sign system of Morse code is further obscured through another capricious encoding, yet despite the additional level of indirection, the mystery solver understood that there should be meaning in the noise and sought it (even though its discovery was accidental). The mechanisms for recording and reproducing sound always constitute a disruption, a chain of contingent signs that needs to be deciphered to get to the meaning in the jumble. In Lacanian terms, the answer of the real fulfills an illusion that there is coherence out there, one that was always already there, even if it doesn't seem so at first (Zizek 1991: 33).

The degree of disruption is an effect of a recording's *definition;* the higher the definition, the lower the disruption. As described by Michel Chion, definition has two parts.[25] First, it is the correlation of a sound recording's acuity and precision in terms of measurable sonic characteristics (wow and flutter, frequency response, signal-to-noise ratio, etc.) that are compared against other (usually previous) audio technologies. For example, the magnetic tape recorder might be said to reproduce recordings with higher definition than its predecessor, the phonograph. In contrast, fidelity is based on an ideological assumption that there should, or even could, be a direct correspondence between a live and a reproduced sound—the desire to believe that recording isn't a symptom. Second, definition is dependent upon the listening audience's familiarity with those norms. Every established audio reproduction technology is considered to be completely adequate to the task of representing sound, at least until it is displaced by "something better." Long-playing records were considered a high definition medium, and the inherent surface noise of LPs disturbed few listeners, until they were measured against compact discs. A commonly held point of reference is necessary, and this is true of the predecessor to magnetic tape as well.

Wire and steel-band recorders, along with long-playing records, were one of the norms against which magnetic tape was measured, though the definition of magnetic systems was a quickly moving target—they were continuously developed and used through the 1950s. The following are examples of improvements made to Poulsen's basic recorder design, all introduced between 1900 and the end of World War II: DC bias noise reduction system by Poulsen and his chief engineer, P. O. Pedersen;[26] AC bias developed by W. L. Carlson and G. W. Carpenter of the U.S. Naval Research Laboratory (U.S. patent granted August 30, 1927); and Dr. Kurt Stille's use of electronic amplification in his improved telegraphone (which used a cartridge-loaded tape medium) (Tall 1958: 1–22).[27] During the 1920s, Stille organized the Telegraphie Patent Syndikat to license the use of the magnetic recording patents that the syndicate had acquired. Along with companies in Germany and America, he sold rights to Louis Blattner, who formed a motion-picture company in England and developed magnetic recorders to use instead of phonograph records for synchronization. The Blattnerphone (later bought by the American Marconi Company) was used beginning in 1929 by the BBC in collaboration with Dr. Heising of Stille Labs, and it incorporated frequency-correcting circuits to improve definition. It was used for broadcasts such as King George V's New Year's Day address of 1932, featured on the new BBC Empire shortwave radio service. Wire recorders were also produced in America through the 1950s.[28] Throughout the war American firms such as General Electric continued to improve and supply wire recorders for military use. Armour Research Foundation alone supplied the U.S. Navy with approximately ten thousand wire recorders between 1942 and 1945.

The medium, wire and eventually tape, improved during this time as well. Poulsen's original machine used wire media with a 1/100-inch diameter that traveled across the recording and playback heads at a speed of eighty-four inches per second. In 1937, C. N. Hickman (an acoustical researcher at Bell Telephone Labs) developed Vicalloy, a material used in a solid metal band form. Using a transport speed of sixteen inches per second, it was said to have an audio quality equal to Poulsen's wire, which moved five times faster. The Blattnerphone used six-millimeter-wide steel bands, which could, with great care, exertion, and danger, be edited with metal shears, the bands then welded back together and reused in subsequent recordings. Before World War II, the Echophone Company, formed by Illinois Institute of Technology graduate Marvin Camras, improved upon the unwieldy spools of wire with a cartridge-loaded machine, the textaphone. This company was bought by the German firm of C. Lorenz Company in 1933, and the Nazi party and secret police acquired the machines in large numbers during the war (Holmes 1985).

Until 1927 all magnetic recording media were solid or plated bands or wires. J. A. O'Neil received the first U.S. patent in 1927 for a flexible paper backing coated with a ferrofluid that was then dried. Magnetic tape was first produced beginning in 1928 in Germany by independent inventor Fritz Pfleumer. Instead of magnetically reorienting the iron particles within the wire or steel band, paper or plastic ribbons were coated with iron oxide particles and laminated onto the medium. The results created a better quality recording medium that was lighter in weight, less expensive, easier to manufacture, and more manageable. As Morton notes, the idea of a coated magnetic surface is still instrumental in all modern media: audio and video tape, computer storage disks, and credit card stripes (Morton 2000: 58). During the war, wire and steel band recorders were in wide production and was a known quantity. However, the advances that made magnetic tape superior to wire media were a clandestine German wartime development. This audio technology, as yet unknown in America, still had no definition—but a high potential for disruption.

During World War II, the Germans improved the formula for magnetic tape further still. In 1932, the German industrial giants BASF (Badische Anilin und Soda Fabrik, a division of I.G. Farbenindustrie chemical combine) and AEG (Allgemeine Electriziäts Gesellschaft) developed a magnetic tape system based on Pfleumer's model for a machine they called the magnetophon. Incorporating American technological innovations such as the AC bias technique, the vacuum tube amplifier, and the condenser microphone, the Germans designed a machine that used coated tape as the recording medium, one that was far more accurate than the older wire recorders—that is, one that recorded a higher degree of definition in terms of frequency response, signal-to-noise ratio, and transport speed consistency. This innovation also combined the capacity to record with an improved ability to reuse the medium and to edit it by splicing. Though this new technology was not entirely unknown to the Allies, their lack of understanding of definition in the new system produced a disproportionate disruption. For example, when Adolf Hitler broadcast radio speeches that were prerecorded using the new magnetic tape recorders, it was mistakenly assumed that he was in the announced location at the time of the broadcast (Gelatt 1954: 286–87). In monitoring radio broadcasts, American audio experts expected to be able to discern the difference between a "live" broadcast and a prerecorded one. Formerly, the characteristic imperfections of transcription disc and earlier magnetic recording systems—that is, their definition—had revealed the presence of reproduction and allowed the Allies to track Hitler's movements. Though the specific results of this deception are unknown, the material losses caused by Hitler's "invisibility" are easy to imagine.

Memorex tape commercials staged a similar scene based on the improved definition offered by cassette tape. In the mid-1980s, when cassette tapes occupied a technological and marketing position roughly equivalent to reel-to-reel tape in the 1950s—that is, they allowed mass culture participation in the recording process—the Memorex Corporation ran a series of television advertisements that portrayed a "blindfold" test. Set in a recording studio (that originatory temple of reproduced audio), well-known musicians such as Ella Fitzgerald were asked differentiate a "live" performance from a recording of the same thing: "Is it live or is it Memorex?"[29] After watching Chuck Mangione play a final few notes for the microphone, Fitzgerald tries to tell through audition (not vision) whether she is listening to Mangione play again or to his reproduction. Presumably, her labors are themselves repeated, first by the audience of the commercial, then by consumers and listeners using the cassettes being advertised.[30]

Through reproduction, Mangione's sound, as a presymbolic substance, returns to Fitzgerald from the void, out of body, divorced from its real space and time. Experiencing the world via our senses, we generally use sound to supply, in part, a sense of reality. Yet apparently, like a clone, the recorded Mangione is the same as its original in all aspects accessible to Fitzgerald, and it is just as unsettling. Even with her considerable experience and skill as an analyst of both live and technologically mediated listening, she is baffled and must be rescued by the band. Though historically conditioned in many regards, reproductive techniques have consistently caused sound to return in the form of a disembodied unconditional demand; reproduced sound has represented the listening subject's frustrated desire for coherence, and its subsequent rationalization.[31]

Why can't Ella tell whether Chuck Mangione is live or on tape? Like the return of the living dead in Hollywood horror films, this is a fantasy of "unreal" sound coming back to traumatize us, to make us question what we should accept as reality. Is that walking person alive or dead? Are we hearing a real trumpet or merely its specter? The return of the real is experienced as a challenge, a disturbance of our established understanding and categories (Zizek 1991: 21–32). Like a Hollywood screen, the recording process acts as a space for the projection of the desire to make sense out of our surroundings. But ultimately those projected fantasies must reaffirm our previous understanding. Our listening is contingent upon placing the sounds we hear in the familiar context of the known definition of the reproductive system, as one element in a chain of signifiers. Recorded sound returns with its own answer, encased in the protective armor of technology, the ideology of fidelity.

Unfortunately, comparisons seem unavoidable. Were Ussachevsky's disfigured and barely recognizable instrumental sounds like zombies, back from the dead to terrorize and confuse us? Are Ussachevsky and Murch modern-day Dr. Frankensteins who have reanimated the sonorous corpses of familiar objects merely to remind us of the value of the "real" sound of an untransposed piano note or of a voice not heard through a vacuum cleaner tube? What of the sense of experimentation and enjoyment suggested by the opening anecdotes?

It Is Rendered

There is a third way to encounter the real that is neither the (imaginary) *simulacrum* nor the (symbolic) code. Evident in many cinematic examples, and discussed in the theoretical literature of film sound and music, *rendu,* or rendering, goes beyond the fulfillment of fidelity evident in the Memorex example, beyond a simple, exact reproduction of an "original" sound. Sound is rendered through recording when it functions as a reinforcement, interpreting details of an original that wouldn't have been evident even if we were present at the "reality" of the performance. As an example of rendering, Zizek offers the scenes in David Lynch's film *Elephant Man* in which the sonic perspective is "inside" the elephant man's subjective experience. "The matrix of the 'external,' 'real' sounds and noises is suspended or at least appeased, pushed to the background; all we hear is a rhythmic beat, the status of which is uncertain, somewhere between a heartbeat and the regular rhythm of a machine" (Zizek 1991: 40–41). Murch used the tape recorder to explore that quality of sound in which "recording magically lifted the shadow away from the object and stood it on its own, giving it a miraculous and sometimes frightening substantiality" (Murch 1994: xvi). For Ussachevsky, with the tape recorder, "[E]ach detail, formerly compressed in the one impression of a whole (or into a single sound) began to be interesting in itself . . ." (Ussachevsky 1977: 4). Bypassing the imposed order of the symbolic realm and the prediscursive chaos of the imaginary, rendering is a state of sublime enjoyment. But this unfettered enjoyment is necessarily ephemeral, as Murch recalls:

One evening, though, I returned home from school, turned on the radio in the middle of a program, and couldn't believe my ears: sounds were being broadcast the likes of which I had only heard in the secrecy of my own little laboratory. As quickly as possible, I connected the recorder to the radio and sat there listening, rapt, as the reels turned and the sounds became increasingly strange and wonderful.

It turned out to be the *Premier Panorama de Musique Concréte*, a record by the French composers Pierre Schaeffer and Pierre Henry, and the incomplete tape of it became a sort of Bible of Sound for me. Or rather a Rosetta stone, because the vibrations chiseled into its iron oxide were the mysteriously significant and powerful hicroglyphs of a language that I did not yet understand but whose voice nonetheless spoke to me compellingly. And above all told me that I was not alone in my endeavors. (Chion 1994: xiv)

This passage describes perfectly the first steps from youthful, isolated experimentation toward a career as a Hollywood sound designer, as Murch's play necessarily becomes inscribed within the context of a language, a semiotic formation of sounds that was always already there. Paradigmatically, Murch discovered his Rosetta stone, and the cultural logic disorganized by the increased technological definition was quickly recovered, the changes reinscribed in the discursive network.[32]

Putting the Real Back into Place

Consumer Culture: Realism and Romanticism

In the case of Murch and, by extension, in mass culture, the practices engendered with the magnetic tape recorder were quickly subsumed into quotidian practices, reasserting a semiotic meaning to recorded sounds.

We can see this in the description of these practices. By the end of the 1930s, electricity was introduced into recording and reproduction practices, paving the way for the subsequent improvements that came with magnetic tape recording. Recordings made electrically differed from the discs recorded by the older acoustic methods in three ways, each with an attending audible impact. First, the range of frequencies that could be recorded was expanded, resulting in a noticeable timbral difference. Definition and details previously missing, such as vocal sibilants, were now recordable for the first time (Gelatt 1954: 223). The frequency range of electrically recorded music was improved from 250–2,500 Hz to 50–6,000 Hz, almost four times the range of acoustically recorded sound. Ironically, and not at all dissimilar to the reaction that accompanied the introduction of compact discs, the initial public perception of the quality of electrical recordings was that they sounded harsh and artificial (Read 1959: 373). Second, the condenser microphone drastically altered recording studio practice. When recording with an acoustic horn, musicians were typically forced to crowd together, arranged as closely as possible around the recording horn in order of quietest to loudest with the soloist at the center, who ducked out of the way for ensemble passages (Gammond 1980: 20). Furthermore, to be heard at all, the quieter string instruments had to be reinforced with a "Stroh" attachment—a diaphragm and a horn mounted on the bridge, which added considerable distortion (McGinn 1983: 42). By contrast, with electrical recording practices performers could be arranged in more comfortable and conventional ensemble set-ups, and the microphone placement would compensate. This meant more than simply added convenience. Instrumental balances and spatial effects were dramatically improved, and the acoustics of the recording space, the "room tone," was now audible on the recording (Gelatt 1954: 223; McGinn 1983: 42). Finally, electrical recording practices resulted in recordings with greater dynamic range. Especially with the introduction of amplified phonographs, the sound level of the original performance could be approximated at playback for the first time (Gelatt 1954: 223). In addition, other technological advances of this time include the beginnings of stereophonic reproduction for spatialization, scratch filters to reduce a recording's surface noise, and the vertical-cut cellulose acetate disc records that could be played with a lightweight moving coil pickup (McGinn 1983: 39–50).

These technological advances were subsumed, by contemporary critics, into two competing philosophical approaches: "realism" and "romanticism" (Read 1959: 373–89). Realism sought to bring the listener to a concert experience by capturing the hall ambiance and introducing live performances of larger ensembles into the repertory. Freed from the constraints of the recording studio, live recordings in large auditoriums regularly included the acoustic effects of the performance space upon the music as a discernable feature of the recording. Radio broadcasts and recordings of symphony orchestras, operas, and choral work became more viable, and new artists such as Stokowski, Beecham, and Weingartner rose to prominence by producing recordings that were, in part, made conceivable by the aesthetics of realism.[33] If realism implied an external, objective reality, romanticism seems designed to contain rendered sound, the subjective experience made manifest audibly. Associated with the affect of an intimate musical experience, romanticism in recording sought to bring the performance into the listener's living room. In "romantic" recordings, small groups of singers or instrumental ensembles would use a small, non-reverberant studio and work close to the microphone, effectively eliminating any trace of room ambiance. Out of this technique developed the musical style of the popular "crooner" vocalist, where the music as well as the technique strove to reduce the distance between listener and performer. An influx of new popular artists were associated with this technique: Billy Holiday, Louis Armstrong, Fred Astaire, and especially Bing Crosby in the Paul Whiteman orchestra were among the first to establish "intimate" recorded musical identities.[34] But it was through more than his "romantic" recording that Crosby encouraged the popular acceptance of the tape recorder.

The American acquisition of the improved tape recorder from the Germans began immediately after World War II.[35] There seem to be two interlocking histories of how the technology was transferred; these might be called the institutional as opposed to the personal accounts. At the end of the war, the U.S. Army Signal Corps recovered German magnetophons and magnetic tape from the facilities of Radio Frankfort and Luxembourg and the tape manufacturing plant at Wald Mittelbach, and the United States claimed rights to this technology under the War Reparations Act (Gimbel 1990).[36] The U.S. government assumed all American patents on the magnetophon and offered the license to any American company. Initially, three took up production in earnest: Magnecord, Rangertone, and Ampex Electric Company.[37] The low-grade tape used by the Germans was soon improved upon by the Minnesota Mining and Manufacturing Company (3M), with its "Scotch" brand of paper tape, introduced in 1949.[38] By 1948 high-quality magnetic tape recorders such as the Ampex 200 and magnetic tape were commercially available in America and were used for measurement, data acquisition and storage, conversion, and analysis in fields such as aerospace and rocket telemetry research, undersea exploration, industrial machine control, medical testing, information storage and processing, and communications (Lowman 1972: 1–7).[39]

In contrasting accounts, the technology of the magnetophon was introduced to the United States through the initiative of individual G.I.'s John Mullin and John Orr. Both were attached to the Army Signal Corps, which was charged with collecting electronic innovations from the defeated Germans, and each, independently, managed to bring home something for themselves. Orr was involved with reactivating a German tape manufacturing plant after the war and befriended the German head chemist, Pfleumer. As a parting gift, the chemist gave Orr a brown paper bag filled with the iron-oxide formulation BASF had used to make its tape. Orr returned to his native Alabama and started Irish Tape, one of the first American tape manufacturing facilities in the states. Orr reportedly told a committee of the Alabama state legislature that he had initially succeeded in duplicating the German formula by coating a strip of plastic with red barn paint (Angus 1984).

John Mullin disassembled and shipped two complete magnetophons back to the states through the mail—a feat only slightly surpassed by the apocryphal stories of G.I.'s mailing themselves entire Jeeps (Mullin 1976). While presenting his booty to a meeting of the Institute of Radio Engineers in May 1946, Mullin attracted the interest of a small company that had previously supplied precision motors to the navy, the Ampex Electric Company. Mullin soon put his magnetaphons (later Americanized to magnetophones) to work in radio production. Ampex, meanwhile, with Mullin as a consultant, set to work to improve upon Mullin's German model and produced one of the first American professional magnetic tape recorders.

Bing Crosby's influence here extended beyond the mass appeal of his music to the control he asserted as a producer and financier. Through his intervention, the Crosby radio show was the first major network program that was not broadcast live. In 1946, at the height of his career and exhausted by the stress of live radio scheduling, Crosby moved from NBC to the struggling ABC, lured by the new network's promise to allow him to prerecord his show for later broadcast (Morton 2000: 67). As with the examples of Nazi wartime broadcasts and the Memorex test, Crosby's shows, "transcribed" with the tape recorder, were initially considered to be indistinguishable from "live" ones. Crosby himself saw and pursued the potential offered by prerecording as a way to control precisely both the content and the length of his show. Mullin was hired to record and edit the first twenty-six shows of the 1947 season and used the original German tape machines and tape the entire time, resplicing the tape back together after each show. His original German equipment was replaced by some of the earliest available Ampex equipment because, in the interim, Bing Crosby had acquired a financial interest in Ampex Electric and had encouraged research, development, and distribution of the tape recorder for use in broadcast. Due in large part to Crosby's lead in developing the practice of prerecording and editing, professional tape recording equipment was in use in a majority of the commercial radio stations by 1950.[40] The unfamiliar degree of definition in the recording initially allowed audiences listening to Crosby's broadcasts with "old" ears to assume that they themselves were, as they had been before, a remote, technologically produced extension of the "live" studio audience. The process of knowing the difference was one of enculturation, of understanding the "answer." That came with the mass popular commercial distribution of the machine.

An enormous number of recorders had been sold and were in use by the mid-1950s, and those consumers used these machines in quite varied manners. Harry Olson estimated that by 1952 there were two million tape recorders plus many more wire recorders in use in the United States (Olson 1954: 637–43). With the introduction of stereo in 1955, tape recorder sales benefited from the most dramatic increase ever recorded for a single product in home audio entertainment (Read 1959: 427–28). The education of these home tape recorder enthusiasts was achieved through several cooperative channels. People learned how and what to record by the examples created by professional producers, which were presented to the public and analyzed by the media and equipment manufacturers.

Several forms of print media helped to educate the public as to what was being done professionally and encouraged individual experimentation. Periodicals such as *The Phonograph Monthly Review, Musical America,* and *The American Record Guide* had existed almost from the invention of the record itself, but the postwar era saw a boom in the number, and a change in the focus, of these journals. Publications that catered to the concerns of recorded music listeners committed significant space to reports on developing technology, and to editorials with particular emphasis on the tape recorder. Columnists and critics such as Edward Tatnall Canby of the *Saturday Review* and recording engineer Joel Tall published equipment and "how-to" guides that focused on how this new technology was utilized professionally and how it could be used at home (Canby 1952; Tall 1958). Soon even books explaining tape recorders to children were commercially available (Krishef 1962).

In 1951 a new magazine began that was dedicated specifically to the interests of a newly identified group: audiophiles. The appearance of *High Fidelity* signaled a critical mass, at least in terms of market targeting, of people interested in the idea of high-definition sound reproduction, since the magazine's professed goal was the wide dissemination of technical information and advice and equipment reviews, in addition to consumer information such as model numbers and prices.[41] Articles often provided detailed equipment specifications and information on engineering topics such as impedance matching, grounding, monitoring, and feedback. In addition to warning hobbyists of potential recording problems, articles often encouraged experimentation. For example, in Alan Macy's 1952 *High Fidelity* article on tape recording, he describes the danger of creating feedback by listening to what was being recorded (monitoring through the tape recorder's playback head). But he also encouraged his readers to experiment, to attempt to create a more controlled version of this phenomenon—an echo effect—themselves.[42] Participation in hi-fi for the individual hobbyist was predicated on the knowledge of and ability to replicate, in one's own home, prevailing recording aesthetics, and *High Fidelity* magazine and other publications were a primary source for information on what the professionals did.

For the hobbyist, the tape recorder was the audio equivalent of the photo album; it could be used to preserve and comment upon both special events and daily living, so that they could be relived, scrutinized, or enjoyed later. People used the tape machine to record their children's performances, to make soundtracks for home movies,[43] and to record music from the radio.[44] Audio fairs, regular local conventions that already drew New York crowds of as many as eight thousand audiophiles by 1951, communicated the inside information on the latest equipment innovations and served as a meeting ground for hobbyists and professionals [*High Fidelity* 1 (1952): 3]. Nationwide clubs for tape recordists were advertised in which enthusiasts would record local events, such as concerts, and exchange tapes by mail (Geraci 1961). Hobbyists were also encouraged and instructed how to seek out new environments to record, in imitation of professional ornithologists and ethnographers. Seemingly an international army of amateur recordists was mobilized to preserve indigenous bird songs and folk musics.[45] But recording was a way of performing a reality check for oneself too. For example, a captionless cartoon accompanying a *High Fidelity* article on tape editing showed a grimacing woman in a nightgown holding a microphone over her snoring husband. Finally, academic composers in the United States began to experiment with the tape recorder in the early 1950s, in order to turn the technology to their own particular creative and professional ends. At stake was, as with Murch, the proper way to see their creative efforts made sense of, coded as if on a Rosetta Stone. However, equally important was the effort to communicate that code: to find the right "museum" in which to store their Rosetta Stone.

"The old and the new are the same": The Structure of Concert Life

> This music [both continental and American electroacoustic concert music of the 1950s] was utterly unprecedented. Mostly, these works seemed to escape all conventions of communicability and to operate without any recognizable codes, like a completely hermetic artificial language which was programmed to generate itself. A somewhat baffling condition for semiology and music critic alike—although perhaps no more so than music has always been. (Chanan 1994: 269)

To look at reviews of the Museum of Modern Art (MOMA) performance of October 1952, the concert that programmed the "premieres" of four musical compositions by Ussachevsky and Luening, one might be inclined to agree with Chanan that this early music produced with tape recorders was unlike anything previously composed and therefore inaccessible and incomprehensible.[46] However, as in the realm of mass culture, the appearance of tape music on this program signaled that the product of a tape recorder could be understood—or, at least, it was beginning to be understood. This concert stands as an early indication that Ussachevsky's experiments could be accepted as musical compositions and included in the structure of American concert life. Rather than open a chasm of incomprehensibility, these compositions helped to affirm that the tape recorder, as a compositional medium, was not antithetical to traditional formal concerns and aesthetic values, and that the electronic music studio was to become a site of cultural maintenance.[47]

The concerts at MOMA of October 26 and 28, 1952, were organized by the American Composer's Alliance (ACA) under the auspices (and baton) of conductor Leopold Stokowski, that renowned and tireless champion of "modern" music, and a pioneer in music technology.[48] Beginning in 1945, Columbia University, with funding from the Alice M. Ditson Fund, had been sponsoring "mini" festivals of new music (Daniel 1982: 561). The 1952 festival honored Stokowski, who led the CBS Orchestra in the climactic program, which included orchestral works by Alexei Haieff and Roger Goeb. Two days later, the Goeb symphony was recorded by Stokowski and the Orchestra for RCA Victor, as were compositions by Weber and Harrison from the MOMA concerts.[49] Oliver Daniel, then the coordinating manager of the ACA, had been recruited by Columbia faculty and ACA members Otto Luening and Normand Lockwood, and it was in the successful aftermath of these Ditson Fund concerts that he and Stokowski began planning further collaborations, first the concert series at MOMA, and then the Contemporary Music Society.

The ACA and its associated Arrow Music Press, a composers' membership organization and a music publishing company, were formed in 1937 by a group of notable New York composers: Copland, Thomson, Harris, Blitzstein, Piston, and Sessions (Sanjek 1988: 98). Born at the height of the federal ASCAP antitrust probe, the ACA was established to watchdog the performance rights interests of its membership of "serious" concert music composers.[50] At issue were royalty rights from the traditional sources (concert licensing and the sale of musical scores) and—increasingly important—from recordings and radio broadcasts. Between 1938 and late 1940, ACA was courted by both ASCAP and the newly formed performances rights organization BMI.[51] By October 1952, with the advent of a new distribution system at ASCAP that (based upon its surveys of radio and television performances) disadvantaged "standard-music," BMI solidified its agreement with the ACA and its members, took over all coverage of ACA's performance rights interests, and housed the ACA library and its Composers Facsimile editions.

Seeking to produce a series of concerts modeled on Stokowski's International Composers' Guild, and the League of Composers concerts of the 1930s, Daniel and Stokowski met with the director of MOMA, René d'Harnoncourt, who enthusiastically agreed to host the concerts (Daniel 1982: 562). Daniel quickly secured funding for the series from BMI, and it was agreed that the ACA would produce them with Stokowski on the podium. While an ambitious start, these concerts were seen by the pair as a stepping stone to a more permanent structure to encourage and support new music: the Contemporary Music Society. As Stokowski said in a press release preceding the October concert, "What the Museum of Modern Art has done for painting and sculpture must be done for music all over the country" (Daniel 1982: 563). Shortly before the concert at MOMA, on October 21, 1952, at the Ford Foundation offices in New York City, Leopold Stokowski met with representatives of fifteen major musical institutions to offer his services as a conductor and champion for new American musical arts (Daniel 1952; Daniel 1982: 575–76). The organizations at this meeting were ASCAP, BMI, the Musicians' Union, the National Music Council, the League of Composers (represented by Claire Reis), the International Society for Contemporary Music, UNESCO, Columbia University, New York University, Julliard, Eastman-Rochester, the Voice of America, the Musician's Trust Fund, the Society for the Publication of American Music, the American Composers Alliance, and representatives of the press. BMI, through its "enlightened" president, Carl Haverlin, pledged $5,000 in support of the venture, while Stokowski offered another $1,000 as a personal contribution—nothing else was forthcoming at that meeting. The resulting collaborations, though, were significant, because, as Howard Taubman of the *New York Times* noted at the time, "it may be the start of a concerted effort by various organizations concerned with the support of the music of our own time to hammer out a united approach" (Daniel 1952: 2–3). Under the auspices of the Contemporary Music Society, in addition to the October 1952 concerts there was a series of four more concerts presented at MOMA as well as two recordings by Stokowski of compositions drawn from these concerts, which were initially released by RCA Victor, then by CRI.

The programs for the October MOMA concerts were largely drawn from the library of compositions by ACA members.[52] While settling the details of the programs in a planning meeting in Zurich while Stokowski was on tour, Daniel and Stokowski, aware of developments in Europe, agreed that one of the concerts at MOMA would introduce electronic music to American audiences (Daniel 1982: 565–66). These innovations seemed to be everywhere at once—in America and throughout Europe simultaneously. Pierre Schaeffer, a radio engineer at Radiodiffusion Française (RF) in Paris, had already coined his term *"musique concrete"* in 1948 and, working with Pierre Boulez, Pierre Henry, and others, had produced concerts and radio broadcasts of tape recorder "music" in France (Chadabe 1997: 26–44). While lectures on electronic music by Werner Meyer-Eppler began in 1948, construction of the studio at Westdeutscher Rundfunk (WDR) in Cologne, initially built around the newly developed vocoder from Bell Telephone Labs, started in late 1951 and lasted into 1952. The first compositions of *elektronische Musik* follow shortly thereafter. In America, one of the first musical projects using the magnetic tape recorder began in 1951 as the Project for Music for Magnetic Tape. Organized under John Cage's initiative, the project proceeded for several years with the support of David Tutor, Bebe and Louis Barron, and Paul Williams, with occasional access to Richard Ranger's studio of Rangertone magnetic tape recorders in New Jersey (Chadabe 1997: 56). Daniel initially approached Cage to request a composition for the MOMA concert. However, as the concert date neared, it became clear that Cage would not be ready. (*Williams Mix,* Cage's first tape piece, was finished in late 1952.) Instead, Daniel approached Luening and Ussachevsky. While the concerts as a whole were apparently well received by both the public and the press, the works of the "tapesichordists" were clearly the evening's novelty. So, even though tape music was flourishing in the home market, the MOMA concert is often acknowledged as the successful inauguration of tape music in America.

So how did these recordings themselves differ from what the young Murch produced, or even from the work of contemporary peers of Ussachevsky? The move from "experiments" to "compositions" is audible in these objects themselves, even while, at the level of manipulation of the tape medium, there is an unavoidable resemblance between the techniques of Ussachevsky, Murch, *musique concrete,* and the "Music for Magnetic Tape" crowd. As Otto Luening explained,

> I think from the first moment that Vladimir Ussachevsky and I began working on our first short pieces, although using machines and later developing, finding and using new electronic means of tone production or manipulation, we were interested primarily in making this part of the great stream of music that we had inherited. Our standard did not depend on how much equipment was at our disposal but how well we used what we had. I could never forget that Busoni had said, "The old and the new are the same," and that he wrote in 1907 that a long period of ear training would be necessary to make electronic sounds useful for artistic purposes. (Ussachevsky 1977: 42)

In *Sonic Contours,* Ussachevsky balanced a professed fascination with the novelty of the tape recorder with a classically trained composer's concern for structural coherence, balance, and contrast. The sound sources were local and familiar sounds, mainly recorded piano, supplemented toward the end of the piece with snippets from recordings of voices. The recorded piano and voices underwent the basic manipulations of playing a segment back at altered speeds and lengths, removing the very onset of the sound (its attack), regenerating the recorded signal as a controllable echo, and splicing the whole together from smaller snippets of tape. But unlike other hobbyist's experiments, this piece was constructed *motivically,* where the manipulation of a few recorded sounds fed a larger organizing principle: a canon in five voices at three different speeds.[53] Such concerns were summed up also in Ussachevsky's description of his compositional process of audition and decision, what he called "the record-play-listen-examine routine." In contrast, Cage's *Williams Mix* was constructed using chance procedures generated with the *I Ching* and assembled from haphazardly collected (but assiduously catalogued) pieces of recording tape (Cage and company didn't even have regular access to a tape recorder as the piece was fabricated). *Sonic Contours* was assembled from Ussachevsky's experiments with the tape recorder and microphone; its materials include his improvisations at the piano, some serendipitous recording of human speech and laughter (part of a recording made while demonstrating the feedback effect for his wife), and passages deliberately composed to go with those that were experimentally derived.[54]

Given the conventional nature of their construction, perhaps the most radical element of these early compositions is their apparent evasion of written form. Through the first half of the century, the role of composer has been, by convention, constrained through standard musical notation (the twelve note per octave tuning system, an additive temporal grid, and timbral limits described through instrumentation, for example). At first glance, then, these compositions seem to owe a greater than acknowledged conceptual debt to musical forms predicated upon improvisation and performance—such as rock 'n' roll, for example, a contemporary development. However, the limits of musical notation are infinitely elastic in the later twentieth century. What has remained immutable, on the other hand, is the conviction that the critical aspects of a musical idea may be contained and preserved through notation. It is telling, then, that, in an effort to secure legal protection from the U.S. Copyright Office, Ussachevsky laboriously transcribed his recordings into musical notation after the fact, extending conventional notation, for example, by writing notes with wavy stems to indicate sounds created through feedback. If musical compositions are vessels that contain knowledge, then works such as *Sonic Contours* are predicated on the Enlightenment ideals of structural listening, firmly grounded in accepted musical reality, and part of its ideological apparatus.[55] The creators of these sonic experiments were similarly reinscribed.

In the aftermath of the MOMA concert and the Ford Foundation meeting, a small grant from the Rockefeller Foundation facilitated the purchase of equipment specifically for electronic music production. Ussachevsky's work *Incantation for Tapesichord* was commissioned by Stokowski in the fall of 1953 to be included in a CBS radio program under his direction (Ussachevsky 1977: 9). (The piece was described by one reviewer for the *New York World-Telegram and Sun* as "[t]he strangest music this side of Paranoia.") The Louisville and Los Angeles orchestras later provided further performance opportunities. Ussachevsky traveled to Paris and Cologne, supported by BMI, to speak on American tape music in 1953 (Ussachevsky 1953). Again in 1955 Ussachevsky and Luening received a grant from the Rockefeller Foundation to investigate the state of studio facilities both at home and abroad. Finally, the works presented at MOMA were release on phonograph record on the Innovations label in 1955.[56] These efforts are indicative of the institutional effort supporting new music in general and electronic music in particular. In response to a plea to the university president, the Columbia electronic music studio was moved (partly from Ussachevsky's living room) to more appropriate quarters in 1954—the "Charles Addams" house located on the site of the former Bloomingdale Insane Asylum (Chadabe 1997: 45). RCA demonstrated the Olson-Belar Sound Synthesizer to the director of the School of Arts at Columbia in 1955, and in January 1959 this seed of institutional sponsorship had fully flowered to produce the Columbia-Princeton Electronic Music Center under the auspices of the two universities, the Rockefeller Foundation, and the RCA Corporation.

Conclusion: It Contains Knowledge

The introduction of the magnetic recorder and tape represented a moment of limitless auditory possibilities. However, that emancipatory moment was brief. Where once the aspirations of the historical avant-garde carried the utopian hope for an emancipatory mass culture (under socialism), this was, in fact, preempted by the rise of mass-mediated culture and its supporting industries and institutions (Huyssen 1986: 15). So too with the tape recorder in America after the war. Murch ultimately found a job that contained his pleasurable activities with sound. As one of Hollywood's preeminent sound designers, he has produced some of its most unusual and adventurous soundtracks. We need only look to *The Conversation* for evidence of his continued engagement with rendered sound and technology. But there his play is safely circumscribed by its mass culture frame, the Hollywood film. Conversely, Ussachevsky and Luening, through the assertion of their play as "composition," were reinscribed in the (slightly modified) established social orders of "music." Given the acceptance of their "compositions" in the concert hall, their sounds must ultimately be understood in terms of the modernist category of high autonomous works of art. Using a new technology, both were able to imagine "noises" that (only momentarily) evaded the cultural logic of recorded sound.

But we are back where we started—Murch and Ussachevsky on opposite sides of the "great divide." This essay has examined the historical moment at which a form of free sonorous enjoyment was made possible through the magnetic tape recorder and then brought under control by it. That control was expressed through the supporting industries and institutions of not only mass culture but high art as well. Or, to use Marcuse's terms, the reality principle necessarily gives rise to affirmative culture. These two anecdotes might be viewed as evidence of a utopian moment followed by order reasserted, by an encounter with the real and the necessary (i.e. healthy) reaffirmation of reality. By imagining creative, singular uses for a new technology, the magnetic tape recorder, both Ussachevsky and Murch were able to "liberate technology momentarily from its instrumental aspects, and thus [they] undermined both bourgeois notions of technology as progress and art as 'natural,' 'autonomous,' and 'organic'" (Huyssen 1986: 11). Murch and Ussachevsky were able to express what Huyssen has called the hidden dialectic, to articulate the relationship between technology, art, and mass culture that also engaged the historical avant-garde until World War II.[57]

Notes

1. Earlier and abbreviated drafts of this paper were presented at a meeting of the International Association for the Study of Popular Music (IASPM-US) at UCLA, at a northeast regional meeting of the Society of Composers, Inc., at Connecticut College, and at the Western Illinois University New Music Festival. My thanks to everyone at these gatherings who offered constructive criticisms, and especially to Nancy Newman.

2. As a sound mixer and designer, Murch is well known for his contributions to films such as *The Godfather, The Conversation,* and *Apocalypse Now.* As an explicator of film sound, he is articulate, insightful, and prolific. See Murch 1994; "Interview with Walter Murch," *Positif* (1989), no. 338; "Stretching Sound to Help the Mind See," *New York Times,* October 1, 2000, Section 2, p. 1; "Sound Design: The Dancing Shadow," in J. Borrman, T. Luddy, D. Thomson, and W. W. Donohue, eds., *Projections 4: Film-makers on Film-making* (London: Faber and Faber, 1995); "Sound Mixing and *Apocalypse Now,*" in J. Belton and E. Weis, eds., *Film Sound: Theory and Practice* (New York: Columbia University Press, 1985); and "The Sound Designer" in R. Madsen, *Working Cinema—Learning from the Masters* (Belmont, Calif.: Wadsworth Publishing, 1990). In an industry of specialization, Murch is unique—a respected specialist in two distinct areas. As a film editor, see his extraordinary *In the Blink of an Eye: A Perspective on Film Editing* (Los Angeles: Silman-James Press, 1995).

3. There is an extensive, though somewhat diffused, literature on the development of the magnetic tape recorder. See Angus 1984; Belton 1992; Chanan 1994; Gelatt 1954; Haynes 1949; Jones 1992; Millard 1995; Morton 2000; Read and Welch 1959; and Sanjek 1988.

4. See Day 2000 for an extensive study of the effects of reproductive audio technology on musical repertoire.

5. See Riesman 1953 and Riesman 1950 on the role of popular music (in combination with other mass media) in the socialization of youth.

6. See Marshall McLuhan, *Understanding Media: The Extensions of Man* (New York: Penguin Books, 1964) for a discussion of the role of technology as an extension of the human senses.

7. The equipment consisted initially of a single Ampex model 400 and a Western Electric 639 microphone; see Ussachevsky 1977: 5–6. Ussachevsky has said: "I was the most junior of the music faculty. It was my job to look after all the audio equipment. I was fascinated with the tape recorder—I don't know why. I immediately did a lot of recording" (Darter 1984: 149). For more biographical information and Ussachevsky's professional credentials, see Darter 1984: 148–49.

8. Speed variation, or transpositions, meant playing the tape at the speed other than that of the original recording, usually 7.5 inches of tape per second. Not only the pitch but the timbral quality of the recorded sound was affected; the greater the transposition, the more noticeable the apparent change in timbre. However, Ussachevsky couched the musical utility of this technique in terms of its acoustic consistency: the ratios between partials of the overtone series remain unaffected (Ussachevsky 1977). The desired parallel of this spectral alteration to conventional motivic usage and development is apparent.

9. In Ussachevsky's own words, "[F]eedback is an automatic but controllable repetition of any sound or sounds being recorded on magnetic tape. In the normal magnetic head configuration on a professional tape recorder the tape passes by [the] erase head, then [the] record head and then the playback head. A sound is first recorded and then heard a fraction of a second later through [the] playback head. If

the output of the playback head is immediately shuttled back to [the] record head, everything that is being recorded will be immediately repeated. If, as is predominantly the case, the sound pattern is longer than the rate of repetition, then, obviously, overlapping of the original and the subsequent repetitions will take place. The number of repetitions can be regulated but the quality of the recording deteriorates" (Ussachevsky 1977: 5).

10. Also see reviews and articles in *Vogue* magazine (July 1953), "Tapesichordists" by Peggy Glanville-Hicks, and *Downbeat* (Chicago, July 29, 1953), "Counterpoint" column by Nat Hentoff.

11. The question of their "official" premiere might be contended because of the Composers' Forum program, which was reviewed by Henry Cowell in *Musical Quarterly* (Manning 1993). See, however, the discussion below on the larger institutional significance of their MOMA presentation.

12. See David Laing, *One Chord Wonders: Power and Meaning in Punk Rock* (Milton Keynes: Open University Press, 1985); Sarah Thornton, "Strategies for Reconstructing the Popular Past," *Popular Music* 9/1 (1995): 87–95; and Paul Théberge, *Any Sound You Can Imagine: Making Music/Consuming Technology* (Hanover, N.H.: University Press of New England: 1997) for the use of music periodicals in "reading" the popular past.

13. See Schwarz 1997 for a description of the listening gaze and its relationship to subjectivity.

14. For Lacan, it is Marx who "invented" the symptom. The Lacanian notion of "symptom" is derived from what Zizek calls "a fundamental homology between the interpretive procedures of Marx and Freud—more precisely, between their analysis of commodity and dreams." Alternatively, for Freud, the essential psychoanalytic understanding of dreams aims not at the *manifest dream-text* or the *latent dream-content* but at the unconscious desire articulated through the dream's formal dimensions, or the structures that translate latent into manifest, "consisting entirely of the signifier's mechanisms, of the treatment to which the latent thought is submitted" (Zizek 1989: 13).

Beyond the psychoanalytic notion of case study, I am thinking also of the field of sociology of technology, which has recently developed a number of analytic tactics to write technological case studies concerned principally with the social nature of technical innovation. Moving away from a focus on an individual inventor as a central explanatory concept, these analytic approaches emphasize social construction in an attempt to avoid technological determinism and to give equal weight to technical, social, economic, and political aspects of technological development. This attitude has been summarized with the anthropological metaphor of a "seamless web" (Geertz 1973; Pinch 1985; Bijker 1990), where both the impact of technology on society and the influence of society on technology are considered. In this "seamless web" the term "technology" is used broadly, and three levels of meaning in the word have been distinguished (MacKenzie and Wajcman 1985). First, there is the level of *physical objects or artifacts;* second, *activities or processes;* third, what people *know* as well as what they do—the *know-how* of creative application and design. In terms of the technology of the tape recorder, descriptions might include the machine itself; techniques of use, such as editing or recording practices; and the background, ambitions, and attitudes of its designers, builders, and users.

15. The broad organization of this paper, as well as my initial access to these Lacanian concepts, comes from Zizek 1991. In addition to the above, Zizek has a fourth category: the real constitutes *knowledge in itself.*

16. As Zizek notes, "Lacan's 'return to Freud' is usually associated with his motto 'the unconscious is structured like a language,' i.e. with an effort to unmask imaginary fascination and reveal the symbolic law that governs it. In the last years of

Lacan's teaching, however, the accent was shifted from the split between the imaginary and the symbolic to the barrier separating the real from (symbolically structured) reality." (Zizek 1991: viii).

17. See Zizek 1991: 18–20 for a discussion of psychosis in relation to failed attempts to deal with the Lacanian real.

18. See Zizek 1997 for a systematic exploration of the Lacanian category of fantasy.

19. The Lacanian category of fantasy is critical to the structure of desire, and the objects through which those desires are embodied. "[W]hat fantasy stages is not a scene in which our desire is fulfilled, fully satisfied, but on the contrary, a scene that realizes, stages, the desire as such. The fundamental point of psychoanalysis is that desire is not something given in advance, but something that has to be constructed—and it is precisely the role of fantasy to give the coordinates of the subject's desire, to specify its object, to locate the position the subject assumes in it. It is only through fantasy that the subject is constituted as desiring: through fantasy, we learn how to desire" (Zizek 1991: 6). In that fantasy never fully consumes desire, it is self-perpetuating and thus, in the surplus it creates, circular.

20. See MacKenzie 1985 and Bijker 1990 for a theoretical discussion of the social constructivist view of the sociology of technology.

21. The wire recorder was an application of already well established and understood theories of magnetism applied to the new area of sound recording. It developed nearly simultaneously in a variety of places, in manner reminiscent of Thomas Kuhn's paradigm for "normal science."

22. See Reynold Weidenaar, *Magic Music from the Teleharmonium* (Metuchen, N.J.: Scarecrow Press, 1995), for a complete history of this fascinating instrument.

23. Angus reports that the enormous shortwave transmission and antenna station at Sayville, Long Island, was established by the Atlantic Communication Company and was linked to a German receiver in Nauen, Germany, in order to relay commercial messages between America and Europe. This sophisticated system used wire recorders to automate and speed transatlantic transmissions. There was a contemporary suspicion that the company was concealing its true German ownership, and, in fact, espionage was subsequently discovered (see Angus 1985).

24. The U.S. Navy quickly closed down the shortwave transmitter and antenna, secreting the entire episode. Only recently did the Freedom of Information Act (1971) enable the National Archives to make the original cylinders and other documents available. The incident seems to have been suppressed for reasons of national security.

25. See Chion 1994: 98–99 for a further discussion of *definition* and *fidelity*.

26. Poulsen's article in *The Electrician* of November 30, 1900, alluded to another of their innovations of this time: multiplexing, a technique of critical importance to later communications technology. "An elegant method of compensation has been invented by the engineer P. O. Pedersen and allows several speeches to be intermingled so that they can afterwards be reproduced separately. As it is not feasible to describe this method satisfactorily in a few words, I shall not speak further of it here" (quoted in Tall 1958: 7).

27. Lee DeForest, inventor of the Audion tube amplifier, used a telegraphone in San Francisco around 1912 to demonstrate a high-speed telephone repeater system, though this innovation never went into production (Tall 1958: 9).

28. See the following for details: Haynes 1949; Dale 1952; Gelatt 1954: 284–289; Read 1959: 426–428; Angus 1984; and Holmes 1985.

29. See John Mowitt's extensive analysis of this example in Mowitt 1987.

30. A wonderfully expressive example of the same challenge can be found in a publicity photography of Frieda Helpel from an Edison Company sound test, c. 1916–25 (reproduced on the back cover of Day 2000). On the left side of this photograph,

Hempel is standing beside a phonograph with an acoustic recording horn visible in the background. Five blindfolded men are seated with pensive expressions while Hempel demurely covers her mouth to suppress a giggle. If nothing else, evident here is the "face" at stake when the gender roles of the test are exchanged.

The literature that presents and analyzes the original/copy problem in recording is quite extensive, particularly from the perspective of film sound. For the most comprehensive treatment see J. Lastra, *Sound Technology and the American Cinema: Perception, Representation, Modernity* (New York: Columbia University Press, 2000). Also see Theodor Adorno, "The Radio Symphony: An Experiment in Theory," in *Radio Research 1941*, Paul Lazarsfeld and Frank Stanton, eds. (New York: Columbia University Office of Radio Research, 1941: 110–139); Rick Altman, "The Material Heterogeneity of Recorded Sound," in *Sound Theory Sound Practice*, Rick Altman, ed. (New York: Routledge, 1992: 15–34); Mary Ann Doane, "Ideology and the Practice of Sound Editing and Mixing," in *The Cinematic Apparatus*, Teresa de Lauretis and Stephen Heath, eds. (New York: St. Martin's Press, 1980: 47–56); James Lastra, "Reading, Writing, and Representing Sound," in *Sound Theory Sound Practice*, 65–86; Thomas Y. Levin "The Acoustic Dimension: Notes on Film Sound," *Screen* 25/3 (May–June 1984): 55–68; Thomas Y. Levin, "For the Record: Adorno on Music in the Age of Its Technological Reproducibility," *October* 55 (Winter 1991): 23–47; Christian Metz, "Aural Objects," *Yale French Studies* 60 (1980): 24–32; A. Williams, "Is Sound Recording Like A Language?" *Yale French Studies* 60 (1980): 51–60.

31. For a discussion of the listening subject—particularly in relationship to the acoustic mirror phase and the sonorous envelope theories—see Michel Chion, *The Voice in Cinema*, Claudia Gorbman, trans. (New York: Columbia University Press, 1999); Philip Brett, "Musicality, Essentialism, and the Closet," in *Queering the Pitch: The New Gay and Lesbian Musicology*, Philip Brett, Elizabeth Wood, Gary C. Thomas, eds. (New York: Routledge, 1994); Guy Rosolato, "La Voix," in *Essais sur le symbolique* (Paris: Gallimard, 1969); Schwarz 1997; Kaja Silverman, *The Subject of Semiotics* (New York: Oxford University Press, 1983); and Silverman, *The Acoustic Mirror: The Female Voice in Psychoanalysis and Cinema* (Bloomington: Indiana University Press, 1988).

32. A discourse network is an expanding and contracting system of inscription by which the position of subjects is determined in relation to contemporary means of communication, along with the shifting patterns of human sociability. See Friedrich A. Kittler, *Discourse Networks 1800/1900* (Stanford: Stanford University Press, 1985).

33. For a further description and analysis of the rise of broadcast and recorded concert music within the context of the music appreciation movement, see Joseph Horowitz, *Understanding Toscanini* (London: Faber and Faber, 1987).

34. Often noted as an invention critical to the history of recorded sound, the electrical microphone is a device that is sufficiently sensitive so as to decouple the volume required of live performance from the loudness needed to make a recording. See Evan Eisenberg, *The Recording Angel: The Experience of Music from Aristotle to Zappa* (New York: Penguin Group, 1987), 156; Simon Frith, *Music for Pleasure: Essays in the Sociology of Pop* (New York: Routledge, 1988) 18–19; and Peter Gammond, and Raymond Horricks, *The Music Goes Round and Round* (London: Quartet Books, 1980) 23.

35. There were several companies that continued to produce and improve upon magnetic recording during the war, though none improved the system nearly as much as the Germans. The most successful of these were the Brush Development Company (later renamed Brush Electronics Co.), which beginning in 1939 researched magnetic recording systems under the supervision of S. J. Begun. During

the war, they improved the coating of paper-backed magnetic tape. From late in 1941, Marvin Camras, a researcher at the Armour Research Foundation of the Illinois Institute of Technology, patented and produced wire recorders that employed AC erasing and biasing. Both of these companies sold machines used by U.S. armed forces during the war.

36. Along with a multitude of other wartime German innovations, tape recorder technology became the focus of U.S. government scrutiny after the war. The results of reports, investigations and government-supported research were made public through a multitude of Congressional committee hearings and defense department symposium proceedings. Unfortunately, the bulk of this material, while it probably still exists, is almost totally inaccessible. The government system of document identification changed during the 1960s, and no index was created to translate the older PB designation to the current Superintendent of Documents Numbering system. With only contemporary citations of the older PB numbers, most of these documents are unlocatable. Some examples are *Report to the British Intelligence Objectives Subcommittee of 1945* by M. J. L. Pulling, entitled "The Magnetophone Sound Recording and Reproduction System," which was available through the Department of Commerce, Office of Technical Services, government document P.B. 60899; the same office produced *Reports Resulting from Investigation of German Technology 1945–46*, "The 'Magnetophon' of A.E.G.," 150 Hohenzollern Damm, Berlin, Grunewald, B.I.O.S. Report 207, H.M. Stationery Office and U.S. Department of Commerce, 4 pp. (1946), "Magnetophon Sound Recorder and Reproducer 1939–1946, A.E.G. Berlin and Kiel," Office of Publication Board Report, P.B. 95210, 711 pp. (in German and English), and *Reports: An Index to Bibliography on Magnetic Recording and Reproduction 1947;* also the *Proceedings of the Department of Defense Symposium on Magnetic Recording* of March 1954, Government Publication P.B. 118037, part of the Fifth Symposium on Acoustic-In-Air Research and Design.

The U.S. government and industry's interest in the technology of the tape recorder grew with its expanded field of application. Immediately upon its introduction in America, the tape recorder was a focus of scientific, military, and medical research, as well as of the music industry.

37. Before 1948 there were no professional magnetic tape machines available in the United States. Some broadcasters used wire recorders, such as the General Electric Model 50 in the studio, and the Pierce Wire Recorder for field recording. In 1946 and 1947, the Brush BK 401 was introduced for home use and subsequently adopted for use in many broadcast studios.

38. The 3M Labs under the direction of W. W. Wetzel improved upon the German formula with acicular red oxide (Tall 1958: 34). The eventual solution was to produce a durable base for tape from a petroleum-based plastic (in place of paper) and to impregnate that tape with a different magnetic material: magnetite, a gamma ferric oxide. 3M, with its Scotch 111 product, set the standard for all American recording tape until the introduction of chromium dioxide-based tape in 1969.

39. The periodical writing about magnetic tape recorders, in both the scientific and popular press, began immediately after the war and quickly became a regular magazine feature. Here is a representative bibliography: "German Magnetic Tape," *Science* 48 (December 1945): 399; "It Pays to Listen," *Scientific American* 173 (July 1945): 18–20; "The Magnetic Tape Recorder," *Radio News* 55 (June 1945): 32; "Sound on Paper," *Scientific American* 174 (April 1946): 156; "Recent Developments in the Field of Magnetic Recording," *Journal of the SMPTE* (January 1947); Joel Tall, "The Art of Tape Recording," *Audio* (May 1950); "Inventory Taking Speeded by Wire Recorders," *Electrical World* (December 18, 1948); "Magnetic

Tape and the CBC," *Radio News* (June 1952); "Using the Tape Recorder," *Curriculum and Materials* (New York Board of Education) 7/1 (September 1952): 4–5; "Sound Inscribed on Paper Tape," *Business Week* (January 26, 1946); "Tape for the Networks," *Newsweek* 52 (May 3, 1948).

40. Soon after, under the auspices of Bing Crosby Enterprises, John Mullin and Wayne Johnson demonstrated the fruits of their further research in magnetic media: the first video tape recorder was demonstrated in 1952.

41. The concept of high fidelity seems to have begun circulating in 1935 as a record marketing ploy to advertise the improved quality of electric phonograph records, but it quickly became associated with a "modern" listener's concern and active participation in the best, most professional reproduction possible. Victor's 1935 recording of Richard Strauss's *Also Sprach Zarathustra* was billed as "high fidelity," with "the greatest crescendo ever recorded" (Gelatt 1954: 269).

42. *High Fidelity* 1 (1952): "From the Editor," 3.

43. Private conversations with Donald Whitfield.

44. In contrast to more recent recording industry tactics regarding MP3 audio files, this last practice was one actively encouraged by manufacturers in the 1950s through their advertising, which advocated the "personalization" of home listening by making one's own tapes. Choosing the repertory and setting the musical sequence of pieces was equated with musical self-determination and creativity. Beginning in the early 1940s, equipment manufacturers began selling home entertainment packages that included both a radio and a recorder (first wire, then, following the war, tape). The design was to facilitate one of the primary uses of the recorder: to record from the radio. See Gelatt 1954.

45. See E. T. Canby, "Recording Nature's Musicians," *High Fidelity* 2/4 (January–February 1953); Peter Paul Kellogg, Ph.D., "Recording Sound in Nature," in Tall 1950; and *The Collecting of Folk Music and Other Ethnomusicological Material: A Manual for Field Workers*, Maud Karpeles, ed., issued by the International Folk Music Council in 1951, as representative examples.

46. See Daniel 1952 for reprints of concert reviews from the *New York Times*, the *New York Herald Tribune*, *Time* magazine, and the *New Republic*. Also see the citations in note 10 and the epigraph that opens this essay.

47. For a booklength study of the logical consequences of this trend, see G. Born, *Rationalizing Culture: IRCAM, Boulez, and the Institutionalization of the Musical Avant-Garde* (Berkeley: University of California Press, 1995).

48. See McGinn 1983 for an expansive history of Stokowski's collaboration with the Bell Telephone research labs.

49. In 1954, with funding from the ACA and the Ditson Fund, Douglas Moore and Otto Luening formed with Oliver Daniel Composers Recordings, Inc. (CRI), a vanity recording label. As Daniel reports, the impetus for this new venture came as a result of RCA Victor's practice of dropping recordings from their catalog because of poor sales. When Victor discontinued Stokowski's recordings of works from the MOMA series by Harrison, Weber, and Goeb, ACA retained "all rights to the works, including tapes, masters, etc." As a result of a clause in the recording contract with Victor, the ACA was in sudden possession of valuable master tapes of recordings by Stokowski, but they had no means to distribute them. CRI was formed as a reliable and enduring means of distributing these and subsequently many other recordings of contemporary music (Daniel 1982: 568).

50. See John Ryan, *The Production of Culture in the Music Industry: The ASCAP-BMI Controversy* (Lanham, Md.: University Press of America, 1985), for more on the breakup of ASCAP and the formation of BMI.

51. Copland was represented by ASCAP, and he negotiated unsuccessfully with ASCAP president Gene Buck to have ASCAP support ACA.

52. The programs at MOMA were as follows: On October 26, 1952, Henry Cowell, *Hymn and Fuguing Tune No. 2* and *Hymn and Fuguing Tune No. 5* (first New York performances); Ulysses Kay, *Suite* for strings (first New York performance); John Lessard, *Cantilena* for oboe and string orchestra; Wallingford Riegger, *Study in Sonority;* and Alan Hovhaness, *30th Ode of Solomon;* on October 28, 1952, Lou Harrison, *Suite* for solo violin and solo piano with small orchestra; Elliot Carter, *Eight Etudes and a Fantasy* for woodwind quartet (first performance); Vladimir Ussachevsky, *Sonic Contours* (first performance); Otto Luening, *Low Speed, Invention,* and *Fantasy in Space* (first performances); and Ben Weber, *Symphony on Poems of William Blake,* opus 33 (first performance).

53. Ussachevsky 1977; for an interesting parallel, see James Tenney, "Conlon Nancarrow's STUDIES for Player Piano," in *Conlon Nancarrow Selected Studies for Player Piano* (Berkeley: Soundings Press, 1977), for discussion of similar formal concerns in the music of Conlon Nancarrow, who was composing for modified player piano.

54. The recordings consist of Ussachevsky, his wife (unnamed), and Peter Mauzey "sharing in verbal comments on the strange sensation of listening with earphones to the feedback of one's own speech. Everything we said and each note that I played as an illustration came back in psychologically disorienting fashion that made one stutter and struggle to continue speaking. The louder one spoke or chuckled, the longer the repetition. The tape was preserved" (Ussachevsky 1977: 9).

55. See Theodor W. Adorno, *Introduction to the Sociology of Music* (New York: Seabury Press, 1976 [1962]), 7; and Rose R. Subotnik, "The Challenge of Contemporary Music," in *Developing Variations: Style and Ideology in Western Music* (Minneapolis: University of Minnesota, 1991), 265–94.

56. The works were reissued in 1991 by CRI as part of their American Masters Series (CRI CD611: *Pioneers of Electronic Music*), along with works by Smiley, Arel, Davidovsky, and Shields.

57. Murch and, in different ways, Ussachevsky might be seen to embody the Americanization of the historical avant-garde's project—transformed by the intervening war.

Since it has become more difficult to share the historical avantgarde's belief that art can be crucial to a transformation of society, the point is not simply to revive the avantgarde. Any such attempt would be doomed, especially in a country such as the United States where the European avantgarde failed to take roots precisely because no belief existed in the power of art to change the world. Nor, however, is it enough to cast a melancholy glance backwards and indulge in nostalgia for the time when the affinity of art to revolution could be taken for granted. The point is rather to take up the historical avantgarde's insistence on the cultural transformation of everyday life and from there to develop strategies for today's cultural and political context." (Huyssen 1985: 7)

References

Angus, Robert. 1984. "History of Magnetic Recording." *Audio* 68. 27–330.
A Tradition of Vision and Innovation. CRI, Inc., webpage (1998). <www.cri.org>.
Beckett, Samuel. 1958. *Krapp's Last Tape. Collected Shorter Plays.* New York: Grove Press. 53–64.
Belton, John. 1992. "1950s Magnetic Sound: The Frozen Revolution." In *Sound Theory Sound Practice.* Rick Altman, ed. New York: Routledge. 154–70.
Bijker, Wiebe, Thomas Hughes, and Trevor Pinch, eds. 1990. *The Social Construction of Technological Systems: New Directions in the Sociology and History of Technology.* Cambridge, Mass.: MIT Press.

Canby, Edward Tatnall, C. G. Burke, and Irving Kolodin. 1952. *The Saturday Review Home Book of Recorded Music and Sound Reproduction*. New Jersey: Prentice-Hall.

Chadabe, Joel. 1997. *Electric Sound: The Past and Promise of Electronic Music*. New Jersey: Prentice-Hall, Inc.

Chanan, Michael. 1994. *Musica Practica: The Social Practice of Western Music from Gregorian Chant to Postmodernism*. London: Verso.

———. 1995. *Repeated Takes: A Short History of Recording and Its Effects on Music*. London: Verso.

Chion, Michel. 1994. *Audio-Vision: Sound on Screen*. New York: Columbia University Press.

Dale, C. B. 1952. "What About Wire Recording?" *High Fidelity*. 2/4 25–26.

Daniel, Oliver. 1952. "A Laurel Leaf for Stokowski." *American Composers Alliance Bulletin* 2/4. This newsletter includes reprints of reviews of the Oct. 26 and Oct. 28 MOMA concerts by Oliver Daniel in *Time* magazine, Nov. 10, 1952; Olin Downes in the *New York Times*, Oct. 27, 1952; Virgil Thompson in the *New York Herald Tribune*, Oct. 27, 1952; Howard Taubman in the *New York Times*, Oct. 29, 1952; Jay S. Harrison in the *New York Herald Tribune*, Oct. 29, 1952; Robert Evett in the *New Republic*, Nov. 10, 1952; and uncredited in *Time* magazine, Nov. 10, 1952.

———. 1982. *Stokowski: A Counterpoint of View*. New York: Dodd, Mead, and Company.

Darter, Tom. 1984. *The Art of Electronic Music*. New York: GPI Publications.

Day, Timothy. 2000. *A Century of Recorded Music: Listening to Musical History*. New Haven: Yale University Press.

Geertz, Clifford. 1973. *The Interpretation of Cultures*. New York: Basic Books.

Gelatt, Roland. 1954. *The Fabulous Phonograph: From Tin Foil to High Fidelity*. Philadelphia: J. B. Lippincott Co.

Geraci, Philip C. 1961. "The Home Recordist . . . Hobbyist or Hoodlum?" *The Stereophile* 1/1 (September 1962): 3–6.

Gimbel, John. 1990. *Science, Technology and Reparations: Exploitation and Plunder in Postwar Germany*. Stanford: Stanford University Press.

Haynes, N. M. 1949. "Modern Achievements and Trends in Magnetic Tape Recording." *The American Record Guide* 15:

Holmes, Thomas B. 1985. *Electronic and Experimental Music*. New York: Charles Scribner's Sons.

Huyssen, Andreas. 1986. "The Hidden Dialectic: Avantgarde—Technology—Mass Culture." In *After the Great Divide: Modernism, Mass Culture, Postmodernism*. Bloomington: University of Indiana Press.

Jones, Steve. 1992. *Rock Formations: Music Technology and Mass Communication*. Newbury Park: Sage Publications.

Krishef, Robert. 1962. *Playback: The Story of Recording Devices*. Minneapolis: Lerner Publications Co.

Lowman, Charles (Manager, Instrumentation Technical Writing, Ampex Corporation). 1972. *Magnetic Recording*. New York: McGraw-Hill.

MacKenzie, D., and J. Wajcman, eds. 1985. *The Social Shaping of Technology*. Milton Keynes: Open University Press.

Manning, Peter. 1993. *Electronic and Computer Music*. Oxford: Clarendon Press.

McGinn, Robert E. 1983. "Stokowski and the Bell Telephone Laboratories: Collaboration in the Development of High-Fidelity Sound Reproduction." *Technology and Culture* (January). 38–75.

Millard, Andre. 1995. *America on Record: A History of Recorded Sound*. Cambridge: Cambridge University Press.

Morton, David. 2000. *Off the Record: The Technology and Culture of Sound Recording in America*. New Brunswick, N.J.: Rutgers University Press.

Mowitt, John. 1987. "The Sound of Music in the Era of Its Electronic Reproducibility." In *Music and Society: The Politics of Composition, Performance and Reception*. Susan McClary and Richard Leppert, eds. Cambridge: Cambridge University Press.

Mullin, John T. 1976. "Creating the Craft of Tape Recording." *High Fidelity* 26/6 (April).

Murch, Walter. 1994. Foreword. *Audio-Vision: Sound on Screen*. New York: Columbia University Press.

Olson, H. F. 1954. "25 Years of Sound Reproduction." *Journal of the Acoustic Society of America* (September). 637–43.

Pinch, T. J. 1985. "Recent Trends in the History of Technology." *BSHS Newsletter* (January 16). 19–21.

Read, Oliver, and Walter Welch. 1959. *From Tin Foil to Stereo*. Indianapolis: H. W. Sams & Co.

Reisman, David. 1990 [1950]. "Listening to Popular Music." In *On Record: Rock, Pop, and the Written Word*. Simon Frith and Andrew Goodwin, eds. New York: Pantheon.

Riesman, David, Nathan Reuel, and Denney Reuel. 1953. *The Lonely Crowd: A Study of the Changing American Character*. New York: Doubleday Anchor.

Sanjek, Russell. 1988. *American Popular Music and Its Business: The First Four Hundred Years*. New York: Oxford University Press.

Schwarz, David. 1997. *Listening Subjects: Music, Psychoanalysis, Culture*. Durham, N.C.: Duke University Press.

Struble, John Warthen. 1995. *The History of American Classical Music: MacDowell through Minimalism*. New York: Facts On File.

Tall, Joel. 1958 (1950). *Techniques of Magnetic Recording*. New York: Macmillan Company.

Ussachevsky, Vladimir. 1953. "As Europe Takes to Tape." *American Composers Alliance Bulletin* 3/3 (autumn): 10–17.

Ussachevsky, Vladimir. 1977. *1952 Electronic Tape Music by Vladimir Ussachevsky and Otto Luening: First Compositions*. New York: Highgate Press.

Zizek, Slavoj. 1989. *The Sublime Object of Ideology*. London: Verso.

———. 1991. *Looking Awry: An Introduction to Jacques Lacan through Popular Culture*. Cambridge: MIT Press.

———. 1997. *The Plague of Fantasies*. London: Verso.

CHAPTER ELEVEN

Tails Out
Social Phenomenology and the Ethnographic Representation of Technology in Music Making

Thomas G. Porcello

Encountering Recording

There is a phenomenon known as print-through that is characteristic of magnetic (analog) audio tape, whereby any stored signal is transferred through adjacent layers when the tape is wound on a reel. According to sound engineer John Woram,

> Since magnetic tape is stored on reels, each segment is wound between two other segments. The tape's magnetic field may be sufficient to partially [*sic*] magnetize these segments, resulting in print-through: an audible pre- and post-echo of the signal on the two tape layers that come in contact with it. On many recordings, the program itself will mask the print-through, especially the post-echoes. However, print-through may be noticeable at the beginning and end of a recording, and during sudden changes in dynamic level, where a quiet passage is not loud enough to mask the echo of a loud passage immediately before or after it.
>
> Since print-through is usually greatest on the outer tape layer it is advisable to store tapes tails out; that is, without rewinding after playing. This way, the worst print-through comes as a post-echo and stands the greatest possibility of being masked by the program itself. (1982: 267)[1]

Audible print-through has both epistemological and phenomenological ramifications for music. It questions the autonomous status that formalist theories (e.g., those of Immanuel Kant, Eduard Hanslick, and Nelson Goodman), through their obsession with musical structure, have granted the musical text, performances of the text, and reproductions of performances of the text. Simultaneously, print-through elasticizes the boundaries drawn around standard conceptions of encounters with music; one's way of experiencing a given musical work needs not—in practice, likely *does not*—begin with the first note and end with the last.

In my teens, I owned a few record albums with extreme print-through at the beginning. Perhaps the master tapes had accidentally been wound heads out; in any event, I would put the needle down and faintly, but very distinctly, hear a perfect, amplitudinally miniaturized replica of what I was *about* to hear, an auditory analog of seeing the sun through thin, high cirrus clouds, before they are blown off and full sunlight ensues. That tiny audio shadow had the power to generate a visceral inner tension; I would hold my breath, waiting for the release that came with the "real" beginning of the song. Even at the time, print-through struck me as a type of effective narrative practice that foreshadowed events in small ways prior to their further revelation or manipulation in the music. And because of the very fact of foreshadowing—the building of anticipation, tension, and desire attendant to the partially known object—the eventual impact of the events was that much more intense. An unknowing pubescent disciple of Roland Barthes, I reveled in the material *plaisir du texte* and the boundaries of time and sensation that can be blurred by intense encounters with music.

Like the phenomenologist, then, I suspected that the ultimate significance of music resides not solely in musical texts per se but rather in social and individual processes of musical encounter. Yet the phenomenology of music has remained largely text-centric, at least to the extent that the particulars of textual structure are implicitly positioned as the agents driving the listener-text relationship (see, for example, Ingarden 1986).[2] To balance this tendency to privilege the text, one must stress the importance of the temporal aspects of experiencing music. Processes of musical encounter are, after all, inscribed in the passage of time, and local epistemologies of time are therefore crucial to understanding how concrete encounters work, and for arriving at a socially informed phenomenology. This is especially true for music, with its duality of time: the temporal relations that are established internal to the musical work by rhythmic and harmonic structures, and the flow of that internal structure through the temporal epistemologies of the social world in which music is performed, listened to, remembered, or otherwise experienced.

In this essay, print-through is offered as a metaphor for cumulative listening experiences engendered in the mediated social spaces of musical encounter, whether such encounters consist of listening, performing, or ethnographic research. My use of the print-through metaphor both draws upon and opens the space for a reconsideration of the work of Alfred

Schutz, who, as part of a larger effort to probe the phenomenology of experience and consciousness, authored two works concerned with music and its implications for a general philosophy of intersubjectivity (1971 [1951] and 1976; see also Wagner 1983).

Schutz attributes the lion's share of the production of intersubjective musical experience to the internal temporal flow of the musical work. His assertion of the possibility of an experiential "togetherness" derived from music's inner time is provocative and insightful for face-to-face musical encounters but becomes more problematic when applied to recorded musical events. To pursue the metaphor, by generating pre- and post-echoes, recorded print-through disrupts the continuity of the musical work's inner time as conceived by the composer or performer. Fragments of a musical sound appear both before and after the sound's "real" placement in the recorded work's inner time, in which case perceptual instantiation no longer corresponds precisely to the inner flow of musical time. As a result, the experience of "togetherness" takes on a character different from the one Schutz postulates.

Similarly, the ability of the metaphor of print-through to raise questions about ethnographic representation is an issue of time and the boundaries around experience, all the more so when the ethnographic project concerns music and the human experiences involved in its creation. Researching, writing, and reading ethnographies are processes inscribed in the forward and backward flows of time, yet most ethnographies submerge the reader in a perpetual ethnographic present that creates a static temporal logic. Recent admissions that the ethnographic present is a potentially fictionalizing trope suggest the utility of writing strategies that—like print-through—disrupt its inner logic, creating reading experiences that construct the implicit "togetherness" of researcher, researched, and reader in new ways. The production of music in the recording studio, with technologies and work strategies that disrupt the linear flows of musical time, is a context particularly well suited to reconsidering the temporal aspects of musical experience and strategies for their ethnographic representation.

Think more deeply for a moment about "pre-echoes," as Woram describes print-through, or "hearing what you are about to hear," as I have suggested. A semiotician such as Barthes would likely characterize them as a Dionysian striptease (evoking associations among music, the body, dance, and sex), a miniaturized, eroticized, veiled glimpse of the musical text to come that generates affect as powerfully as the music itself. For recording engineer Woram, however, print-through is a technical problem to be solved by a particular manipulation (tails-out storage) of the recording medium (analog tape) in order to protect the purity and integrity of the music itself. Pre- and post-echoes that to Barthes and my teenage self are sensual become a peril to the *tonmeister;* sound engineers and the industry clients they serve generally abhor this particular form of titillation unless its incorporation can be justified on the grounds of intentional artistic activity. As a researcher, I see print-through as both a phenomenon and an event, a process deeply implicated in the structuring of experiences both musical and extra-musical. In the end, Barthes, my teenage self, the sound engineer, and the academic ethnomusicologist represent competing epistemologies of the recorded musical text in the particular temporal flows generated by the echoes of print-through, and for each such epistemology there exists a different phenomenon of encounter. Print-through thus suggests the need for new examinations of the nature and description of intersubjective musical experience.

Encountering Ethnographic Research

The relevance of print-through for re-examining the social phenomenology of music became evident during research on the interpenetration of music, performance, discourse, and technology that I conducted by working as a sound engineer in a professional recording studio. In relation to many technical issues, including print-through, this production role often structurally embodies the discrepancy between Barthes' subjective, sensual *plaisir du texte* and Woram's objectivist, technical *analyse du texte*. Throughout my research, such tensions are deeply structured into a tripartite subject positioning: as an engineer and researcher in the studio, and as a music aficionado both in and out of the studio. To be a proficient engineer is to balance the technically effective with the musically affective; to be an insightful researcher is to understand how the social actors involved discursively define and negotiate that balance within local and global industry, aesthetic, and ideological systems; to be an engaged listener is less scripted but surely involves appreciating the results of that balance, without it intruding to the point where the effective dislocates or overwhelms the affective.

The roles of engineer, listener, and researcher were, of course, not entirely distinct; nor were they mutually exclusive or autonomous from larger social positionings such as race, class, and gender. One never dons a single identity to the total exclusion of others. Rather, I envisioned myself working through a series of cohabiting discourses within the physical space of the studio, discourses that aligned with various degrees of fixity to multiple professional roles and social positionings. Much like the musical texts being built layer by layer, day by day, performance by performance on multitrack tape, the relative mix among these discourses and positions built up, unfolded, and changed throughout the temporal flow of given recording sessions.

When I was most actively wrapped up in engineering duties, my discursive positioning and daily practice—what I thought, heeded, said, and did—would generally mirror that of the other engineers. But even then, a vital component of engineering is the ability to listen from multiple subject positions. For example, one must listen with technical ears for acoustic phenomena such as phase cancellation, 60 Hz radio frequency hum from crossed audio and lighting circuits or ground loops, tape saturation, distortion, and so on. Yet the ultimate goal of such technical listening is to project oneself into the space of the eventual consumer, making judgments about what is or is not an aesthetically (and, often, commercially) viable sound. In this role, the engineer listens as a consumer, a music fan, a radio station program director, the owner of a high-end stereo system or boombox, a drive-time commuter, a club owner, a talent scout, and so on. In other words, the engineer (along with producers, musicians, and everyone else in the studio) projects as many listening situations and experiences onto the musical text as possible, or, one might say, experiments with multiple phenomenologies of the musical work that correspond to projected subject positions.

Beyond the multiple listening positions embodied in the role of engineer, I sought out, as frequently as possible, because my research was motivated by a set of intellectual questions about making highly mediated music, the additional listening position of academic researcher. When the session slowed down, the musicians took breaks, and there were no technical problems to solve, or while driving home at night, I would stress this third kind of listening and thinking, the discursive space of the ethnographer. Like Bobby, the studio's chief engineer, moving the faders up and down on the console while working on his musical mix, I manipulated the balance of my subject positioning throughout the research process, constantly searching for the best possible mix of my own.

To characterize ethnographic research by means of taxonomizing, monolithic terminology that fixes the researcher's subject position (e.g., participant-observer, observer-participant, and so on [see Junker 1960; Henderson 1994]) is to ignore the fact that, much like the phenomenological characterization of musical experience, the ethnographic experience is built around encounter. Further, the ethnographic encounter is highly dynamic, unfolding in time as does the experience of performing or hearing a piece of music. Like multiple performances of a given musical work in which numerous elements remain the same across each performance (elements that make it recognizably the same work), numerous aspects of the ethnographic encounter remain constant from day to day and help define it as recognizably the same experience. But if ethnography is reminiscent of music in this respect, it might best be likened to an improvisational genre like jazz. The melody of encounter is not necessarily the same from day to day; the groove of participation (Keil and Feld 1994) has an enormous range of possible variations, some good, some bad, some productive, others not.

Like jazz improvisation, the success of participant ethnography is a matter of interaction and communication, shifting patterns of strangeness and familiarity, even practice. This very play across sameness and difference is perhaps the most instructive aspect of ethnographic work, revealing much about what is being researched and about the research process itself. Adopting the wrong mix of subject positions described above often led me to the research equivalent of an improvisation with no groove, that is, a tension-filled, frustrating day in the field. But as with a good performer, the key is to learn from one's mistakes, to replay the tape of the gig and decipher which conditions are being violated and thus preventing the successful establishment of the groove.

My suggestion of a parallel between music and ethnography, then, is not simply by way of remarking that both similarly unfold and emerge through time. I am also suggesting that the process of learning to do effective participant research and ethnographic writing can be similar to the process of learning to make music in an ensemble.[3] Both require a certain technical familiarity at the outset (knowing basic research methods, being competent on an instrument or with one's voice) but are ultimately dependent upon subsequent learning of effective, relevant, and locally meaningful patterns of coded social interaction—that is, performance skills. For the musician, such coded interaction might consist of, for example, knowing appropriate and inappropriate performance techniques for a given musical style or acquiring "the ability to recognize, distinguish, and deploy the musical possibilities organized in styles or genres by various communities" (Walser 1993: xii; see also McClary 1991: 27). For the ethnomusicologist or anthropologist, it may involve moving beyond the abstract level of "knowing how to do field research" into the concrete specificity of knowing, for instance, that acceptable behavior in one situation may be completely out of bounds in another, or in the presence of certain persons.[4] Learning those codes occurs not only during public interaction but also in moments of private introspection, in reflective time away from that interaction. The musician practices with the group and alone; the ethnographer, after time with informants, reviews and rewrites field notes. In such moments, both engage in individual analytic interpretive processes that complement social, dialogic, ensemble-centered interpretation, to aid in developing communication and improvisational skills appropriate to the performance/research context.

Encountering Writing

It's very late at night, almost 1 A.M., I think, but I don't have a clock on my dash. The interstate heading north from San Marcos is nearly deserted, and I settle into a comfortable ten-miles-an-hour over the speed limit in the middle lane. As I'm driving home, I slip one of today's cassettes into the car's tape player and head quickly for the spots I had mentally flagged during the session for an immediate re-listen. What triggered those two sour hours in the early evening when the tension was so palpable that I almost turned off my microphone? Who made that crack about drummers' IQs that had us all on the floor? What was Bobby saying about the Lexicon's left channel? Was that split-second dropout in the rhythm guitar track during the third verse of "Pretty Little Rain" really that noticeable? I speed backward and forward through the tape. I listen for a second, then impatiently jab the rewind button. Niederwald slips by, and I'll be home in thirty minutes, so I'd better find what I'm looking for soon. Wrong tape? I punch the eject button, pull out a different cassette, and continue hunting and pecking. The fast forward is in cue mode, so the day squeals by, forward and backward, a high pitched burbling of music and voices. Birds under water, I think to myself. The material remains of my day, a flock of magnetized, subaquatic birds trapped on acetate, squawking as I punch the buttons on my cassette deck.

I find the first spot I'm after and listen through it. I hit rewind and play it again. And again. Now I'm in control of today's recording session. I define what was significant and what was not. I can make those significant events happen as many times as I want them to. Later, on my transcribing machine, I will even be able to speed them up and slow them down. But for now, I listen to them and think. Stop. Rewind. I play what happens for five minutes before and after each event to get a better sense of context. Here in the front seat of my car, I've taken control over my informants, making them repeat themselves over and over—what did you say? what was that? come again?—until I see patterns, or imagine I'm seeing them. Stop. Rewind. Play. At home, if I'm not too tired, I'll jot down some thoughts and questions. Things to ask Bobby tomorrow if we have time. Things to think about. Places on the tape to listen to a year from now when I start writing. I pull into the parking lot, eject the tape, and break the record-safety tabs to prevent accidental erasure. This was a good day. Remember to make a backup copy of tape 7, side A.

As I'm heading up the interstate, Bobby is still sitting behind the Otari mixing console. Usually, if we drive separately, Bobby and I lock the studio and head up the road together, until he gets impatient and punches his accelerator, pulling ahead of me on his way up to Austin. But tonight, he begs off. There's that problem with the master tape, a sudden dropout in the rhythm guitar track in the third verse of "Pretty Little Rain." It's one of those things that passes by so fast that the first time you wonder if you're imagining it. But no, it's very real. An eighth note, just gone. If you don't know it's missing, it's nearly seamless. But once you're aware of its absence, it looms like an enormous pothole you've seen too late to swerve away from, and you cringe, waiting to smash into it. Of course you hear the dropout before the client does, so every time you're playing through the third verse, you're desperately looking for an excuse to stop the tape before you get there, glancing nervously over your shoulder, dreading the word, "Hey!"

Bobby rewinds the tape to the critical spot, hits stop, then play. The guitar churns along, "ch ka ch ka ch ka ch ka." Then, "ch ka ch ka . . . ka ch ka." Stop. Rewind. Play, "ch ka ch ka . . . . . . ka ch ka." Stop. Rewind. Play, "ch ka ch ka . . . . ka ch ka." Each time, the space looms larger, until its silence becomes a black hole threatening to consume the entire song, "ch ka ch ka . . . . ka ch ka." Stop. Rewind. Play, "ch ka ch ka"—here-it-is-big-enough-to-drive-a-truck-through—"ka ch ka." Bobby stops the tape, punches rewind, and lets the tape wind back to thirty seconds before the dropout. Play. The song slides by, and Bobby lets it play on, getting a better sense of context. He programs this long minute into memory repeat and plays the tape over and over again, listening for the patterns in the music, the attack and sustain, listening hard for clues on how to patch this hole, how to stretch a tarp over the Grand Canyon. Stop. Bobby seizes control of the song and jots down ideas for how to work through the problem. Maybe he'll have time to discuss strategies with Mark tomorrow before the band arrives. A sample from the second verse might do the trick. Or maybe just bring the rhythm guitar down in the mix. Some delay on the track might also work. Rewind. Stop. Play. Fast forward, finally, until the end of the tape slap-slaps on the spinning take-up reel. Bobby puts the tape in its box, tails out, shuts off all the outboard tube gear, hits the lights and the alarm, and heads to his car.

Heading south on the interstate, Jon slips a cassette of today's rough mixes into the stereo in his van. Tomorrow he plans on rerecording his guitar solo in

"Pretty Little Rain." Today's solo was a complete bust, he thinks; the notes were there, but it sounded like he was playing along with a record, not jamming with the band. Before leaving, he asked Bobby to make two copies of the song: one with today's solo, one without. "Pretty little rain / can't wash me clean . . ." As the second chorus slips into the solo, Jon leans to the right, his head hanging over the stick shift, so he can hear the stereo image of the guitar. As the bend up to the F-sharp skids harshly into a G, he makes a mental note to be more careful tomorrow. Less finger pressure on the string, stay away from the fret, less bite with the pick. But it's the groove that bothers him most. He just wasn't in the pocket today. Rewind. Stop. Play, "Pretty little rain/can't wash me clean . . ." One more time through his solo. Selma slips by in the rear view mirror. Stop. Fast forward. Stop. Play, "wash me clean . . ." No solo now, just the bass and drums. Concentrate on the groove. Vessie's ahead of the beat, pushing Alex. Stop. Birds burbling under water. Play, "clean . . ." Listening. Leaning over the stick shift. Pushing Alex. Remember that for tomorrow. Stop. Jon flips the stereo off before the third verse starts and drives in silence toward downtown San Antonio.

Rewind. Stop. Play. With our tapes, Jon, Bobby and I delinearize our experiences in the creation of the musical artifact known as the recording and reflect on issues of participation in quiet isolation. How can Jon make his solo better; how can Bobby make the recording better; how can I make my research better? We each use our audio tapes strategically to carve up and rearrange time as we work on our respective grooves. We are able to do so as the result of interpenetrating technologically, socially, and physiologically mediated processes that meet on the medium of analog tape.

A Phenomenology of Encountering

Situated in this interplay of shared and individuated moments of recorded musical experience is, I would argue, a modified version of what Alfred Schutz refers to as "musical tuning-in," the living through of a "vivid present" by experiencing togetherness as a "We" (1977 [1951]).[5] In attempting to unpack the social significance of music, Schutz adopts a phenomenological stance favoring the socially informed experience of live performances and suggests that one should focus on performers' and listeners' interpretations of the "signs" of a historical musical culture out of which musical works arise: "Any work of art, once accomplished, exists as a meaningful entity independent of the personal life of its creator. The social relationship between composer and beholder as it is understood here is established exclusively by the fact that a beholder of a piece of music participates in and to a certain extent re-creates the experiences of the—let us suppose, anonymous—fellow-man who created this work not only as an expression of his musical thoughts but with communicative intent" (Schutz 1977 [1951]: 113).

It is worth noting that Schutz occupies a position in transition from author-centered textual criticism to the poststructuralism of Michel Foucault and Jacques Derrida that trumpets the death of the author. Schutz stresses the independence of the work from the author once that work has been launched into the social world in which it is apprehended, but this does not lead him to claim that authorial intent is irrelevant. Rather, textual meaning is truly dialogic: musical experience is created in the space between the form-in-time created by the composer (the "communicative intent") and the processes of participation and experience activated in the presence of the beholder. That the social relationship between composer and beholder is comprised of the participatory engagement involved in beholding the work should not be taken to mean that extra-musical social knowledge is irrelevant, however. Rather, Schutz is suggesting that social knowledge brought to such encounters constitutes merely a ground against which participation and communication via the musical encounter are figured.

I suggested above that most phenomenological accounts of music implicitly position the text as the agent driving each encounter. Schutz is no exception here, adopting the concept of *durée*—the notion of an inner time structured differently from outer or objective time—and applying it to the arrangement and flux of tones within a musical work (c.f. Qureshi 1994). The movement of the musical work through these tones in inner time draws the beholder into "an interplay of recollections, retentions, protensions, and anticipations which interrelate the successive elements" (Schutz 1977 [1951]: 113).[6]

Schutz argues that while musical inner time moves irreversibly forward in the flow of outer time, the composer can control this movement in ways that refer the listener backwards: "The consciousness of the beholder is led to refer what he actually hears to what he anticipates will follow and also to what he has just been hearing and what he has heard ever since this piece of music began. The hearer, therefore, listens to the ongoing flux of music, so to speak, not only in the direction from the first to last bar but simultaneously in a reverse direction back to the first one" (ibid.). Thus the inner time of the musical work, largely attributed to the textual manifestations of the composer's agency, draws the beholder into participation with the composer's stream of thought in a "polythetic" fashion, that is, as a step-by-step process.

Finally, Schutz posits that "this sharing of the other's flux of experiences in inner time, this living through a vivid present in common, constitutes . . . the mutual tuning-in relationship, the experience of the "We," which is at the foundation of all possible communication" (ibid.: 115), and he concludes:

> This social relationship is founded upon the partaking in common of different dimensions of time simultaneously lived through by the participants. On the one hand, there is the inner time in which the flux of the musical events unfolds, a dimension in which each performer re-creates in polythetic steps the musical thought of the (possibly anonymous) composer by which he is also connected with the listener. On the other hand, making music together is an event in outer time, presupposing also a face-to-face relationship, that is, a community of space, and it is this dimension which unifies the fluxes of inner time and warrants their synchronization into a vivid present. (ibid.: 118)

Beholders may share the inner flux of tones in the common passage of outer time, but in his ultimate privileging of inner over outer time—in locating musical experience so firmly in the tonal flows of the musical work—Schutz retains, despite his desire to accommodate external social knowledge, a strongly textual bias in describing the nature of musical encounters.

Schutz's phenomenology remains a remarkable effort in describing the mechanisms by which music is capable of generating shared, intersubjective, affective experiences. However, his suggestions that outer time is necessary to unify the fluxes of inner time, that inner time unfolds polythetically, and that unification must occur within the context of face-to-face interaction are, I believe, problematized by the way that Bobby, Jon, and I use our audio recordings to fragment inner and outer time and thereby eradicate the posited irreversibility of their flow.

Stop. Rewind. Play. In a radical reorganization of the progression of "Pretty Little Rain's" internal temporal flow through outer time, each of us wrestles control of "we-ness" away from composer and musical text by disrupting the polythetic flow of the musical work's inner time. Jon's concern in his van with the ensemble performance we-ness of his guitar solo, Bobby's concern in the studio with the service-oriented we-ness of a successful recording session, and my twin concerns in my car with how we-ness is constructed discursively and musically from the larger social processes involved in making and recording music, and with how we-ness mediates my ethnographic and musical experiences, all suggest that Schutz's location of we-ness in the inner time of the musical work unnecessarily restricts our ability to discern the multiple ways in which musical experiences are actively manipulated by beholders to create multiple we-ness*es*.

At first glance, this would seem to contradict Schutz's belief in the possibility of shared musical experience. However, if we dislocate we-ness from the linear flow of music's inner time by disruptive use in outer time of the buttons on our tape machines, and in so doing wrestle polythetic agency away from the composer, we are presented *not* with a complete absence of what Schutz argues constitutes we-ness (tuning in, co-performance, vivid presents/presence) but rather with a malleable series of we-presences built more equally from the internal *and* external flux of musical experience. Because of our ability to manipulate the flow of music, we can selectively and intentionally create multiple variations—partially shared improvisations—on we-ness, as opposed to the monolithic vision implicit in Schutz's description. The vivid present of Jon's guitar solo and the tuning-in to the process of making his record did not simply disappear when we disbanded for the night and listened to our respective tapes alone. Instead, each of us constructed what might be called an individuated or individually experienced we-ness that emerged from a set of previously shared, and now relived, musical and social experiences.[7]

The existence of a plurality of we-nesses that play among the spaces of multiple individuated and shared musical experiences suggests that we-ness is better described as a fluid movement between social coalescence and fragmentation, shared and individual modes of apprehension and tuning-in, and public and private beholdings of music. In contrast to Schutz's seemingly monolithic characterization of we-ness, in which musical experience seems relatively undifferentiated among those present, the we-ness achieved in the recording studio often appears more tentative, experimental, and distanced: one perhaps characteristic of individual epistemologies brought into contact by audio technologies that make it easy to manipulate temporal boundaries of music and that are being used to create a shared experience from joint, though spatially and temporally fragmented, musical encounters.

Further, Schutz restricts his discussion of face-to-face musical encounters to the Western art music tradition in the concert hall setting: with its conventions of silent listening in a darkened space, and with well-marked distinctions between performers and audience, between musical production and musical consumption. In the recording studio, however, musical experience is shared simultaneously—if unevenly—as music, as motion, *and* as discourse about music and musical experience. Moments of individuation and sharedness emerge out of the interpenetration of talk, musical performance, and performative talk specifically about music, often punctuated by expressive, performed bodily motion. And especially in popular music sessions, the lines of speaking and singing get blurred and slurred as voices and instruments rip out riffs in dense layers of collaborative—or contesting and competing—ever-escalating, gyrating performances of musical, verbal, and popular cultural competence. Jamming and singing and punning and joking often flow seamlessly into one another and build thickly textured expressive texts, layer upon layer, performance upon performance, like the musical tracks being laid on tape. Days later these heightened moments may be recalled and reanimated in talk. And often they produce specific inspirations that wind up on tape as part of the final recording. As such, they become portable and renewable experiences of we-ness, rejuvenated in subsequent performances of the songs involved, or talked about long after the session is over.[8] Unlike the live concerts discussed by Schutz, however, these studio performances are encoded onto a replayable medium; social memory is magnetized and digitized. CDs, records, and tapes are "burned," "cut," "pressed," "etched," and "baked," all of which are processes that congeal the multiplicity of experiences leading up to the commodity state. The delinearized experience of music in the studio is thus partially relinearized in its transformation into the cultural commodity that the consumer ultimately purchases as the final musical text—the recording.

It would be easy, at this point, to adopt a theoretical stance that argues that the we-experiences just described exhibit a characteristic postmodern fragmentation of signs and apprehension about positing fixed, unified referents of social experience: that these musicians and engineers have substituted the shared ownership of multiple copies of a nonexistent musical original—a simulacrum of a shared performative experience made possible by postindustrial technology—for that shared performative experience itself. And, in fact, the musical experiences that I was surrounded by during my studio research strongly exhibited such features. Nonetheless, the vast majority of musicians and engineers I have worked with strongly believe in and articulate ideas of shared musical experience virtually identical to those described by Schutz. That is, the emic discourse is firmly rooted in notions of tuning-in to shared experiences built out of musical expression, a seemingly romantic stance in light of the multiple layers of technological, social, and discursive mediation involved in building that tuning-in. Neither discursive stance should be accepted uncritically, though both are construed as equally real interpretations for their respective adherents.

Instead, I would suggest that it is more profitable to characterize the tension between the romanticized emic and the overly cynical postmodernist theoretical discourses as a simultaneous problem of epistemology and representational practice. Understanding why Schutz would frame we-ness as an undifferentiated experience is not a difficult task if one recalls his explicit indebtedness to G. H. Mead and other writers of a generation for whom it was relatively easy to imagine homogenized cultural wholes. Schutz's "Making Music Together" was originally published in 1951, when anthropologists were doing ethnological research that sought to explain and integrate all domains of experience under the master rubric of "Culture" when unified, undifferentiated, non-stratified, raw, small-scale societies still appeared to exist in remote uncooked corners of the globe. One might argue that finding we-ness in musical experience was no more difficult in 1951 than seeing it in any other cultural domain studied by anthropologists, such as kinship, religion, or economic reciprocity systems.

Encountering Representation

The representational practices of anthropologists who painted such neatly sewn-up, totalized portraits of social experience—usually based on ideological claims to objectivity that have since been demonstrated to be no less socially contingent than those of any other culture—have come under increasingly explicit scrutiny in the past twenty years, but especially since George Marcus and Michael Fischer's *Anthropology as Cultural Critique* and James Clifford and Marcus's *Writing Culture,* both published in 1986. These works strongly problematize the relationship between research and writing, reminding us that anthropology—and certainly ethnomusicology—are as much the latter as the former, are representational practices as well as strategies for investigating Others' understandings of the social worlds they construct and inhabit.

One recent strategy for creating viable alternatives to totalizing narrative representations of cultural experience has been the adoption of more fragmented, experimental, exploratory, and tentative ways of writing about culture, methods that background attempts at making objectivist truth claims, stressing instead the central importance of multiple local cultural "reals."[9] Kathleen Stewart's ethnographic account of "just talk" in West Virginia coal camps, for instance, adopts an avowedly "nervous" shifting between story and analysis that evokes not only a local cultural poetic but also her congruent desire to "reopen stories, and spaces of cultural critique, that are . . . continuously being slammed shut with every new 'solution' to the problem of culture and theory" (1996: 40, 6). In Stewart's definition, culture is a process, a continual working-through, enacted in multiple performances that, rather than creating fixed meanings and epistemologies, continuously problematize, pile up, densify, accumulate, and *work on* local ways of knowing. "Roaming" among texted genres (ibid.: 210), the processes of narrative representation can take on a fluid, processual character not unlike that of a culture as seen through its multiple forms of mediation and performance. Instead of seeking to resolve the dialogics inherent in local epistemologies and ethnographic research, written representation can seek to highlight them, or, at the very least, to present their complexity and texture in new ways, and perhaps therein to convey their lived experience as a cultural poetic.

Similarly, though reacting against a different set of theoretical issues (specifically, recent suggestions that all representation of Others is a form of real or symbolic violence) and with a sense of teleology not present in Stewart's account, Mark Whitaker offers that ethnographic writing should adopt the form of a series of "'tries,' acted experiments of gradually increasing complexity that move one ever closer to simply living a form of life" (1996: 8).[10] Derived from Ludwig Wittgenstein's solution to the paradox in which it becomes impossible to make meaningful statements about given epistemologies without either using—and thereby validating—them or creating new—and equally contingent—ones, Whitaker suggests that the scholar's focus should be less on the finalized, neat epistemological constructions conveyed in published works and more on the learning processes that accompany ethnographic research. Such tries, which substitute the uncertainty, mistakes, and tentativeness that characterize early phases of field research for the (relatively more) tidy explanations of cultural order that we often claim to have arrived at by the end of our research, leave epistemological and representational doors cracked open, actively resisting totalizing narratives of cultural experience.

A recent collection of manuscripts that apply such critiques of field research and representation directly to ethnomusicology specifically highlights an experiential perspective that emerges in musical ethnographic research and that should, it is argued throughout, be more openly encoded in ethnographic monographs (Barz and Cooley, 1997). Most of the contributors explicitly address field research rather than writing practices, but Michelle Kisliuk foregrounds their interpenetration, suggesting that "there is no definable border between the field and the space of writing—we write when we are doing research, and we research while we write. An awareness, therefore, that field experience and ethnography are inseparable must infuse both" (ibid.: 41). Kisliuk's point is well taken, especially with regard to field notes and other forms of *in situ* writing. Yet the practicalities of academic authorship (often very real distances in time and space from the physical site of field research and the people we work with) generally insure that the majority of our *published* ethnographic material will be produced in contexts well separated from our research, when the borders between the field and the space of writing have once again grown all too defined.

In his contribution to the volume, Jeff Titon argues that "knowledge is experiential and the intersubjective product of our social interactions," and that fieldwork experiences are, as a result, "intensely lived" (ibid.: 95); phenomenological accountability demands, Titon suggests, that such experiences be included in representations of field research. Acknowledging the spatial and temporal distances inherent in writing an ethnographic monograph, however, he points to a very real problem that arises in seeking to encode experience through the use of documents (texts, photographs, and so on) created *in* the field during research:

> When we are with our friends, these documents appear—at best, and when they do not get in the way—not so much as objectifications but as extensions of our relationships. But when we get back from the field, in the university, in the library, or study, alone, particularly if our friends are far away, these field artifacts take on a very different cast. They substitute for experience by evoking our memories of it. Like a photograph taken or a brochure brought back from a holiday abroad, they are documentary and evocative at the same time. They traffic in nostalgia. (ibid.)

In other words, away from the field, the documents too easily turn into objectifications, and too easily become icons of the very objectivist tropes of social science that experiential phenomenology seeks to counteract. Titon suggests the need for representational strategies that, at the very least, *resist* such objectification, strategies such as narrative nonfiction writing, interactive and reflexive film, and weblike interactive multimedia (ibid.: 98).

The recording studio is a particularly appropriate research site in which to work through such experiments in representation, because it is a music-making context that is virtually defined by a refusal of narrativizing (musicalizing) closure. In the recording studio, the continual work performed on constructing moments of tuning-in, the densely layered performances of musical, verbal, and cultural play and competence enacted in a clearly circumscribed space of intense artistic collaboration, where music launches talk and talk launches music, and the fluidity of inner and outer time, coupled with the consequent movement between the individuated and distanced we-nesses that Jon, Bobby, and I have enacted by manipulating our audio tapes, are all lived "tries" in Whitaker's, Stewart's, Kisliuk's, and Titon's deeper sense of refusing to totalize experience. Such "tries" are not mere artifacts of the postmodern fragmentation of experience, to be read as a cultural text of pastiche, leaving in their wake a shallow surface of free-floating social signifiers glued together only by a publicly displayed historical narrative of what their significance once was (see Baudrillard 1968 and Nöth 1990: 444). Rather, they are strategic, intentional, deeply felt forms of performed cultural activity, and living embodiments of multiple local epistemologies enacted in the flow of internal and external time in and out of the recording studio.

The presence of such lived tries in the form of multiple musical "takes" in the studio suggests the appropriateness of a writing strategy that incorporates similar tentative, experimental tries, or roams among representational genres of writing; the learning process that Kisliuk encodes in her ethnographic writing parallels her experience as a researcher, while the notion of "tries" encodes both my learning experience *and* the actual work processes and human experiences involved in the making of technologically mediated music in a recording studio. "Roaming" and "tries" as forms of experimental writing acknowledge the researcher's presence in the experience and representation of local epistemologies and cultural forms, at least to the extent that the learning process in the field is acknowledged as a viable, indeed necessary, subject of discourse. My presence in this text, then, where it intrudes, is a type of print-through, mediating my own learning about, experiencing, and subsequent representations of the building of collective and individuated we-nesses in the flow of internal and external discursive and musical time.

In elasticizing the boundaries around the musical text, print-through calls attention to

In elasticizing the boundaries around the musical text, print-through calls attention to the temporal unfolding of music. Like Jon, Bobby, and I punching the fast forward and rewind buttons on our respective tape decks, print-through also challenges a simple, linear temporal model of musical ontology and experience. Albert Einstein and others have argued that space is curved; Schutz suggests the possibility of a curvature of musical experience in which tuning-in is built from a series of forward- and backward-looking tonal moves operationalized in the space of face-to-face interaction; audio tape allows temporal transformations of both inner and outer time in which spatial, musical, and temporal experience can alternately be distanced or narrowed, individuated or collectivized, forwarded or backgrounded; ethnographic representational tries suggest that the echoes of tentative epistemologies arising during cultural research should be audible in the final mix of ethnographic writing; and magnetic print-through suggests the possibility of an electroacoustic, temporally equivalent curvature, which puts the beginning before the beginning, and the end after the end.
after the end.

In his *Outline of a Theory of Practice,* Pierre Bourdieu suggests the need to reintroduce the concept of time into the study of practice: "The detemporalizing effect . . . that science produces when it forgets the transformation it imposes on practices inscribed in the current of time . . . is never more pernicious than when exerted on practices defined by the fact that their temporal structure, direction, and rhythm are constitutive of their meaning" (1977: 9). His caution is equally true whether one studies chemical reactions or musical performances; it is equally true for the act of writing and for that of researching, both of which are acts firmly "inscribed in the current of time."

The pernicious effect that Bourdieu seems most concerned with is a misapprehension of the nature of the phenomenon under investigation. That is, he suggests that to adopt a research lens that detemporalizes a practice constituted by its very temporality is to establish an irrevocable barrier to arriving at a valid understanding of the conditions of its existence. As ethnomusicology has moved away from comparative, historical, and primarily textual analysis into a more direct investigation of music as a dynamic, emergent cultural process (see Titon 1997), the risk of representational distortions that accrue to misapprehending musical experience with atemporal models increases. In this sense, Bourdieu provides ethnomusicology with a return to the phenomenologists' emphasis on incorporating processes of interaction, a position more open to acknowledging temporality than is the epistemological reflection often characteristic of the hard sciences and much social science, which, whether quantitative or qualitative, position themselves as essentially objective in nature.

Bourdieu's insight thus forges a link between the social phenomenology of music and questions of ethnographic representations of musical experiences. If, as I have been asserting, ethnographic research and writing are both fundamentally caught up in flows of time, then it is not only the researcher's apprehension of the conditions of existence of an event that may be adversely affected by the detemporalization process; it is equally their eventual representation in written interpretation. To ignore or background time can thereby result in a cumulative effect in which representation magnifies the atemporality of epistemological reflection.

I do not wish to suggest that the complexities of incorporating the temporal flow of practices or events into research and writing strategies can be entirely resolved through an oversimplified strategy such as making one's interpretive text somehow mimetic of the temporality of that which is under investigation. Too many layers of practice and process intervene between event and representation (minimally those of researching, writing, and reading) for one to assert blithely even the possible accuracy of such a mimetic attempt. Further, I would argue that the very notion of mimesis is predicated upon a fundamental separation of practice from its later representation; that is, mimesis presupposes two distinct objects characterized by a common set of traits. A less absolutist position suggests the possibility of creating representational structures that are, at best, strategically evocative of the particular temporal flows involved in a given practice. Such structures undermine the fundamental distinction between practice and its representation by questioning the boundaries between event and text. If these representational structures involve the evocation and incorporation in the text of the temporal structures of the event or practice, then the boundaries between text and event have been rendered permeable. The event may be over, it may have happened two years ago in a different geographical location, but in the sense that it is present as an echo in this text, it is in fact not over. Its beginning was before the beginning of what I have written, and its end is now being replicated long after its end in outer time. That event, those practices, have printed-through onto this document you now hold in your hands.

Encountering Performance and Production

Imagine for a moment that you are the first note of Jon's failed guitar solo on "Pretty Little Rain." You are being listened to simultaneously in my car, in Jon's van, and back in the studio with Bobby. Your existence at this moment is the result of a complex series of electronic, acoustic, and social transformations.

Your genesis is an idea in Jon's mind, an idea about your attack, articulation, intensity, duration. Electric currents in Jon's brain are transmitted to the cerebral cortex, then travel to the nerves and muscles in his hands and arms. Neural electrical energy is converted to muscular impulse, and your acoustic embodiment begins when the plastic pick in Jon's right hand strikes the metal string of his electric guitar, exciting a particular frequency that is a result of the thickness and tension of the string as well as its length, which is controlled by the placement of the fingers of Jon's left hand on the guitar neck. Your timbre varies with the type of string (metal composition and gauge), the material of the pick (plastic, metal, fingernail) and its thickness, and the composition of the neck and body of the guitar (maple, ash, or ebony woods).

Struck in a meeting of mind, flesh, metal, wood, electricity, and tissue motion, your organic existence is now transformed. Electronics take over where flesh and wood and metal leave off. The vibrating string lies above a magnetic pickup, and its movement through this magnet's force field generates an electric current. Jon's double-humbucker Fender Stratocaster thus turns your acoustic energy (the vibration of the string) into an electric current that carries you via a cable to his amplifier. Run through a series of tubes or transistors, resistors, capacitors, transformers, or circuit boards, you are modified by the amplifier's circuitry before being routed to a loudspeaker embedded in a wood or fiberglass cabinet. Your electric current arrives at a large magnet, and this "voice coil" moves a nearby paper or thin metal diaphragm, exciting the surrounding air molecules and thereby converting you from electric signal back into a modified version of your original acoustic form.

In front of Jon's speaker cabinet, Bobby and I have placed two microphones, waiting to capture you. We have carefully selected them from among the numerous types of microphones (dynamic, ribbon, condenser, tube, FET), polar patterns (cardioid, bipolar, supercardioid, omnidirectional), and brand names (Telefunken, AKG, Beyer, Shure) available. We have talked with Jon about what he thinks you should sound like, we have listened to him creating all your siblings in earlier rehearsals, and we have formed our own opinions about your strengths and weaknesses. The microphones we eventually chose to represent you were positioned with excruciating patience in order to capture your most flattering attributes. Bobby in the control room listened through his monitors, while I was out in the reverberant hallway with headphones on over my earplugs. Jon flailed away on dry runs of his solo, and the sound pressure was strong enough to push uncomfortably on my chest as I crouched by the speaker, straining to hear Bobby's shout through the console's talkback microphone, "About two centimeters toward the edge of the cone!" The microphones we chose earlier just for you will convert your acoustic form originating from the loudspeakers back into an electric current that will send you via cables through the wall of the hallway into the console in the control room, where we are now waiting for you, all ears.

Once you are in the console, we may channel you in any number of ways. Today, we have decided to send you immediately out of the console to the outboard effects rack. We divert you to the patch bay, and then you zip behind us, under the floor, into a solid-state parametric EQ, where we enhance your good points and dampen your weaknesses. Listen to us talk candidly about you: we want you brighter, but with a rounder bottom. We joke about you, anthropomorphize you, sexualize you, humiliate you, praise you, caress you, make you a slave serving our aesthetic ends. Now, in addition to being flesh, metal, wood, electricity, magnets, and tissue motion, you are constituted discursively, socially.

When we're happy, you get sent back to the console's patch bay and then out to the analog multitrack tape machine. Here, your electric current arrives at the recording head and causes you to excite yet another electromagnet. Tape with magnetic particles coated in an acetate backing runs at fifteen inches per second past this electromagnet, and as the particles pass this head, you magnetize them—force them into particular alignments that are magnetic representations analogous to your prior existence as the electronic current that has so transfixed them.

Approximately seven hundredths of a second later, when the magnetized tape passes the playback head, the inverse process occurs; the magnetic particles generate a new electric current and you are sent to a set of amplifiers, then to the console, and then you are finally split into two paths. The first goes to the control room loudspeakers where Bobby listens to you, and the second is sent to Jon's headphones. In both cases, your electric current is passed once again to an electromagnet that excites a diaphragm that pushes air and thus converts you back into the acoustic energy that we now experience as the recorded sound of the first note of Jon's guitar solo on "Pretty Little Rain."

Encountering Mediation

The technical transformations and mediations that you were subjected to along the way are, of course, inscribed within a series of social mediations, many of which themselves revolve around issues of technology and the technological mediation of musical sounds and experiences. These social mediations pertain to musical and stylistic concerns that may go well beyond the specifics of this recording session; everything from the particular instruments and peripheral technologies used by each band member to the politics of composition, performance, and decision-making rights within the band and between the band (and its management and producer) socially mediates the sounds and the music.

But the social mediations of sound also reach far beyond the internal politics of Jon's band and its stylistic affiliations. For instance, Bobby and I know that in this session no one else possesses the technical knowledge that we do, which gives us certain rights to speak, judge, and act on particular realms of Jon's music—especially those involving the multiple layers of transduction from acoustic to electric to magnetic representations of musical sound. We can exert these rights because we have exclusive knowledge of signal flow, signal processing, and, perhaps most importantly, the language of audio production. With this exclusive knowledge comes power and a certain degree of aesthetic autonomy over the highly technical means of music production. In stark contrast to our position of relative strength is Vessie, Jon's drummer. The lone woman in the studio throughout the session, and the least active member of the band in arranging Jon's songs, she is rarely looked to for advice. In the highly male world of popular music, the studio and its sophisticated technologies are forcefully constructed as male domains; when women are present, it is usually as singers (background singers at that), and they are generally expected to take directions, not give them.

Thus, the social relations internal to the band are themselves inscribed within a larger context of musical and nonmusical social relations. Despite Bobby's exclusive knowledge of recording technology in this session, his agency is constrained by the fact that studio engineers work in a service industry; they are present to assist a paying client. Thus Bobby may have leeway in a session like this to make technical decisions that will have serious effects on the sound of the client's music, but the client will always be the final arbiter of those decisions. (The client may be an individual, a group of individuals, or a corporate entity. It may be a band leader like Jon, a producer, an A&R rep, or a record label owner. Some will have a great deal of musical knowledge, and some very little. Some will be technically fluent, and others may see the music simply as an investment opportunity.) Just as Bobby's agency is ultimately constrained by this larger industry structure, Vessie's agency operates within a set of expectations about the role of women both in the music industry and more broadly in Western society. In the studio (and throughout the popular music industry), historical ideologies of women's technical incompetence are deeply entrenched. While a different woman in this band, or Vessie perhaps in a different band, might have had a somewhat more active decision-making position than was the case here, my larger point is to suggest that all musical performances, and all music production, are implicated in a social arena beyond the musical event per se. In the end, the way you sounded as that first guitar solo note in "Pretty Little Rain," your internal flux of musical tones and the corresponding polythetic flow that governs our living through of a vivid present by experiencing togetherness as a "We," resulted from more than *just* Jon's compositional idea, materialized on his guitar, and recorded on tape.

Music is framed by technological, social, and physiological mediations, but individual and collective experiences of music are not reducible to these mediations. Tonight—around 1 A.M., I think, but I don't have a clock on my dash—Jon experiences music as an issue of time, texture, timbre, groove, finger pressure, flesh on metal and wood, the recorded acoustic and physical materiality of his individual performance within the ensemble's larger groove. Tonight, Bobby lives it as a technical problem, but one that must be addressed at that level in order to prevent it from overwhelming the aesthetic experience of the final recorded product—balancing the technically effective with the musically affective. And I listen to my tape, attending to the complex moves among musical and discursive events, cohesion and dissension, work and play. With our tapes, Jon, Bobby, and I pass fluidly, seamlessly, between shared and individuated moments of musical experience, highlighting them differently as suits our needs and wishes. Tomorrow we go back to work together.

As I'm heading up the interstate, Bobby is still sitting behind the Otari mixing console. Usually Bobby and I lock the studio and head up the road together, until he gets impatient and punches his accelerator, pulling ahead of me on his way up to Austin. But tonight, he begs off. There's that problem with the master tape, a sudden dropout in the rhythm guitar track in the third verse of "Pretty Little Rain." It's one of those things that passes by so fast the first time, you wonder if you're imagining it. But no, it's very real. An eighth note, just gone.

Bobby cues up the tape and listens to the rhythm guitar track, "ch ka ch ka ch ka ch ka, ch ka ch ka . . . ka ch ka." Stop. Rewind. Play, listening, listening. He rewinds the tape to the second verse, turns up the monitors, centers his head in the nearfields, stares absently at the control room glass, listening. Forward to the third verse, listening. Back to the second. Focusing on performance and the sound, listening for similarities and differences in attack, decay, intensity, timbre.

Suddenly he springs out of his chair, grabs two patch cables, and routes the rhythm guitar tracks into the Lexicon 3500, a digital effects processor/sampler in the outboard rack. He starts up the tape machine, swivels around to face the Lexicon, and spends a brief moment adjusting levels and punching buttons on the face of the small box. Still facing it, he gropes his right hand out and rewinds the tape machine, then starts the second verse. The guitar track churns on, "ch ka ch ka ch ka ch ka . . ." He rapidly punches twice on the face of the Lexicon, snagging a perfect "ch" out of the stream and sending it deep into the Lexicon's memory. Bobby quickly forwards the tape to the third verse, and as the machine shuttles ahead, he punches a button on the face of the Lexicon repeatedly. Each time, a perfect "ch" comes out of the nearfields.

Turning quickly back to the patch bay, Bobby sends the output of the Lexicon to the input of the rhythm guitar tracks and puts them in "record ready" mode. Now for the delicate part. Bobby starts the tape and listens through the dropout: "ch ka ch ka . . . ka ch ka." Now that he's about to execute the punch, the Grand Canyon has shrunk to the size of a small irrigation culvert, but no matter. Rewind. Play. Bobby holds his left middle finger on the "play" button and waits for the dropout. At the exact moment it passes the recording head, he punches the orange "record rehearse" button—in, out.[11] *A half second. He listens back. A perfectly placed silence greets his ears. The tape machine is now programmed to record at the right spot.*

Turning back to face the Lexicon, he again gropes with his right hand to hit play/record on the tape machine. At the precise moment that the tape machine activates "record," Bobby punches the "play" button on the Lexicon, and the sampled "ch" is dropped into the rhythm guitar track. Stop. Rewind. Play. Bobby turns up the monitors and again faces the control room glass, listening: "ch ka ch kash . . . ka ch ka." Too quick with his finger: the attack of the punch cut off, the decay ending too soon. Bobby resets the tape machine and the Lexicon and starts the process over: "ch ka ch ka—punch—ka ch ka." Stop. Rewind. Listen. This time, Bobby, the digital sampler, and the tape machine have performed Jon's rhythm guitar part perfectly.

Bobby slaps his hands together, congratulating himself. With the precision of a skilled surgeon performing a skin graft, Bobby has dropped the sampled "ch" from the second verse into the third, and no one except he and I will ever know the difference. In time, we'll probably both forget it ever happened, or at the very least be unable to say for sure which "ch" is live and which was sampled and transplanted. Bobby fast-forwards the tape until it slap-slaps on the spinning take-up reel, slides it tails out into its box, turns off the tube gear, hits the lights and alarm, and heads down the fire escape to his car, in the oppressively hot, silent, Texas summer night.

Notes

1. Storing tape tails out is such common practice that digital tape—which bears no such magnetic properties—is often wound this way too, not out of necessity but habit, especially by older engineers and producers who grew accustomed to the perils of print-through on analog masters.

2. Although he is generally considered a phenomenologist, Ingarden's *The Work of Music and the Problem of Its Identity* (1986) is concerned primarily with eidetic questions about the ontological "essentials" of a musical work that affect the concretizations of that work by listeners (Reiser 1986 [1971]; see also Falk 1981 for a more complete analysis of Ingarden's phenomenological stance).

3. Kisliuk likens ethnography to performance in this respect (1997). I would offer that, especially in its early stages, it more closely resembles a musical rehearsal.

4. Structuralist anthropology, deeply influenced by Saussurean structural semiotics that saw language as composed of meaningful contrasts, tended to carve local epistemologies up along just such lines of difference. But in structuralist ethnographic writing, the processual aspects of learning systems of contrast are often backgrounded to description of the systems themselves (see Marcus and Fischer 1986, especially chapters 1 and 2, for a critique).

5. Schutz's view sharply contrasts that of Pierre Bourdieu, who sees music as the ultimate marker of social distinction: "nothing more clearly affirms one's 'class,' nothing more infallibly classifies, than tastes in music" (1984: 18).

6. The importance of movement through tonal and rhythmic time within the work of music is discussed further in Zuckerkandl (1973 [1956]), Keil (1966, 1987, 1995), Keil and Feld (1994: especially 151–80), Prögler (1995), and Alén (1995).

7. The ability to relive musical experiences points to a further shortcoming in Schutz's theory: how to account for the effects of cumulative listening experience.

What the beholder "anticipates will follow" is not simply a question of tonal flux in the inner time of the musical work but is also socially accrued from years of listening experience. For example, one's ability to anticipate the placement of a bridge or a key change in a given country song is owed in part to accumulated exposure to popular music conventions, not solely to musical events occurring in conjunction with the present encounter with that given work. That is, Schutz's textualism blinds him to the ways in which the experience attendant to a particular musical encounter is conditioned in part by previous similar (and different) listening experiences.

8. The distinction I am making between we-ness and sharedness is located in the realm of discourse. The "moments of heightened experience" are shared through talk as well as performance or recording practices. Sharedness is used to indicate moments more tightly bound to and experienced through discourse, while we-ness suggests moments, more closely bound to the music *per se*, that are emically considered to transcend discourse, leaving participants quite literally speechless.

9. Two points deserve further mention in this respect. First, the notion of "multiple local cultural reals" has a parallel at the level of individual experience in Schutz's own writing, especially in the notion of the "multiple realities" as chronicled in Wagner 1983: 90–91, 225–26. Second, some recent anthropological literature on ethnographic writing is so skeptical of objectivist truth claims that the notion of "writing about culture" has been supplanted by that of actually "writing culture." The implication of the latter phrase is that the truth claims of the anthropologist are less a reproduction of cultural knowledge than the creation of new forms of culture itself.

10. The larger thrust of Whitaker's argument suggests less that writing can move one closer to a way of *living* the Other's life than toward better representations of how that life is lived. Thus, Whitaker refuses, like Stewart, to claim that representations can lead readers to experience the Other in a transparent way.

11. "Record rehearse" is a feature that allows for a dry run of a tight punch. In essence, it cuts the playback signal from the track to be overdubbed without actually erasing the information on that track. With this feature, an engineer can check the timing and feasibility of a punch without erasing what already exists on tape during the space of the punch. Further, the beginning and end points of the punch are stored in the tape machine's memory, so they can be repeated automatically by the machine. This feature allows the engineer to free his hands from the tape machine during the punch if another operation needs to be performed simultaneously.

References

Alén, Olavo. 1995. "Rhythm as Duration of Sounds in *Tumba Francesca*." *Ethnomusicology* 39 (1): 55–71.

Attali, Jacques. 1985. *Noise: The Political Economy of Music*. Minneapolis: University of Minnesota Press.

Barz, Gregory, and Timothy Cooley, eds. 1997. *Shadows in the Field: New Perspectives for Fieldwork in Ethnomusicology*. New York: Oxford University Press.

Baudrillard, J. 1968. *Le système des objets*. Paris: Editions Gallimard.

Bourdieu, Pierre. 1977. *Outline of a Theory of Practice*, trans. Richard Nice. Cambridge: Cambridge University Press.

———. 1984. *Distinction: A Social Critique of the Judgment of Taste*, trans. Richard Nice. Cambridge: Harvard University Press.

Clifford, James, and George Marcus. 1986. *Writing Culture: The Poetics and Politics of Ethnography*. Berkeley: University of California Press.

Falk, Eugene. 1981. *The Poetics of Roman Ingarden*. Chapel Hill: University of North Carolina Press.

Henderson, Lisa. 1990. "Cinematic Competence and Directorial Persona in Film School: A Study of Socialization and Cultural Production." Ph.D. dissertation: University of Pennsylvania.

Ingarden, Roman. 1986. *The Work of Music and the Problem of Its Identity*, trans. Adam Czerniawski. Berkeley: University of California Press.

Junker, B. 1960. *Field Work: An Introduction to the Social Sciences*. Chicago: University of Chicago Press.

Keil, Charles. 1966. *Urban Blues*. Chicago: University of Chicago Press.

———. 1987. "Participatory Discrepancies and the Power of Music." *Cultural Anthropology* 2 (3): 275–83. Reprinted in Keil and Feld 1994, pp. 96–108.

———. 1995. "The Theory of Participatory Discrepancies: A Progress Report." *Ethnomusicology* 39 (1): 1–19.

Keil, Charles, and Steven Feld. 1994. *Music Grooves*. Chicago: University of Chicago Press.

Kingsbury, Henry. 1988. *Music, Talent, and Performance: A Conservatory Cultural System*. Philadelphia: Temple University Press.

Kisliuk, Michelle. 1997. "(Un)doing Fieldwork: Sharing Songs, Sharing Lives." In *Shadows in the Field: New Perspectives for Fieldwork in Ethnomusicology*, G. Barz and T. Cooley, eds., pp. 23–44. New York: Oxford University Press.

Marcus, George, and Michael Fischer. 1986. *Anthropology as Cultural Critique: An Experimental Moment in the Human Sciences*. Chicago: University of Chicago Press.

McClary, Susan. 1991. *Feminine Endings: Music, Gender, and Sexuality*. Minneapolis: University of Minnesota Press.

Nöth, W. 1990. *Handbook of Semiotics*. Bloomington: Indiana University Press.

Prögler, J. A. 1995. "Searching for Swing: Participatory Discrepancies in the Jazz Rhythm Section." *Ethnomusicology* 39 (1): 21–54.

Qureshi, Regula. 1994. "Exploring Time Cross-Culturally: Ideology and Performance of Time in the Sufi 'Qawwali.'" *Journal of Musicology* 12 (4): 491–528.

Reiser, M. 1986 [1971]. "Roman Ingarden and His Time." In Ingarden 1986, pp. 159–173. Reprinted from *Journal of Aesthetics and Art Criticism* 39 (4).

Schutz, Alfred. 1976. "Fragments on the Phenomenology of Music," edited and with a preface by Fred Kersten. *Music and Man* 2: 5–71.

———. 1977 [1951]. "Making Music Together: A Study in Social Relationship." Reprinted in *Symbolic Anthropology*, ed. J. Dolgin, D. Kemnitzer, and D. Schneider, pp. 106–119. New York: Columbia University Press.

Stewart, Kathleen. 1996. *A Space on the Side of the Road: Cultural Poetics in an "Other" America*. Princeton: Princeton University Press.

Titon, Jeff Todd. 1997. "Knowing Fieldwork." In Barz and Cooley 1997, pp. 87–100.

Wagner, Helmut. 1983. *Alfred Schutz: An Intellectual Biography*. Chicago: University of Chicago Press.

Walser, Robert. 1993. *Running with the Devil: Power, Gender, and Madness in Heavy Metal Music*. Hanover, New Hampshire: Wesleyan University Press / University Press of New England.

Whitaker, Mark. 1996. "Ethnography as Learning: A Wittgensteinian Approach to Writing Ethnographic Accounts." *Anthropological Quarterly* 69 (1): 1–13.

Woram, John. 1982. *The Recording Studio Handbook*. Plainview, New York: ELAR Publishing Co.

Zuckerkandl, Victor. 1973 [1956]. *Sound and Symbol*, trans. W. Trask and N. Guterman. Princeton: Princeton University Press.

CHAPTER TWELVE

"*There's not a problem I can't fix, 'cause I can do it in the mix*"
On the Performative Technology of 12-Inch Vinyl

Kai Fikentscher

This essay focuses on the relationship between music as realized in performance[1] and music as tangible, authoritative text, and how this relationship has been impacted, if not defined, by music technologies over the course of roughly one century.[2] Specifically, I wish to examine the disc jockey (deejay), who, starting in the early days of radio in the 1920s, and especially since the disco era of the 1970s, has had a crucial part in a process of redefining the role of music technologies. This process has been marked by a shift of emphasis away from the creation of authoritative musical texts and toward the creation of new performance modes. While we are awaiting the full impact of "interactive" musical interfaces (thanks to the explosion of digital technologies and the growth of the internet) on the ways we have grown accustomed to conceptualizing music, we can examine the club deejay[3] in a role that so far has received little attention: as a visionary figure who, in the world of dance music, has helped to bridge the transition from the analog to the digital era by redefining the turntable and sampler as performance instruments.

To appreciate the club deejay as a pioneering force transforming the relationship between music as defined by performance and music conceptualized as authoritative text, I will outline a brief history of recording technologies and their impact on musical performance.[4] This will be followed by a discussion of three aspects of deejay musicianship: technique, technology, and musical concepts.

Introduction: Before Recording Technology

In the Western world, before the arrival of early music technology—e.g., the cylinder, the gramophone, the graphophone, the phonograph, or the player piano—music was conceptually associated with two primary concrete forms: an audible event (a performance) or a written abstraction of—or prescription for—such an event (a score, a lead sheet, or some other kind of manuscript containing musical notation). In the case of performance, each manifestation of a musical work varied with the number and type of instruments used, the acoustics of the location where the performance took place, and the degree of training of the musicians. Musical performances were thus unique to time and space, since they could not be documented and reproduced in an identical form. The ways we have come to conceive of music from that era are primarily based on written documentation, including scores, sketch books, eyewitness accounts, travel logs, drawings, and early photographs. As sounded performances, music from that era does not exist. Not until the arrival of recording technology did musical sound become divorceable from the particularities of time and space.[5]

The Impact of Early Recording Technology

The advent of recording technology meant that musical performances ceased to be unique to time and place.[6] To judge by the numbers of early commercial recordings sold, this aspect more than outweighed whatever limitations in timbral fidelity the new technology had. The fact that a 78rpm recording of Enrico Caruso's voice did not quite sound like a concert performance by Caruso was deemed less important than the fact that the Victrola used to play this recording could be taken along to a picnic in the park.

The instant ability to reproduce a musical performance made the record a potent rival to sheet music (and the piano roll), even if that relative potency was only slowly recognized.[7] Eventually, however, the record became the music industry's most important commodity. Whenever a new recording technology emerged (e.g., audiomagnetic tape in the 1940s,[8] microgroove vinyl in the 1950s, compact discs in the 1980s), the new recording format necessitated the development and marketing of new playback equipment, which in turn translated into a considerable portion of the music industry's earnings.

Since recordings could serve as a form of documentation of musical performances, record libraries and archives emerged in the first decades of the twentieth century. This helped to put scientific weight behind a fledgling new academic field, comparative musicology. Types of music that hitherto had been cultivated entirely as oral traditions by their respective practitioners became available to interested outsiders on a scale previously unimaginable. Working in Berlin, Erich M. von Hornbostel was able to hypothesize on the nature of music in Asia and Africa[9] based on the availability of recordings, while at about the same time, savvy entrepreneurs in New York became interested in marketing hillbilly and blues, at the time both marginal musical traditions.

As radio and sound film technologies evolved in the 1920s and 1930s, recorded music gradually became a mass phenomenon in the United States. Still, both environments presented music in formats that dated back to the pre-recording era, in that radio shows, when not actual live broadcasts, were formatted as make-believe concerts, and American films were often structurally based on the format of theatric vaudeville and musical variety shows. However, as early as the late 1920s, enterprising radio disc jockeys such as Jack Cooper and Martin Block began to regard records not merely as documents of musical performance but as the basic elements for performance on their radio shows. Initially contrary to radio station management policy, some radio deejays began to broadcast music from recordings, thereby giving birth to the concept of musical performance as illusion (what you hear is not what is going on while you're hearing it) and to the concept of the deejay as creative authorial agent.[10] Still, during the first half of the twentieth century, recording technology served largely as a means to approximate, simulate, or document musical performance, and the actual performance served as the authoritative principle by which the recording was evaluated. This changed with the arrival after WW II of a new recording technology, audiomagnetic tape.

Performance as Illusion: Enter the Multi-track Recorder

In addition to technical shortcomings such as a limited frequency range, early recordings of music reflect the acoustic qualities of the recording locations, which—literally and figuratively—left little room for sound engineers to maneuver in. We know that in the 1920s, Louis Armstrong was placed at a greater distance from the recording megaphone than the rest of his band, since there was no other way to balance the various amplitudes produced by the members of his Hot 5 and Hot 7 ensembles, Armstrong's cornet consistently being the loudest.[11] With the gradual improvement of recording studio equipment, however, the role of the recording or sound engineer increased. While the inventions of electric amplification and the microphone were important, the greatest impact on recorded music, and our general concept of it, was the arrival of the multitrack tape recorder in 1958. This technology enabled the production of simulated performances that were similar in concept to those of prewar radio deejays such as Block and Cooper. The crucial difference was that recorded music could now be a product of illusionary performance. For example, ensemble instruments could now be recorded individually, in sequence, and then be balanced with each other in a final mix, which sonically represented a collective performance that had never really taken place. Thus, from this point on, a recording no longer necessarily documented a performance at all. Instead, it was a new type of authoritative text, a sonic score. For example, the material on the Beatles' landmark album *Sgt. Pepper's Lonely Hearts Club Band*, created on two 4-track recorders, was unperformable onstage by its authors after its release in 1967. Since about that time, especially in the context of Western popular music, the record has become the measuring stick for musical "live" performance. Also in the postwar era, sound recording has come to represent a new art form, the recording studio has become a new instrument, and producers ranging from Phil Spector and George Martin to Holger Czukay and Trevor Horn, together with their sound engineers, have not hesitated to make full use of the new universe of sonic options as they become available. As a result, ever since the end of the 1960s, for the purposes of recording music, acoustic characteristics or conditions have not necessarily been linked to any particular performance location. Rather, they can be—and frequently are—simulated and modified at will and the touch of a button.

Recording Meets Performance: Enter the Club Deejay

The elasticity of the medium of audio tape, combined with the technological sophistication of the recording studios centered around multitrack tape recorders, has revolutionized Western concepts of music as much as the invention of recording had done previously. In the 1950s and 1960s, avant garde composers such as Edgar Varèse, Karl-Heinz Stockhausen, and Milton Babbitt composed pieces based on audio tape technology, splicing, reversing, and editing the medium to suit their creative endeavors. In the post–Sgt. Pepper period of the 1970s, the making of albums by rock artists such as Emerson, Lake, and Palmer or Pink Floyd could take months, and recording studios—and recording budgets—in rock and pop music grew accordingly. At the same time, and for the first time since the invention of recording technology, disco deejays, at the time a not-yet-established category of musician, began to experiment with audio technologies in an attempt to shrink the distance that had been created between performers of music and audiences across space and time through the pervasive use of records and radio. Initially, these deejays did not work in recording, radio, or television studios but in bars, clubs, and discotheques, where they were in direct contact with their patrons, who came not only to listen but to dance. Unlike the invisible if personable radio deejay of the preceding decades, and more like the swing band leaders of the 1930s, disco deejays engaged with their dancing clientele directly and dynamically. But instead of playing saxophones or brass riffs on the bandstand, they played records and used turntables, mixers,[12] and amplifiers to create music that, although based on sounds created, arranged, and recorded by others, became ultimately "theirs." Like radio deejays, disco deejays became identifiable through their programming and treatment of recorded musical material. Unlike radio, which simulated musical performances featuring invisible performers and, frequently, physically passive listeners, the discotheque combined at least two levels of performance, deejaying and dancing. Dancing, the response to successful deejaying, involved the "listener" actively. Dancing transforms a "listener" into a "dancer," whose performance has the often realized potential of influencing the deejay's programming, thereby creating a feedback loop between deejay booth and dance floor. In this way, since the disco era, recorded music has been reintegrated into the process of musical performance, rather than just serving as commodified, authoritative text.

Vinyl Records: Mediated Music in Performance

The performance of disco and club music takes place in a setting in which a deejay creates a musical program based on the use of music and sounds recorded onto vinyl discs. While this sound is fixed, the deejay has a variety of means to manipulate that sound in creative ways so as to render his or her[13] performance unique to time and place. Among the creative choices and variables available to deejays are the musical repertoire; the technology used to play music for dancing; the techniques used to play, mix, and remix records into the flow of one musical performance; and the rapport and interaction between deejay and dancers. To deejay is to make mediated music immediate, using recordings, turntables, mixers, and sound reinforcement technology—originally designed for recording and playback purposes only—in creative ways. To understand club[14] deejays as musicians, it is necessary to discuss not only their craft and their musical equipment but also their approach to musical artistry.

The use of records as sound sources of music for social dancing has a relatively short history that begins in the late 1920s with the increasing popularity of the electrically amplified jukebox.[15] In the context of deejaying, vinyl recordings are not merely played for dancing; rather they form the basis of creative individual musical expression. They are as indispensable to the deejay's musical instruments (turntables, mixers, equalizers) as strings are to violinists, harpists, and guitarists. Underlying this approach to recorded music is what I call the "disco concept."[16] This concept fuses mediated music with musical immediacy for the purpose of a musical performance in which a deejay interacts with a body of dancers at a specific location for a specific amount of time. The essence of the disco concept translates into "spinning," "mixing," or "working" records on the part of the deejay and into "dancing" or "working your body" on the part of the dancers. The history of this concept is inextricably linked to New York City and its recent sociocultural history.[17]

Disco: History and Concept

The term "disco" has many definitions, most of which are associated with a period during the 1970s when it referred to a type of music, a musical environment, a style of dress, and a variety of leisure time activities all marked by a penchant for extravagance, hedonism, and even decadence, summarily known as "disco boom" or "disco fever."[18] However, I am concerned instead with disco as a musical technology whose chief characteristic is its mediated yet performative character.

With mediation as a central aspect of its musical performance, disco likens the discotheque to a type of institution that was established well before the 1970s (Clark 1974; Goldman 1978; Sukenick 1987). As such, a discotheque may be a circumscribed location, usually an indoor setting such as a basement, a loft, or a converted garage or warehouse, where an acoustic/spatial environment, an illusionary "world" contrasting with the "real" world outside, is created through the transformation of aural and visual perception. Part of the definition of the disco concept is the disco environment,[19] which prioritizes the aural realm at the expense of the visual. Inside a discotheque, visual perception is deemphasized through the use of (relative) darkness and lighting effects, while auditory perception is drastically heightened through the constant presence of music at a high volume and with a wide range of frequencies (which may at times cross the boundary between music as sonic versus physical sensation). Typically, this music is performed not by musicians but by a deejay, who determines its intensity, continuity, tempo, and style through choices of repertoire and program (i.e., the structure within a selected musical repertoire). Typically, a disco is visited at nighttime by a clientele that comes most visibly to interact with each other and the deejay through dance. The clientele of a deejay, often referred to as "his following," is crucial to the success of the disco concept, as it has an active role in the creation of the "otherworldly" character of the disco environment. The disco experience is thoroughly interactive.

In Western societies, gathering for the purpose of social dancing has a long history. While the origins of disco can be found in the context of the austerities and prohibitions imposed on wartime Paris by its Nazi conquerors,[20] in the United States social dancing to mediated music dates back to the 1930s, when, following the repeal of Prohibition on January 5, 1933, many bars, taverns, and similar establishments reopened with jukeboxes.[21] As a precursor of the discotheque, the jukebox-equipped tavern was an important factor both in the emerging American music industry and in the shaping of the acoustic habits and aesthetic sensibilities of the American public. The most notable consequence of the jukebox is that, in the Western hemisphere after World War II, the idea of dancing to technologically mediated music became a familiar one.

In postwar Europe, particularly in Paris and London, discotheques had developed as the gathering places of the jet set. Paris discotheques such as Chez Castel, Chez Regine and New Jimmy, for example, were social institutions that set a standard soon imitated in other Western urban centers. Disco historian Steven Harvey makes the connection between the French model and its imitations in New York City and contrasts both with a later form of discotheque that became the creative forum for 1970s disco culture:

> The first discos to open in New York were Le Club in 1960, followed by Arthur . . . in 1965. These were the earliest manifestations of the disco as upper-class watering holes where the rich could be seen in a colorful setting. This syndrome continued into the Seventies through Studio 54 and Xenon. Under these circumstances the music has to be considered secondary—a backdrop. It was not these clubs that sustained the new music, even with Studio [54]'s major impact and the great DJs who played there. Rather, it was the underground clubs which catered to Blacks, Latins, and gays.[22]

Among the first underground clubs in New York City were Salvation and Sanctuary, which opened in 1969 and 1970, respectively. Sanctuary, located on West 43rd Street between Ninth and Tenth Avenues, was not only the "first totally uninhibited homosexual discotheque in America,"[23] it had "the most celebrated disc jockey of that early disco era,"[24] a Brooklyn-born Italian American named Francis Grasso. Both sources quoted here agree on Grasso's influential role as resident club deejay, using slightly different terminology to describe Grasso's uniqueness. Goldman credits Grasso with the invention of the technique of "slip-cueing," which allows for the construction of an uninterrupted musical program through the blending together of individual records to form one seemingly unending musical soundscape,[25] while Joe describes Grasso's merits as the perfection of "the technique of 'mixing' records (a system of dovetailing one tune into another . . . designed to reduce disconcerting breaks in the music and [for] the momentum it creates."[26] Grasso appears to be one of the pioneers of a new way of playing records that was perfected by New York underground club deejays of the 1980s and 1990s. At the time, though, this approach to records was not only a sign of a new dance culture, it redefined records as being central to a new type of musical performance. Consequently, the concept of musical performance itself needed to be expanded to include the creative use of pre-existent recorded music.

During the first half of the twentieth century, American society experienced several dance crazes.[27] By World War II, jazz, under the name "swing," was transformed from a primarily African American musical genre into American mainstream entertainment. Since then, the war has come to be considered the line of demarcation between two periods of social dancing, based on their principal mode of musical delivery. Despite the availability of jukeboxes in bars, jook joints, and honky tonks, the standard by which the quality of dance music was measured prior to World War II was the performance of a dance band or orchestra playing "live" on a podium, stage, or bandstand while the dancers would take to the adjacent floor.

After World War II, particularly in the 1950s and 1960s, two factors combined to diminish the role of this form of dance music performance. One was the increase in the funding needed to sustain dance orchestras; the other factor was the rapidly developing recording industry, accompanied by significant advances in recording and playback technologies. Stereophonic sound became a standard soon after its development in the 1950s, and radio, gramophone, and television technology became quickly available to the consumer.[28] This combination of cost efficiency and a high degree of technological sophistication led to the emergence of discotheques in the United States in the 1960s. By that time, an industry had developed that supplied a steady flow of musical product to an audience whose ears had come to expect a high level of acoustic fidelity in the areas of musical recording and playback. Stereophonic and hi-fi sound, 7-inch singles and 12-inch LPs, high-output amplification, and sophisticated loudspeaker design all became part of the disco technoscape.

Pioneers working in the art world[29] of disco, such as Francis Grasso and Walter Gibbons (who remixed the first commercially available 12-inch single for Salsoul Records), combined these elements and thereby contributed to the development of a new art form, the art of deejaying, or "spinning."[30] Spinning "wax," as vinyl recordings are affectionately called by club deejays, developed to a point where the technology of mediated music—originally used for passive consumption only—was transformed into both active and creative forms of musical production, performance, and interaction. This transformation took place in the disco environment of New York City in the 1970s, and it provides the basis for the standards of dance music performance and production of the 1980s and 1990s.

Working Wax: Deejay Technique and Deejay Technology

Modern-day club deejays typically use upwards of two turntables, an audio mixer, and two separate amplification systems to address the dance floor and the deejay booth, respectively. Depending on the sophistication of the P.A. system,[31] a crossover or another equalization unit may also be used.[32] Once the sound amplification system is set to the deejay's liking, he will focus on his main instruments for the rest of the evening, the twin turntable and the audio mixer. Together, these form one unit that is referred to as the "console" or "set."[33]

Unlike the mobile deejay, a club deejay usually works with stationary technology; comparable to the visual appearance of an airplane cockpit, the deejay booth is the center of the venue's public address system, complete with a system controlling the lighting and other special effects such as smoke and fog machines, possibly video screens, and definitely a sound system with enough wattage to power outdoor events. The turntables are usually installed so that vibrations from the floor or the P.A. system will not negatively affect their operation, and typically the mixer is installed between, in front of, or below the turntables for easy access. While mixers vary greatly in design, features, and mode of operation, the Urei 1620 model, based on a precursor made by Bozak, is a favorite among club deejays.[34] By the same token, turntable technology has hardly changed since the mid-1980s, when Panasonic introduced the Technics SL-1200 model. Versions of this heavy-duty turntable, which features a high-torque, direct-drive motor and a vari-speed mechanism, are installed in most deejay booths and, since the 1980s, have become a worldwide industry standard.[35] This hardware standard has affected the evolution of deejay technique in general, to the extent that the art of spinning and mixing is learned and practiced on basically one type of mixer and turntable.[36]

To the casual observer, the technique of spinning appears too similar to playing records to be considered anything but a fairly straightforward enterprise. This may be one of the reasons why deejaying is sometimes not viewed as a very prestigious activity or profession to be associated with the specific musical skills or techniques that traditional musicians must acquire. This perspective changes once one considers the twin turntable set and the audio mixer as one instrument consisting of three units that have to be operated simultaneously as well as synchronously in order to allow the artistry of deejaying to emerge. The coordination of two or more turntables and a mixer requires intimate knowledge of the appliances, and operation skills that can be acquired only through practice. The extent of this knowledge is most apparent whenever a deejay segues from music playing on one turntable to music playing on the other, which necessitates using the cross-fader control on the mixer to control the flow and balance of two (or more) separate audio signal feeds. At that time, in addition to other optional considerations, the simultaneous control of tempo, volume, and balance of timbres and textures is as crucial as on any other musical instrument. While the tempo of a piece of recorded dance music is fixed, the tempo of the music on a Technics SL-1200 turntable is adjustable from minus 8 percent to plus 8 percent.[37] Tempo changes can also be affected without using the speed control. Slowing down can be accomplished by touching the rotating platter with a thumb or finger and applying the desired amount of friction, whereas an index finger applied toward the outer edge of the record's center label is used to speed up the tempo. Both techniques serve to get one record to sync up with another.

Among club deejays, the most common way of segueing from one record to the next is by slip-cueing. Other common techniques are "fast cuts" and "overlays." Fast cutting involves a rapid, almost instantaneous switch between turntables, usually just before the first downbeat of the section or song about to be played. Overlays are achieved by playing two records at the same time through the P.A. system for an extended period of time, often lasting minutes. The aim is to synchronize two different records so as to make them sound like one piece of music (which they then become in the hands of an accomplished deejay).[38]

While mixing skills have helped many a deejay achieve cult status, especially in the field of hip-hop music, they are not considered as important as programming skills in the context of disco and club music. Tony Humphries, a professional deejay since 1977 who has since become a major influence on his peers in New York City, speaks from a club deejay perspective in the following interview excerpt:

DJs have to understand the concept of programming. How to break a record. How to play with records, repeat intros, lengthen breaks, endings . . . There is an art to programming . . . The DJ who plays all his hottest records in a row is not doing his job right . . . what you do is you play a track, followed by something new and then you back it up with something that they know and like. It's like a train ride. The clubgoers become very trustworthy that you will come back with something they like. It's the 15-minute game. About every third song, you give them a well-known song. After one hour, the crowd has been exposed to ten new records. That way, you please yourself as well as the crowd. Larry Levan was great at this. The most important thing to remember is that musical content, i.e., how you program, is more important than actual mixing skills.³⁹

Someone who has mastered the operation of two or even three turntables and an audio mixer is not necessarily considered a good deejay. Paradise Garage deejay Larry Levan, praised in the preceding quote, explained his own approach: "When I listen to DJs today they don't mean anything to me. Technically some of them are excellent—emotionally they can't do anything for me . . . There is actually a message in the dance, the way you feel, the muscles you use, but only certain records have that."⁴⁰ These "certain records" and how they are employed are the ultimate indications of a deejay's quality, skill, and style, as the preceding comments show. A deejay's technological skill is thus at best valued equally with his knowledge and choice of repertoire.

As mentioned before, the deejay's repertoire travels with the deejay. While some clubs have a (usually limited) record library, most underground deejays bring along several crates of records.⁴¹ These crates are then stacked in the deejay booth, often facing the set, so that the deejay positions himself between the set and the records. Whereas hip-hop deejays may employ a helper to handle and store records as they are switched quickly, club deejays usually do the handling and programming by themselves. The records are organized according to certain criteria, and intimate familiarity with this organization and the records themselves is considered essential to efficient deejaying. The most common division is between new and old records, the latter often being referred to as "classics." Also, deejays differentiate between stylistic categories (such as hip-hop, house, acid house, deep house, techno, or trance) and functional ones (such as vocals, instrumentals—also known as tracks or dubs—acapellas, and sound effects). Some arrange their records by label, or by year, whereas others distinguish between major and indie label releases, as well as between domestic and imported records. Few deejays use bpm (beats per minute) as an organizational principle.⁴²

Deejay David DePino, who learned deejaying under the tutelage of his colleague Larry Levan at the Manhattan club Paradise Garage in the early 1980s, comments on aspects of repertoire selection, programming, and mixing:

[Through] shopping at all different stores and shopping often you find things that make you different [from] everybody else. And that's the magic. I don't go to any other clubs, because I don't want to hear what other deejays are playing... My club [Tracks] is open from ten to four, and nobody even gets there until twelve, one o'clock, and they're gone before four, [so] I've got three, four hours to really let people hear good music—my taste in music, and what they want to hear. Junior [Vasquez, who spins at Sound Factory] is open from twelve to twelve, he's got twelve hours to do it. I got four hours... I can't go through many different moods like I used to. Tracks used to be open at 9 [P.M.], crowded by 9:30, and I used to close at 7 [A.M.]. I had many hours. I could have took them through an oldies hour or two, I could have took them downtempo, uptempo, mood swings, unusual music... Can't do it right now.[43]

Junior Vasquez, at the time resident deejay at the Manhattan club Sound Factory, echoes DePino's concerns:

I have always relied upon having a 10- to 12-hour night in which to play everything I think is important. A long night allows a jock to completely satisfy at least two different kinds of clubgoers. The crowd who dances to you at 6 or 7 AM is pretty much ready for you to lead them wherever you will, which is a very powerful position for a jock to be in. My 4 AM-to-noon crowd could let me know which records the 10 PM-to-4 crowd would like three months down the line. If I have to force those two crowds to party together, I run the risk of pleasing no one—including myself.[44]

Through the influence of early discotheque deejays in New York such as Terry Noel (at Arthur) and Francis Grasso (at Salvation and Sanctuary), the art of spinning was gradually transformed into a new form of American vernacular art—the art of mixing. In the 1970s, the emergence of disco as a mainstream phenomenon owes as much to the creativity of deejays like Francis Grasso as to the development of new sound technologies designed and built for, at times even by, club deejays. One of these technologies was the development of the direct-drive turntable with speed control. The other was that of the 12-inch single, to be discussed in more detail later.

In addition to the standardized formats and conventions of an industry concerned primarily with pop product,[45] early disco deejays were faced with the particularities of the standard playback equipment, especially with the inconsistencies of belt-driven turntables. When switched on, a belt-driven turntable requires a significant amount of time before it reaches either of the two fixed speeds (33⅓ or 45 rpm). With the later development of speed controls at both 33⅓ and 45 rpm and the option of controlling the rotation speed directly via a quartz-controlled motor instead of indirectly via a rubber belt, deejay mixing has become less laborious and more flexible. Still, it takes practice to acquire the manual dexterity and skill to slow down or speed up a spinning record without noticeable pitch or tempo changes that may confuse or annoy the dancers.

As described above, slip-cueing is the basic way of stitching together music, record after record, into one uninterrupted flow. Pieces of songs can be superimposed over each other, creating a new musical effect, and affect, at the same time. "The 'seamless segue' was brought to the level of an art by early deejays, who went on to develop a whole repertoire of effects created by manipulation and interaction of turntables."[46] Walters's account of David Morales's approach to deejaying is typical of club deejay practice:

> Morales doesn't play records so much as he transforms them. He can't let a piece of vinyl simply be. Many of the 12-inch singles Morales mixes are grating and monotonous when left unaltered or heard out of their club context. But layered on top of another, cut up, stretched out, and paced to create an evening of multiple climaxes, these bass-and-drum machine-generated records turn into grand, almost symphonic soundscapes of urban life . . . Morales often takes an instrumental or a dub version and adds vocal fragments from other records. Or he layers one instrumental on top of another. He adjusts his mixer to create unexpected shifts of volume and tone. Sometimes he plays two copies of the same record with the needles in exactly the same place, and the sound whooshes like a jet taking off. Morales knows how to make a record sound live.[47]

What Walters describes here is known as "working a record" among club deejays and dancers. When he works a record, Morales the deejay becomes Morales the performer.

Through creatively exploiting mediated music and the means of playback technology, deejays have acquired the means to become musical authors, producing their own interpretations of prerecorded music, not only through the choice of repertoire but also through style of mixing. In New York, this concept, formulated by deejay pioneers such as Terry Noel and Francis Grasso, was developed by such influential figures as David Mancuso (at Loft), Nicky Siano (at Gallery), Tee Scott (at Better Days), and Larry Levan (at Paradise Garage), and transmitted to the younger deejays of the 1980s and 1990s (among them Frankie Knuckles, David Morales, Basil Thomas, Louie Vega, Danny Tenaglia, Ken Carpenter, Andre Collins, Roger Sanchez, David Camacho, and Kim Lightfoot), establishing an unbroken tradition of this art. Almost half a century after the first discotheque, this tradition still has no textbook or manual, no comprehensive documentation, no established recognition as a form of musical artistry. Rather, it is carried orally from one generation of deejays and dancers to the next.[48]

Part of the relative obscurity of this unbroken tradition can be explained by the fact that it developed largely out of sight of the public eye, in the dance underground of the city. It was not until hip-hop music broke the mass media barrier in 1979 that the concept of the deejay as an artist and culture hero gradually became formulated.[49] As dance or party music, hip-hop developed at the same time as disco but with a different cultural territory and agenda.[50] Still, while the focus in hip-hop (and much hip-hop scholarship) is on its verbal component, and hence on the emcee, the central musician in hip-hop is the deejay, respectfully referred to as the "master on the wheels of steel."[51] As *New York Times* critic Mark Dery observes,

> On the surface, the ascendance of the DJ is a result of rap's pervasiveness. More significantly, the . . . DJ stands at a cultural crossroads—the post-modern intersection of "high" art and "low" folk traditions. By using an everyday appliance to make sound collages that people can dance to, the DJ signifies the fusion of avant-garde and entertainment. [He descends,] on the one hand, from John Cage's "Imaginary Landscape," a 1939 percussion piece whose instrumentation included two phonographs; on the other, from [hip-hop deejay] Grandmaster Flash, whose 1981 audio collage, "The Adventures of Grandmaster Flash on the Wheels of Steel," stands alongside Jimi Hendrix's abstract-expressionist "Star-Spangled Banner" as one of pop music's most dazzling moments.[52]

The forays by hip-hop deejays into the institutions of the entertainment industry (primarily recording, as well as radio and television broadcast studios) were also undertaken by club deejays. By the time Steven Harvey published his survey of the New York "disco underground" in 1983, some of its protagonists weren't mere deejays anymore. Mastering the art of spinning wax on the wheels of steel had afforded many deejays an entry to the recording studio as tape editors, remixers, and producers. In the process, they acquired a new title, "mixer," which, by the end of the decade, changed to "remixer" and "remixer/producer."[53]

From Mix to Remix: The Deejay as Composer

The need to mix originally arose from the limitations of early deejay technology. When using just one turntable to play records for dancing, there would be "dead" time between each record that was contrary to the continuous dynamics of a dance floor atmosphere. Dead time could be avoided when using two turntables, but their outputs had to be fed through the same sound system.[54] Hence the need for a device to alternate between the audio signals from two (or more) turntables. This device, the mixer, originally evolved in the context of multitrack recording technology. A mixer, known in recording studio parlance as a "board" or "console," allows for the control of all audio aspects of individual sound sources.[55] It can therefore be used for both recording and playback purposes.

During the 1970s and 1980s, many club deejays became record producers by involving themselves in the recording process as the flip side of their activities involving primarily the manipulation of musical sound in the process of playback. This involvement meant a transfer of sound manipulation concepts into the realm of recording and caused a blurring of the conceptual divisions between production and reproduction, between sound engineer, producer, and deejay. This in turn started the dance music industry.⁵⁶ Its central musical format is the 12-inch single. A seminal figure in this context is Tom Moulton, a New York disc jockey–turned-producer who conceived of the 12-inch single in 1975, in effect standardizing the sound carrier format of the dance music industry for the next quarter of a century.⁵⁷ Moulton's introduction of the 12-inch single as the ideal disco format coincided with the first establishment (by David Mancuso) of a record pool in New York,⁵⁸ which in turn happened at a time when local radio disc jockeys began to accept club deejays as trendsetters by playing what they had heard in clubs.⁵⁹ In hindsight, it is hard to imagine the start of the disco boom in 1975 without the coincidence of these three developments.

Doug Shannon, a former disc jockey in Cleveland, explains the emergence of the 12-inch format as "the only logical alternative . . . for record companies to place longer versions on special promotional album-size . . . records. These . . . would enable longer cuts and a different studio mix to be used, compared to what was available on an album or 45, and would also assure that a song's reproduction would . . . have better sound quality and greater volume."⁶⁰ These concerns reflect a compromise between what deejays wanted record companies to provide them with and what advantages and risks record companies saw in terms of marketing. Shannon continues, "after the longer cuts . . . on promotional 12-inch singles began to increase in numbers and in popularity through discotheque exposure, record companies realized there was a potential commercial market for 12-inch singles."⁶¹ They subsequently started to issue 12-inch singles for commercial release in the fall of 1976. To differentiate the new format, 12-inch singles were marketed as "Giant 45" or "Giant Single." This was supposed to underscore their nature as singles and avoid confusion with the similar looking album, for which 12-inch singles quickly came to serve as a promotional tool. In addition, the packaging was different. From the beginning, many 12-inch singles were (and still are) most often dressed in a plain one-color record jacket with the center cut out to show the sticker on the record bearing the song titles.

Initially, early 12-inch records contained extended versions of songs that, in most cases, were also available in shorter form on 45s or LPs, with the B-side often providing the same music without vocals. This latter version was called "instrumental" even if remnants of the voice track were present. Over time, the term "mix" became used synonymously with "version" and eventually displaced it in usage and in print on vinyl records and record jackets. Thus, a term used primarily in the recording studio was adopted in the discotheque as well.

In the former context, a "mix" denotes the phase following the multitrack recording, or the product of that phase, during which the final balance of all tracks to each other in terms of volume and timbre is determined. In that sense, a mix is comparable to a score of Western art music: it is an authoritative "text." In the context of a dance venue, however, a mix refers to the programming and blending by a deejay of records and, perhaps, sound effects, accomplished with less sophisticated equipment and, more importantly, in real time. Still, strictly speaking, a deejay's performance constitutes a "remix," since he uses, for his own mix, vinyl recordings that have been previously mixed in the recording studio.

In the course of the 1980s, however, particularly with the advent of sampling technology, remixing made its way into the recording studio, this time as a money-intensive industry. Not only were musical sections stretched through the repetition of material already available on master tapes, but individual tracks were changed, either by using different signal processing technology (e.g., echo or different equalization) or by actual re-recording using additional musicians, often synonymous with, or recruited from, the production team.[62] The results of this approach became known as remixes, to differentiate them from the "original" mixes found on 45s, LPs, or previously released 12-inch singles. As deejays learned to master the individual steps involved in remixing, they began to compose their own music into a music that, by name and title, did not necessarily reflect their input. Still, as remixers, deejays became known for their sound and style beyond the confines of the deejay booth and began to reap the economic benefits of being composers and/or producers.[63]

In the aftermath of the disco craze, through the efforts of club deejays, the late 1970s and early 1980s were a period in which the lines between studio producers, engineers, songwriters, and disc jockeys became increasingly blurred.[64] Many deejays, in addition to spinning records at clubs, ventured into dance music production, bringing many of their workplace concepts and techniques into the recording studio. In the process, the arts of mixing using a multitrack console and recorder and of mixing at a dance venue using two or more turntables and a comparatively unsophisticated audio mixer began to converge. The more savvy deejays fed the know-how thus acquired back to the dance venue. As a result, the number of versions found on a 12-inch single increased from two (A- and B-side) to about four to six. To account for this expansion, different categories of versions or mixes were developed during this period, as deejays became increasingly involved in the composition, production, and engineering of dance music.

The oldest, most established of these (now more or less standard) categories is the dub. Its roots are in the deejay culture of 1960s Jamaica, associated with mobile sound systems run by competitive teams of deejays.

Outstanding traits of the original dub, which has evolved into a subgenre of 1970s reggae, is the treatment of song structure and the addition of sound effects, notably echo. Through the use of two copies of the same record, sections can be not only repeated but shortened or extended, and the vocal part can be replaced with "live" performances by "dub poets," "DJs" or "toasters."[65] A standard feature on 12-inch dance singles since the mid-1980s, dubs have become one of the main categories of dance music remixing. For underground deejays who have ventured into remixing, a dub has become the mix category that allows for the highest degree of individual musical expression. The deejay becomes a composer in the recording studio, as New York deejay/remixer Victor Simonelli explained in an interview.

V. Simonelli: Now a dub gets fun, like, I can experiment a lot. A dub mix is fun, because I'm able to do whatever I want.

K. Fikentscher: As in "Do your thing"? "Go crazy"?

V. S.: Oh yeah. A dub is "do your thing." Arrangement doesn't really make a difference. It's just a "feel" thing, mainly. It's different in each situation, but I'm talking about most of the time.

K. F.: Some people equate dubs with instrumentals.

V. S.: Sure . . . I agree with that.

K. F.: Okay. Although some records have instrumental versions.

V. S.: And also a dub version. And the difference between an instrumental [and a] dub could be a lot of delay, or vocals coming in and out, breakdowns that you wouldn't expect, just interesting stuff that surprises you . . . There is no standard format to a dub in my opinion, or I never found one.

K. F.: Interesting. Because Junior [Vasquez], on the other hand, gets accused by some of playing too many dubs at the Sound Factory.

V. S.: [laughs] To tell you the truth, most of the stuff that gets played in clubs today around the city would be considered dubs.[66]

The various categories of mixes are the result of a gradual process whereby deejays became regarded as the key players in a new type of music industry. This industry is now built upon the concept of remixing as recycling one musical piece in as many appearances as feasible, in order to target a variety of markets distinguished by musical style. It took about a decade for this concept and industry to develop. Initially, deejays such as Vasquez and Simonelli were hired to do "edits," i.e., to extend the song, particularly introductions, breaks, and instrumental sections, for 12-inch versions. They did not otherwise tamper with them. "It was a triumph when you could add a conga," recalls *Billboard*'s dance music editor Larry Flick.[67] However, remixing became established not on its artistic, innovative, or aesthetic merits but for economic reasons.[68] Writing for the *New York Times* in

1992, Rob Tannenbaum quotes Leslie Doyle, at the time national director of dance music at Elektra Records: "'Remixers can salvage records.' For just a few thousand dollars . . . these audio auteurs refashion records to match changing styles. Success has made the practice rampant; one executive estimates that half the singles on the Top 100 chart are remixed."[69]

This estimate comes almost a decade after the practice that led to its formulation was initiated by enterprising deejays, both in the fields of early house music and pre–"Yo! MTV Raps"[70] hip-hop. Tannenbaum credits the evolution of remixing from underground oddity to major label standard practice to pop artist Madonna. "Madonna, the most statistically significant artist of the '80s, carried dance music out of the underground and into the mainstream. [At that time] a cadre of remixers—mostly black and Latino, often gay—came to dominate the field, and dance-oriented acts released entire albums of remixes. Soon the job title was a misnomer: remixers were also rearranging, rewriting and reproducing."[71] Shep Pettibone, mentioned earlier, had abandoned deejaying by 1984 and moved completely into remixing and producing. He has since worked profitably for prominent pop singers such as Michael Jackson and Madonna, who went as far as saying she preferred his remixes to the originals. Their first two collaborations went to number one on the charts and transformed Pettibone's career.[72]

The evolution from spinning to mixing to remixing to producing, as exemplified by Pettibone, has been a gradual and continuous process for most deejays. Over time, through the increasing involvement of deejays in the art of record making, the transfer of technologies and aesthetics between the recording studio and the deejay booth increased to a level so financially profitable that for years it has been the main driving force behind the dance music industry. As a result, representing its characteristic institutions—the independent label, the record pool, the underground club, the specialty retail store—is now a group of deejays who base their activities as remixers, producers, A&R personnel, and recording artists on an ever expanding concept of their musicianship. In addition to enjoying the loyalty of their dance floor clienteles, they wield more influence and money-earning power in the music industry than ever before. The increase in status from record spinner to record producer has made the deejay a cultural hero, even if the culture in question is of marginal, subcultural character.

"You can't work CDs like you can work wax."
On Why 12-Inch Vinyl Is Critical

The development of the 12-inch single and its almost immediate appeal to consumers, but especially to deejays, has played an important role in the emergence and evolution of the dance music industry in America, with New York City as its center. The picture that emerges from the rather tight connections between 12-inch vinyl and deejay technology, deejay technique (spinning, mixing), and deejay style (programming, working, remixing) is one that explains how both the acoustic properties of the 12-inch single as an analog medium and the physicality of the equipment (records, turntables, mixers) have helped shape the art and craft of dance music performance as it developed in New York City over a period of more than twenty-five years. Although during this time other sound carriers were developed and mass-marketed, it is not surprising that most New York deejays, and virtually all those who work in the dance underground, prefer vinyl over any other sound carrier format.

Since the late 1980s, this has led to a situation of conflicting interests within the recording industry as digital audio carriers, most notably the CD, have surpassed analog carriers such as vinyl and, to a lesser extent, cassette tape in terms of sales. Being disproportionately more profitable than vinyl, CD technology has effectively marginalized vinyl in the general music consumer market, which is dominated by large recording companies, known as the "majors." Within the dance music industry, independent companies ("indies") have attempted to fill the vinyl void created by the majors. Since the initial phasing-out of vinyl product by the majors around the turn of the decade, the number of independent labels catering exclusively to the dance music industry has increased. These companies, which initially almost exclusively dealt in vinyl product, were affected by the shrinkage in the vinyl market. Over the last few years, the number of 12-inch single sales has decreased. In response, independent label executives have turned to marketing music that may yield higher sales, such as hip-hop (the labels include Strictly Rhythm and Nervous) or techno (Radikal); they have reissued old catalog repertoire on CD (Easy Street, Cutting); or they have gone out of business altogether (NuGroove, 111 East). Almost without exception, these and similar companies are now selling their catalog on both vinyl and CD.

"The vinyl record is going to be the black-and-white television set of the audio world. It will be around, but in small numbers. Just as the color TV has taken over, the CD eventually will," judged Michael Gussick, president of Easy Street Records, in a personal communication in 1993. Only two years earlier, Mark Finkelstein, president of Strictly Rhythm Records, had taken a slightly more optimistic position:

> Here is my take on the difference between vinyl, CDs, cassettes, and why I don't think vinyl should ever be replaced, albeit will be because it's not as profitable. To get the warmth of the sound you need an analog format. And that leaves you with

vinyl and cassette. Cassette doesn't work, for two reasons: One, you can't mix cassettes, and secondly they have an inherent tape sound, tape hiss . . . the CD is too bright, too thin, too exact. Vinyl has got the warmth of the mechanical absorption of the sound through the vibration of the needle that you cannot otherwise replicate. You can get a depth of sound that you cannot get on CD . . . I mean I can identify a Strictly Rhythm record that's playing in a club because I walk in there and I feel the bass in my chest . . . As the majors pull out [of the vinyl industry], it leaves a void in the market place, and we (at Strictly Rhythm) have learned that where there is a demand, there will be a supply. So as they pull out, that DJ who is now looking for something to play Saturday night will pick up a Strictly Rhythm record.[73]

With few exceptions, two decades after the introduction of the CD, vinyl records remain the norm in New York underground dance clubs. Although most deejays continue to view digital audio as inappropriate for club playing purposes (Cooper 1990; Flick 1990), some have begun to use the CD-R in addition to DAT in order to transfer rough mixes and unreleased tracks from the recording studio to the dance club or radio show. Overall, however, the indications are strong that the vinyl market will be of concern to the dance music industry for quite some time to come.

Conclusion

With the arrival of the disco deejay in the early 1970s, the turntable became an instrument of musical performance in the hands of men whose role gradually changed from programming prerecorded music on LPs and 45s to arranging and editing multitrack recordings for the purposes of producing 12-inch singles, to composing and recording original music modeled on the previous two concepts of music making while mixing all of the available sound sources into hours of uninterrupted dance-music-drama. This type of performance happens in the context of a specific evening and a specific club environment populated by a specific group of dancers. In this environment, the twin turntable decks and the audio mixer are treated as one single musical instrument, the deejay set. In tandem with an amplification system, the deejay set has become a powerful performance instrument. The deejay is thus a performer, a musician, and deejaying or spinning is tantamount to making music intended, in most cases, to turn a passive audience active through dance.

Taking into the account the highly context-sensitive nature of deejaying, we can understand spinning as an act of interactive musicianship. Over time, the evolution of deejay musicianship has led to the emergence of the 12-inch dance single, seen by some as the last commercially viable vinyl format.[74] The history of deejay musicianship is also a history of the cross-fertilization between the aspects of sound manipulation (editing, mixing, arranging, composing, and producing) of the deejay booth and the recording studio (the latter being the environment where digital audio had its first impact). As deejays established reputations as mixers and remixers outside the deejay booth, they negotiated the transition from analog to digital sound, using both for the purpose of realizing their respective arts. While discotheques and dance clubs have long been test sites for new (in more ways than one) music, the deejay has become more than a mere technological pioneer or a cultural hero. At the dawn of the new millennium, he is the quintessential post-modern musician.

Notes

1. I define musical performance as any event in which the sound-creating musician(s) and his/her/their audience share the same acoustic space due to their physical presence in one location, at one time.

2. This quote is taken from the rap segment of In Deep's "Last Night a DJ Saved My Life" (New York Sound Records, 1983).

3. In contrast to the radio deejay, the club deejay is not physically separated from the audience but shares the same physical space. This space may be referred to as a discotheque or club. I tend to use the latter, since it has fewer historically specific connotations than the former. See also Fikentscher 2000.

4. For more detailed and comprehensive treatments of this topic, see Chanan 1995 and Théberge 1997.

5. For a discussion of the political implications of sound recording technology, see Attali 1985: 85–132.

6. Recording technology has always implied two processes, sound storage and sound reproduction. Contemporary terminology refers to these processes as recording and playback. Here, whenever I use the term "recording," the concept of playback is always implied.

7. Remarkably, the sales of records don't surpass those of sheet music as a source of revenue in the American music industry until 1952 (Garofalo 1996: 84).

8. Théberge 1993 dates magnetic tape recording back to 1898, while Rietveld 1998 credits the German Fritz Pleumer with his 1928 invention. The first commercial multitrack tape recorders appeared on the American market in 1948 (ibid.: 136).

9. See, for example, Hornbostel 1928 and Hornbostel and Abraham 1909.

10. Two famous examples, among many others, are Al Jarvis's 1932 "The World's Largest Make-Believe Ballroom" on KFWB in Los Angeles and Martin Block's 1935 "Make-Believe Ballroom" on WNEW in New York, the latter having been inspired by Jarvis. For a detailed discussion, see Passman 1971: 48 and Poschardt 1995: 45.

11. See Tanner, Megill, and Gerow 1997: 57.

12. Out of context, a mixer can be understood to be either a device used to mix signals from several audio sources before sending them to an amplification system, or the person who operates this device in the position of a recording engineer, technician, or deejay (see Kealy 1979; Harvey 1983). Unless indicated otherwise, in this text "mixer" refers to the device, not the person.

13. By my estimation, in New York City male club deejays outnumbered their female colleagues by a ratio greater than ten to one during the 1990s.

14. As part of the term "club deejay", club here refers to the working environment of the deejay, not club music, one type of post-disco dance music (see Thomas

1995; see also Rietveld 1998, who claims that club is "purposefully indefinable" but encompasses many deejay-specific types of dance music, such as disco, club, house, hip hop, garage, techno, and rave, among others [20]).

15. See Goldman 1978: 24–25 and Hazzard-Gordon 1990: 112.

16. Miezitis uses the term "disco experience" to refer to the same musical environment (1980: x).

17. See Goldman 1978 and Poschardt 1995.

18. See, for example, George 1988: 153.

19. I am borrowing this term from Miezitis 1980.

20. Twenty years after the demise of disco music, there is still a noticeable dearth of literature on disco and disco history. This unfortunate circumstance highlights the few extant, albeit out-of-print, works, such as Goldman 1978 and Joe 1980. Both trace the origins of the disco concept to Paris in the 1940s.

21. Jukebox technology had existed since 1899 but, mainly due to severe technological drawbacks, did not see widespread use for the ensuing three decades. Electric amplification, developed in the late 1920s, enabled jukeboxes to cut through the din of social dancing environments, where earlier popular technologies such as the Victrola phonograph and the "player piano" had fallen short. Starting in the 1930s, the jukebox became enormously popular in the United States, and it remains so today.

22. Harvey 1983: 40.

23. Goldman 1978: 114.

24. Joe 1980: 18.

25. Goldman 1978: 115.

26. See Joe 1980: 18. While both Joe and Goldman appear to describe the same phenomenon, slip-cueing is more precisely understood as a specific mixing technique. Another term used at the time to describe the effect of musical flow is "disco blending," applied to Walter Gibbons's mixing style on Salsoul Orchestra's 1978 *Greatest Disco Hits* album, appropriately subtitled "Music For Non-Stop Dancing."

27. For a more detailed analysis of social dance history in the United States, see Clark 1974, Goldman 1978, and Hazzard-Gordon 1990.

28. Breh 1982: 166.

29. The term is borrowed from Becker 1982.

30. Rietveld has defined these terms according to geography. Accordingly, "deejaying" is the British equivalent of the American "spinning" (1998: 149).

31. This is an abbreviation of "public address" system.

32. Equalizers and crossovers allow for the subdivision and fine-tuning of the audio frequency spectrum. In this way, highs, mids, and lows can be balanced against each other. These controls can also be used to emphasize certain frequencies, especially when manipulated in sync with the rhythm of the music.

33. The "set" may also denote the musical program designed by the deejay (Rietveld 1998).

34. Neither the Bozak nor the Urei mixer are manufactured anymore. As a result, they have become sought-after collector's items on the used deejay equipment market.

35. Some New York club deejays, such as Jason Load or Kenny Carpenter, still favor and/or own older turntable models that are no longer commercially available, such as the Technics 1100 or the belt-driven Thorens TK 125.

36. Traveling club deejays might therefore be comparable to concert pianists. When on tour, both leave their instruments at home, taking along only their repertoire. Both the deejay and the pianist can safely assume that a finely tuned instrument will be provided at the performance location.

37. Some older deejays, especially the "disco era" vanguard, developed their speed

control skills on turntables other than the SL 1200 and have stuck to the techniques that work on belt-driven units or those without a built-in speed control mechanism.

38. Evidence suggests that terminology differs geographically. While British observers use the term "slow mix" (Rietveld 1998: 110), American synonyms are "beat mix" and the somewhat older "disco blending," the latter attributed to New York deejay Walter Gibbons (see also note 24).

39. Quoted in Paoletta 1991: 12–13.

40. Quoted in Harvey 1983: 40.

41. In New York, milk crates made of sturdy hard plastic are in widespread use.

42. Still, deejay-oriented publications that organize music according to bpm and/or keys, such as Jester 1990, do exist and are available in some of New York City's retail stores frequented by deejays. Some record companies also print deejay-friendly bpm indications on the record label or sleeve. Even when bpms are not used as an organizational principle, many deejays measure bpms by means of a metronome and affix self-adhesive labels to the record indicating the tempo handwritten in numbers.

43. DePino 1992.

44. Quoted in Cooper 1990: 91.

45. The records known as 45s (initially developed by RCA) share with 78s the limited time for recorded sound of less than four minutes per side. For deejays, an additional disadvantage is the fact that 45s are often made from recycled vinyl (as opposed to "virgin" vinyl for LPs and 12-inch singles), which results in a overall lower sound quality.

46. Klasco and Michael 1992: 61.

47. Walters 1988: 22.

48. This applies as well to British club culture, which is much more mainstream than underground in character (Toynbee 1993: 293).

49. I am excluding the radio deejay from this discussion; his role as cultural hero is discussed in George 1988, Harper 1989, Nolan 1969, and Williams 1986. See also Poschardt 1995: 40–93.

50. See Toop 1991.

51. For a discussion of the hip-hop deejay as composer/performer, see Johnson 1994.

52. Dery 1991.

53. In the original introductory editorial to Harvey's article in the magazine *Collusion*, his interview partners are labeled "New York's top mixers and DJs." The reprint of the article in *DJ Magazine* ten years later substitutes for the older editorial an update that basically paraphrases the older version's content, dropping the use of "mixer:" "Steven Harvey and Patricia Bates spoke with many of the DJ/producers at the helm of the sound of the city" (Harvey 1993: 4).

54. Early versions of this set-up used two separate amplification systems without a mixer. Instead, the deejay controlled the volume settings on two separate amplifiers.

55. See Kealy 1979 and Tankel 1990.

56. The history of the dance music industry is detailed in Straw 1990: 119–195.

57. Joe (1980: 63–64). On the topic of the history of the 12-inch single, see also Fikentscher 1991 and Shannon 1985: 203 ff.

58. A record pool is an organization for and by deejays, who, for a membership fee, receive new product (12-inch singles), mostly from major labels. In return, pools collect feedback sheets from pool members and pass those on to the labels.

59. See Goldman 1978: 131.

60. Shannon 1985: 205.

61. Ibid.: 206.

62. Examples of this approach are C&C Music Factory, Masters at Work, and Mood II Swing (from New York City), and Blaze (from New Jersey). The success of these producer teams was based on the collaboration of deejays and musicians turned producers/remixers whose names read like a list of who's who in the 1990s New York dance music industry: David Cole and Robert Clivilles (C&C Music Factory), Kenny Gonzalez and Louie Vega (Masters at Work), Lem Springsteen and John Ciafone (Mood II Swing), Kevin Hedge and Josh Milan (Blaze).

63. Former deejay Tom Moulton arranged for Salsoul Records to have every remix of his labeled "A Tom Moulton Mix." Shep Pettibone followed this example.

64. Kealy 1979.

65. In Jamaican dance music culture, a "DJ" is what in New York is called a rapper or MC (Master of Ceremonies, emcee). The term for disc jockey in Jamaica is "selector." Not surprisingly, there are many historical and contemporary links between Jamaican dub culture and early hip-hop culture in New York City (Rose 1994; Toop 1991).

66. Yu, who equates instrumentals to dubs, also comments on the appeal of dub versions as played by Larry Levan at the club Paradise Garage (1988: 72).

67. Tannenbaum 1992: 23.

68. See also Tannenbaum 1992; Flick 1991; Chin 1987; Pareles 1990.

69. As quoted in Tannenbaum 1992: 23.

70. This show, broadcast nationwide on cable TV, helped to establish the genre as a part of the mainstream music industry.

71. Tannenbaum 1992: 23.

72. Ibid.

73. Finkelstein 1991.

74. Audiophile LPs and independent 45 rpm singles, the other two existing vinyl formats, have not seen six-figure sales in recent history. Twelve-inch dance hits, however, are known to sell between 50,000 and 500,000 copies.

References

Attali, Jacques. 1985. *Noise: The Political Economy of Music*. Minneapolis: University of Minnesota Press.

Becker, Howard. 1982. *Art Worlds*. Berkeley and Los Angeles: University of California Press.

Breh, Karl. 1982. "High Fidelity, Stereophony, and the Mutation of Musical Communication." In Kurt Blaukopf (ed.), *The Phonogram in Cultural Communication*. New York: Springer, 165–177.

Chanan, Michael. 1995. *Repeated Takes: A Short History of Recording and its Effect on Music*. New York and London: Verso.

Ching, Brian. 1987. "Remix Players Are Unsung Heroes." *Billboard*. 4 July, 58.

Clark, Sharon Leigh. 1974. "Rock Dance in the United States, 1960–1970: Its Origin, Forms and Patterns." Dissertation, New York University.

Cooper, Carol. 1990. "Life of the Party." *Egg* (June/July): 89–98.

DePino, David. 1992. Interview with the author, June 1.

Dery, Mark. 1991. "Now Turning the Tables . . . the DJ as Star." *New York Times*, April 14.

Fikentscher, Kai. 1991. "'Supremely Clubbed, Devastatingly Dubbed': Some Observations on the Nature of 12-inch Dance Singles." *Tracking: Popular Music Studies* 4/1: 9–15.

———. 2000. *"You Better Work!" Underground Dance Music in New York City*. Middletown, Conn.: Wesleyan University Press.

Finkelstein, Mark. 1991. Interview with the author, December 16.

Flicks, Larry. 1991. "Remixers Have Found a New Beat: Major-Label Deals Offer Artistic Credibility." *Billboard*. 26 January, 1 and 106.

Garofalo, Reebee. 1996. *Rockin' Out: Popular Music in the U.S.A.* Saddle River, N.J.: Prentice Hall.

George, Nelson. 1988. *The Death of Rhythm and Blues*. New York: Pantheon.

Goldman, Albert. 1978. *Disco*. New York: Hawthorn.

Harper, Laurie. 1989. *Don Sherwood: The Life and Times of "The World's Greatest Disc Jockey."* Rocklin, Calif.: Prima.

Harvey, Steven. 1983. "Behind the Groove: New York City's Disco Underground." *Collusion* 9: 26–33.

———. 1993. "Behind the Groove." *DJ* (March 11–24): 4–9.

Hazzard-Gordon, Katrina. 1990. *Jookin': The Rise of Social Dance Formations in African-American Culture*. Philadelphia: Temple University Press.

Holden, Stephen. 1986. "Disc jockey"/"Disco." In H. Wiley Hitchcock and Stanley Sadie (eds.), *The New Grove Dictionary of American Music*. London: Macmillan Press, 626–627.

Hornbostel, Erich M. von. 1928. *African Negro Music*. Oxford: Oxford University Press.

Hornbostel, Erich M. von, and Otto Abraham. 1909. "Vorschlag zur Transkription exotischer Melodien." *Sammelbände der Internationalen Musikgesellschaft* 11: 1–29.

Jester, Chris. 1990. *Mixxin' Keys and BPM's '88 & '89 Annual Guide to Dance Music*. Northridge, Calif.: Time Warp Publishing.

Joe, Radcliffe A. 1980. *This Business of Disco*. New York: Billboard Books.

Johnson, Laura. 1994. "The Compositional Techniques of the Hip Hop DJ in Relationship to Performance, Recording, and Dance." Unpubl. Master's thesis, Hunter College of the City University of New York.

Kealy, Edward R. 1979. "From Craft to Art: The Case of Sound Mixers and Popular Music." *Sociology of Work and Occupations* 6, No. 1: 3–29.

Klasco, Mike, and Pamela Michael. 1992. "Crushing Grooves: The Art of Deejay Mixing." *Electronic Musician* 8/10: 58–65.

Miezitis, Vita. 1980. *Night Dancin'*. New York: Ballantine.

Nolan, Tom. 1969. "Underground Radio." In Jonathan Eisen (ed.), *The Age of Rock: Sounds of the American Cultural Revolution*. New York: Vintage, 337–351.

Paoletta, Michael. 1991. "Tony Humphries: In the Mix." *Dance Music Report* 15/5 (March 16–29): 12 and 46.

Pareles, Jon. 1990. "If at First You Do Succeed, Remix and Remix Again." *New York Times* (June 10): 26.

Passman, Arnold. 1971. *The Deejays*. New York: Macmillan.

Poschardt, Ulf. 1995. *DJ Culture*. Hamburg: Rogner & Bernhard.

Rietveld, Hillegonda C. 1998. *This Is Our House*. London: Ashgate.

Rose, Tricia. 1994. *Black Noise: Rap Music and Black Culture in Contemporary America*. Hanover and London: Wesleyan University Press / University Press of New England.

Shannon, Doug. 1985. *Off the Record*. 2nd ed. Cleveland: Pacesetter Publishing House.

Simonelli, Victor. 1992. Interview with the author, November 7.

Straw, William. 1990. "Popular Music as Cultural Commodity: The American Recorded Music Industries 1976–1985. Dissertation, McGill University, Montreal.

Sukenick, Ronald. 1987. *Down and In: Life in the Underground*. New York: Collier Books.

Théberge, Paul. 1993. "Random Access: Music, Technology, Postmodernism." In Simon Miller (ed.), *The Last Post*. Manchester and New York: Manchester University Press.

———. 1997. *Any Sound You Can Imagine: Making Music/Consuming Technology.* Hanover and London: Wesleyan University Press / University Press of New England.

Tankel, Jonathan David. 1990. "The Practise of Recording Music: Remixing as Recording." *Journal of Communications* 40/3: 34–46.

Tannenbaum, Rob. 1992. "Remix, Rematch, Reprofit: Then Dance." *New York Times* (August 30): 23.

Tanner, Paul O. W., David Megill, and Maurice Gerow. 1997. *Jazz.* 8th ed. Dubuque, Iowa: Brown & Benchmark.

Thomas, Anthony. 1995. "The House the Kids Built: The Gay Imprint on American Dance Music." In Corey K. Creekmur and Alexander Doty (eds.), *Out in Culture: Gay, Lesbian and Queer Essays on Popular Culture.* Durham: Duke University Press, 437–448.

Toop, David. 1991. *Rap Attack 2: African Rap to Global Hip Hop.* New York: Serpent's Tail.

Toynbee, Jason. 1993. "Policing Bohemia, Pinning Up Grunge: The Music Press and Generic Change in British Pop and Rock." *Popular Music* 12/3: 289–300.

Walters, Barry. 1988. "Last Night a DJ Saved My Life: David Morales Remakes Dance Music." *Village Voice* (June 7): 21–25, 34.

Williams, Gilbert. 1986. "The Black Disc Jockey as a Cultural Hero." *Popular Music and Society* 10/3: 79–90.

Yu, Arlene. 1988. "'I Was Born This Way': Celebrating Community in a Black Gay Disco." B.A. thesis, Radcliffe College.

CHAPTER THIRTEEN

Sounds Like the Mall of America
Programmed Music and the Architectonics of Commercial Space

Jonathan Sterne

Shopping malls have become icons of consumer society. The prophets of advanced capitalism—whether they be post-Marxist academics or developers—have given us the shopping mall as emblem and microcosm of this cultural epoch (see Morse 1990; Shields 1992; Karasov and Martin 1993). Visions of the shopping mall become social visions. Yet the visual bias in cultural critique tends toward the assumption that all that matters presents itself to be seen. What if we were to *listen* to a shopping mall instead? What could be heard?[1]

At the Mall of America (Bloomington, Minnesota), beneath the crash of a roller coaster, the chatter of shoppers, and the shuffle of feet, one hears music everywhere. Every space in the Mall is hardwired for sound. The apparatus to disseminate music is built into the Mall's infrastructure and is managed as one of several major environmental factors. Music flows through channels parallel to those providing air, electricity, and information to all areas of the Mall. "Facilities Management," the department responsible for maintaining the Mall's power supplies, temperature, and even groundskeeping, also keeps the Mall's varied soundtracks running. Throughout the many stores and hallways, one can see the blonde circular speakers that are programmed music standard. The Mall of America has three main sound systems: a set of speakers in the hallways plays background music quietly; a set of speakers hidden beneath the foliage of Camp

Snoopy (the amusement park built into the Mall's atrium) broadcasts the steady singing of digital crickets; and each store is wired for sound so that it may play tapes or receive a satellite transmission. The Mall of America both presumes in its very structure and requires as part of its maintenance a continuous, nuanced, and highly orchestrated flow of music to all its parts. It is as if a sonorial circulation system keeps the Mall alive.

In places like the Mall of America, music becomes a form of architecture. Rather than simply filling up an empty space, the music becomes part of the consistency of that space. The sound becomes a presence, and as that presence it becomes an essential part of the building's infrastructure. Music is a central—an architectural—part of malls and other semi-public commercial spaces throughout the country, yet for all the literature on spaces of "consumer culture," little or no mention is made of the systematic dissemination of prerecorded music that now pervades these places (a notable exception is Frow and Morris 1993). This article can be thought of as an answer to that absence, but it is really part of a larger experiment: what happens when we begin to think about space in industrialized societies *acoustically*? How is sound organized by social and cultural practice? How does it inflect that practice? These are old questions for ethnomusicologists, yet the field has really just begun exploring music and sound in industrialized and recorded forms. Since Charles Keil's call for studying mass mediated music (1984: 91), there has been a growing field of interest in the circulation and culture of recordings (see, for example, Wallis and Malm 1984; Meintjes 1990; Guilbault 1993; Manuel 1993). Much of this work is concerned with the relationships of performers and audiences, with the manner in which music influences or connects constructions of identity, or with music industries themselves. This article takes these problems as a point of departure but explores them at two different layers: (1) where music and listeners' responses to it are themselves commodities to be bought, sold, and circulated; and (2) where this commoditized music becomes a form of architecture—a way of organizing space in commercial settings.[2]

The centrality of music as an environmental factor in commercial spaces should come as no surprise. To the contrary, the idea of music pervading quasi-public commercial spaces is the height of banality. Programmed music, better known by one of its brand names, "Muzak," is one of the most widely disseminated forms of music in the world. Alex Greene notes that "we take Muzak for granted, the word having transcended its status as a product trademark and entered into the realms of everyday language, as a label of all 'easy' listening music" (1986: 286). Americans take for granted that almost every commercial establishment they enter will offer them an endless serenade during their stay. This banality itself is a cause for reflection: in 1982 it was estimated that one out of every three Americans heard programmed music at some point every day; that number has steadily increased since then. Americans on average hear more hours per capita of programmed music than any other kind of music (Jones and Schumacher 1992: 156). As I will shortly discuss, programmed music now encompasses both "easy listening" music *and* original recordings heard elsewhere. In other words, one cannot tell simply by listening to music whether it is "Muzak" or not—*all* recorded music is at least potentially Muzak. (For consistency, I will refer to the service itself—"programmed music"—rather than adopting this common usage of a specific brand name. Currently, the three largest programmed music services are the *MUZAK* Limited Partnership, 3M, and Audio Environments, Inc. Programmed music has been in practical use since the 1930s, and *MUZAK* remains the predominant service in the industry; it is the model on which other services are based.)[3]

The economics and social organization of programmed music presumes and exists on top of a whole culture and economy of recorded music. In other words, programmed music presumes that music has already become a *thing*—a commodity. This reification is represented in the economics of the service, and in the presumptions on which this economy is based. For instance, programmed music requires the absolute separation of performer and audience fostered by many recording industries, thereby circumscribing the experience of music for the majority of the population to that of listening: "Today's Baby Boom generation grew up with music as an integral part of their lives. From the clock radio to the Hi-Fi to the stereo to the CD player, music has always been present. They expect it everywhere they go. In fact, respondents of all ages in survey after survey unanimously agree they prefer to shop, dine and work where music is present. Music moves people" (*MUZAK* 1992b). Here, musical experience is understood entirely as listening, and cultural value is attributed to the very *presence* of music as a kind of sound. These are two key cultural assumptions underlying the production and deployment of programmed music.

I want to be absolutely clear here: while the capitalist and consumerist market structure of mass mediated music contributes to a larger divide between performer and audience, with fewer performers and a larger audience, this is not necessarily a quality inherent in recording and transmission (mediation) of sound itself. In other words, we should be wary of critiques of mediation *qua* alienation. Also, we should be careful to recognize that the treatment of music as purely a kind of sound (as opposed to a whole ensemble of practices such as dancing, playing, and so on) is a specific cultural construct and not universally valid. However, this construct of music as sound is very much alive, and exerts real effects, as the case of programmed music demonstrates.

If—under certain conditions—music exerts effects primarily or solely as sound, then we have to begin asking questions about the very act of listening under those conditions. In a media-saturated environment, listening designates a whole range of heterogeneous activities involving the perception of sound. Everything from aesthetic contemplation in a concert hall to the mere act of turning on a radio or a sound recording in one's everyday environment can be understood as "listening." Here, I will use the term "listener" to denote a person perceiving sound in either the active or the passive sense, or both. This ambiguity is important in thinking about programmed music, since such music certainly isn't meant for contemplative listening; it also isn't always "heard" in an entirely passive fashion—rather, it tends to pass in and out of the foreground of a listener's consciousness. Thus the necessity for understanding "listener" as an ambiguous term that shuttles between activity and passivity. In part, this ambiguous status of listening—especially as it pertains to programmed music—is an effect of the social organization of music in a capitalist mass media environment. Peter Manuel (1993) and others have suggested that recorded music be considered from a "holistic" vantage point that examines its production, circulation, and consumption. The context of programmed music adds a whole second layer of circulation to this economy: *re*production, *re*distribution, and secondary consumption. The "producers" of programmed music are the programmed music services themselves, who assemble already existing songs into soundtracks. The consumers of programmed music are stores and other businesses that purchase the services.[4] Clients generally subscribe to a programmed music service and pay a small monthly fee. The service will provide the subscriber with a tuner and a choice of approximately twelve satellite channels to choose from, or a special tape player and a catalogue of hundreds of different four-hour programs. Clients with tape subscriptions generally receive new tapes every month, or every few months, depending on the type of music. Thus, as Manuel points out, in a thoroughgoing analysis of mass mediated music the analytical tools of ethnomusicology need to be supplemented with those of communications (1993: 7; see also Wallis and Malm 1984). A detour through the political economy of programmed music will clarify my own analytical orientation.

Essentially, the use of programmed music in a shopping mall is about the production and consumption of consumption. Programmed music in a mall produces consumption because the music works as an architectural element of a built space devoted to consumerism. A store deploys programmed music as part of a fabricated environment aimed at getting visitors to stay longer and buy more. Other commercial establishments may use programmed music to other ends, but in all cases its use is primarily concerned with the construction of built and lived commercial environments. Having deployed the music, subscribers such as a store or a mall consume consumption insofar as they are interested in listener response to the music itself. They are purchasing the music so as to consume listeners' responses to it—for instance, if listener responses to music lead to increased average shopping time, increased sales, and increased number of customers (see *MUZAK* 1990, 1992b). In other words, while the people who go to a mall to shop may hear programmed music, the consumers of that music (and listener responses to it) are actually the stores and the mall itself. A thoroughgoing analysis of these relations requires an adjustment in critical orientation. Rather than focusing purely on listener response—that of people we normally think of as "consumers" in a mall setting—I am primarily concerned here with the production, distribution, and consumption of that listener response (or what I called above the second "layer" of circulation). I have focused on the frames of possible experience and the ways in which those frames are constituted, rather than cataloging all possible listener experiences in the Mall of America. I am less interested in an exhaustive survey of possible meanings listeners (or "hearers") may attribute to programmed music than in the uses to which those attributions may be put.

Two other obvious problems obstruct a proper ethnography of listening. As I discuss below, actual hearing and listening practices are not necessarily at the forefront of participants' consciousness—sounds can be quite ephemeral, and therefore my calling attention to the music would not necessarily elicit responses from people that reflected what would happen in my absence. Furthermore, because the music in the Mall comes from a larger field of circulation, it would be an error to isolate music heard in the Mall from other contexts in which the same music is heard. In other words, if I were to do a proper ethnography of listening, given that my subjects would be visitors and not *dwellers* in the Mall, the Mall ceases to be useful as an exclusive site of inquiry. Finally, ascertaining exactly what music means to listeners in the Mall still begs the question of how that experience is put to use by the Mall itself.

A Suburban Ethnomusicology?

Given these concerns, in this article I examine the deployment of programmed music in the Mall of America. While this mall may be more spectacular than other malls, its spectacle is a self-conscious one: a tourist visit to this mall above all others is about the spectacle of consumption itself. The Mall's utter extremity on one hand and everydayness on the other offer a unique perspective on a place where consumerism is conflated with nationalism, and where a private commercial space can be hailed by developers as an "alternative urbanity." The Mall has been promoted by airlines and travel agencies as a tourist destination to rival downtowns; some local architects have echoed this assessment, suggesting that the Mall is downtown for an outer suburb and that *this* downtown offers what Deborah Karasov and Judith A. Martin call a "facsimile of urban delights with almost no urban responsibilities" (1993: 27). But Karasov and Martin are quick to qualify this assessment: "What this commercial imperative suggests is how poorly we understand our cities, present and future, if we view them as little more than accumulated land uses. There is little expectation that shopping malls will contribute to urban design and social goals, no matter how big these malls become or how many people they attract. In the end, the Mall of America is no more a city than Sea World is an ocean" (ibid.).

In this facsimile of urbanity, speakers in the ceilings and walls cascade travelers with an endless flow of music. Perhaps, then, as a footnote to Bruno Nettl's call for an "urban ethnomusicology" (1978: 13), this essay could be understood as a suburban ethnomusicology. Although they are a distinguishing aspect of the space, a mall's acoustical features cannot be understood apart from its other general thematic and structural features. Acoustical space is an integrated and substantial element of cultural practice, not an autonomous sphere.

The Mall of America is the largest mall in the United States; it is second only to Canada's West Edmonton Mall. (For a discussion of the West Edmonton Mall, currently the largest in the world, see Crawford 1992.) Also known as the "megamall," it has become a major tourist and leisure site in the area; it has attracted over 10,000 bus tours (each averaging fifty people) since its opening in August 1992, and the average visit to the mall for all customers is three hours, close to triple the industry average. In addition, adult shoppers spend approximately $84 per visit, which is almost double the industry average (based on Nordberg 1993: 1). You can find billboards advertising the megamall—"the place for fun in your life"—at least as far away as the middle of South Dakota.

Aside from its size, the Mall's most unique feature is its national theme (Petchler 1993). It was built with a self-consciousness about its cultural purpose: while striving valiantly to be "all things to all people," the Mall cultivates itself to simulate a whole range of generic "American" experiences that will appear nonthreatening to its desired middle-class clientele (Karasov and

Martin 1993: 19, 25). While most malls serve as regional centers, the megamall attempts to present itself as a center of national culture. Its tenant stores are well-known national chains, and the four department stores—Sears, Macy's, Bloomingdale's, and Nordstrom's—combined, represent the paradigm of national department stores (it is worth noting in this respect that neither Boomingdale's nor Nordstrom's had any locations in Minnesota prior to the construction of the Mall). Even the Mall's home state is demoted to one region among many, as evidenced by stores with "Minnesota" themes. Considering that the West Edmonton Mall claims to be "the world" in a shopping mall, that the second largest mall in the world should devote itself to an entirely "American" theme illustrates the self-importance of American nationalist ideology.

The Mall's national identity requires a very narrow conception of the nation, centered on the mainstream of retail marketing. There is a great deal of product duplication, and with a few exceptions (like souvenirs), one could find almost all of the products available in the megamall at many smaller malls. Similarly, although there are a great number of specialty stores, all specialties are geared toward an assumed mainstream population. Ideologically, the Mall of America adds an explicit national theme to the usual consumerist and white middle-class worldviews represented in mall design.

While the Mall of America derives special significance from its size and theme, it also represents the cumulative wisdom of almost forty years of mall design and management. Architecturally, the Mall embellishes on industry standards, but not much. Any difference in scope between the Mall of America and other malls is a result of scale. In contrast to the traditional suburban shopping mall surrounded by smaller strip malls, movie theaters, bars, and fast food joints, the Mall of America has simply enclosed all the surrounding activities under one giant roof. As Karasov and Martin put it, the spaces separating the shopping mall from its surroundings have been transformed from "highway to hallway" in the Mall of America (ibid.: 23). The Mall thus foregrounds the connections between consumption and leisure so prevalent in American culture, while keeping each activity in its place: to wit, the Mall has an unusually large entertainment complex—the amusement park is joined by a "Lego Imagination Center" and an indoor mini-golf course. As Rob Shields has remarked, there is a critical interdependence among private subjectivity, media and commodity consumption, and privately owned semi-public spaces like shopping malls (1993: 1). In the Mall of America, they can feed off one another.

Great care was taken to produce an "urban shopping district" sensibility for the interior. Each major corridor of the Mall is called an "avenue" and is painted, carpeted, lit, and named differently from the others. The "entertainment districts" are isolated from the shopping areas—Camp Snoopy is located in the Mall's gigantic atrium, while the movie theaters, bars, and an arcade are located on a separate floor, away from retailing. The Mall compartmentalizes specialized functions like eating and entertainment and retains a general cohesion of design throughout its interior.

In theory and execution, the Mall's soundscape is entirely consonant with other design goals, in part because programmed music is a phenomenon divided according to the same logic as the other commercial enterprises in the Mall: according to a reduction of identity to consumer taste and a universe of taste that rotates on the axis of a consumer class. The music also works because programmed music has become a design feature integral to *any* mall, and therefore doesn't seem the least bit out of place. The acoustical design of the Mall is a result of similar philosophies to those underlying other design features, but it is structured to somewhat different ends.

Sounding Out the Mall

A social space is as much defined by its constant influx and expenditure of energies, by the movements which maintain it, as it is by any stable or structural construct (Lefebvre 1992: 92). Music can therefore be considered as one of those energy flows (such as electricity or air) which continually produce the Mall of America as a social space. In this way, programmed music is both an environmental and an architectural element of the Mall. The acoustical space of the Mall is structured around a central musical tension: the quiet, nondescript music in the hallways contrasting with louder, more easily recognizable and more boisterous music in the stores.

Background Music

The 3M Corporation provides quiet background music for the Mall's common spaces. According to their programming director, Tom Pelisero, 3M "did nothing unusual for the Mall of America" (1993). Pelisero himself suggested that the megamall was a hostile environment for background music, because the common hallways are all filled with the din of Camp Snoopy. So what would the Mall want with a standardized form of background music that is barely audible?

The music for the hallways is known within the industry as "environmental" or "background music." This is the kind of music usually brought to mind by references to "Muzak": symphonic arrangements of well-known tunes, both contemporary and traditional, that make prolific use of stringed instruments but stay away from brass, voice, and percussion. In the last ten years, programmed music providers have begun updating their collections. Instead of hearing "Lucy in the Sky with Diamonds" performed by the Czechoslovakian State Orchestra, one is now likely to hear it adapted for a four- or five-piece jazz group (McDermott 1990: 72). These relatively generic ensembles are chosen in service of background music's ultimate design goal: anonymity. The quest for anonymous or "unobtrusive yet familiar" music animates the entire production process.

All environmental music has certain essential characteristics. All vocals and those instruments considered by programmers to be abrasive are eliminated; both would call attention to themselves and thereby disturb the backgroundness of the music. As several authors have noted, mass-mediated music tends to focus more on the performer than on the song (see Chopyak 1987: 441). In the case of programmed music, this tendency must be countered by stripping the music of any distinctive elements. Background music strives toward anonymity and can thus be understood as the inverse of most industrially recorded and disseminated music. Arrangements of popular and traditional songs are thus performed in "a style devoid of surprise" (Radano 1989: 450) in an effort to render the music familiar and unthreatening—and nondescript.

Background music programming operates according to a technique called "stimulus progression," where each musical selection is rated on a scale from one to six and arranged with other songs in an ascending or descending order to evoke certain emotional responses in listeners. Although *MUZAK* does not share their criteria for stimulus ratings, it is clear that the differences between low and high stimulus ratings are based primarily on rhythm, tempo, and melody. The more upbeat a song, the higher its stimulus rating. (Of course, all background music already operates within a limited range in these respects, given the constraints on musical content and style mentioned above.) Stimulus progression was invented to combat worker fatigue in weapons plants during World War II, functioning on a principle of maintaining a stable stimulus state in listeners at all times. Programming is designed to slow people down after exciting parts of the day and speed them up during sluggish parts of the day. It is an aesthetics of the moderate: not too exciting, not too sedate. While environmental music is no longer used exclusively in factories and production centers, it is still programmed along this line of thinking. In a shopping center setting, stimulus progression could be justified—to pick up visitor movement during the middle of the morning and afternoon, and to slow people down after lunch and at the end of the day.

Thus, background music in the hallways has many possible uses, despite its precarious audibility. To paraphrase a corporate slogan, the music in question is not meant to be listened to, but to be heard. A great deal of market research shows that the presence of quiet, leisurely music increases the duration of shoppers' visits (see Bruner 1990). But even if the din of Camp Snoopy counteracts the kind of psychological effects Gordon Bruner seeks, the simple presence of the music itself—when it can be heard—does carry some significance. For instance, it constructs a continuity among the hallways, bathrooms, and entrances. These spaces are somewhat distinguished by architectural motif, but the background music reinforces their common characteristics through its own non-distinctive and generalized character (Greene 1986: 288). Background music is not devoid of meaning, but its meaning is entirely located in its presence, rather than in the songs in the soundtrack. Even controversial songs that contain a catchy tune may still wind up in an environmental program: for example, one may hear quiet jazz arrangements like Madonna's "Like a Virgin" or Nirvana's "All Apologies" with a piano or saxophone playing the vocal melody. This practice is quite common, although it invites a kind of recognition on the part of listeners that may actually be disruptive—the song, if recognized, could still call attention to itself, even in an anesthetized version, thus contradicting the "backgroundness" of the environmental program.

This soundtracking further serves to structure the hallway as a transitional space, a space for movement. Besides the nondescript (or vaguely familiar) environmental music and the echoing amusement park, the hallway has no markers of its own identity other than a vague architectural theme. The mall management does not intend the hallways as destinations for Mall visitors. (Despite the management's intentions, these transitional spaces often serve as a place for youth to congregate and socialize—to hang out. As this has become a point of contention, I will consider the issue in further detail below.) Through the contrast of clearly identified architectural, visual, and music markers, stores construct themselves as the identifiable localities within the Mall.

The tensions within the acoustical space both affect and reflect the contradictory flows of movement throughout the Mall—into and out of stores, through hallways, among levels, and into and out of the parking lots. Musical programs constantly produce the space; their continuous presence is an insistence or reminder to listeners. Programmed music can be said to territorialize the Mall: it builds and encloses the acoustical space and manages the transitions from one location to another; it not only divides space but also coordinates the relations among subdivisions. As they divide and demarcate, sonorities create "a wall of sound, or at least a wall with some sonic bricks in it" (Deleuze and Guattari 1987: 311). To get anywhere in the Mall of America, one must pass through music and through changes in musical sound. As it territorializes, music gives the subdivided acoustical space a contour, offering an opportunity for its listeners to experience space in a particular way: "music . . . calls forth our investments and hence, our affective anchors into reality" (Grossberg 1991: 364).[5] It also constructs the limits of that experience.

Foreground Music

In contrast to the halls' quiet, sometimes inaudible soundtrack, stores may have varying volumes of foreground music. Foreground music is the industry name for music programming that consists of songs in their original form, as recorded by the original artist. The music itself is still meant to serve as a background wherever it plays, but it is "foreground" in that it can draw attention to itself in ways that background music cannot. So while on the environmental channel we might hear a jazz group quietly working out "Faithfully" by the band Journey, on a foreground channel we'd hear Journey playing the song themselves, complete with wailing vocals and soaring guitar solos. This louder and more audacious foreground music emanating from stores works in tension with the background music in the hallways. If the store is open to the mall (rather than being closed off by a front wall with a door), music distinguishes the store's interior from the exterior hallway. If the volume of the store's music is moderate, the placement of the speakers within the store will determine a sonic threshold: on one side the ambiance of the hallway is primary in a listener's auditory field, and on the other side the sounds of the store will be primary in a listener's ear. This sonic threshold, often a discernible physical point, behaves as a store's front wall. Through clear acoustical delineation, the music produces a sense of inside and outside.

If, on the other hand, the volume of the store's music is high enough, the music will spill out into the hallway. In this way programmed music produces a transitional space from outside in the hallway to inside the store, much as stairs up from the street or a canopy and carpet on the sidewalk would do. Its louder relative volume also directly hails people in the hallway in an attempt to get their attention — it more or less invites them inside the store. Thus, from the hallways, stores can become identifiable by how they sound; this sonic quality is the central preoccupation of foreground music programming.

Foreground music sounds like radio: it "broadcasts" already existing recordings but is itself carefully programmed according to a logic called "quantum modulation." If background music strives toward anonymity and gradual changes in mood, foreground music strives for an absolutely consistent identity and unchanging mood. Quantum modulation produces continuity and maintains flow in the overall soundtrack through assigning each song a composite numerical value based upon a variety of criteria: rhythm, tempo, title, artist, era, genre, instrumentation, and popularity. A flow of music is established through song compatibility and cross-fading so that all transitions from song to song are seamless. This stress on maintaining a flow that does not vary in "intensity" is again based upon the posited listener of the music: it is assumed that the person will hear the music for a shorter duration of time (for instance, while browsing in a store). Therefore, rather than try to gradually alter the listener's mood over time, the music remains at a consistent value (*MUZAK* 1992b; Ritter 1993).

Unlike in environmental music, where services will provide only one program choice (usually just called "the environmental channel" or some such), there are a multitude of foreground programs. Foreground music operates at the levels of taste and distinction, differentiation and association. The standard satellite programs are based on the categories of *Billboard* charts, adjusted for certain demographics like age and gender. Rather than organizing music according to style categories, foreground music organizes it according to marketing categories like "top 40" or "adult contemporary"—which is similar, but not identical, to the way commercial radio stations organize their playlists. Foreground music programs available on tape are more specialized, according to genres such as "classic jazz" or even "Hawaiian," or they are programmed for other specialized uses, such as "holiday music" (*MUZAK* 1992a). Two illustrations will clarify the continuities and discontinuities of foreground music programming.

In programmed music, the recording medium is inseparable from its message: music acquires its value as much from the manner in which it is recorded as the supposed "content" of the recording. The following is part of a "Mixed Tempo Classic Pop" tape in a *MUZAK* catalog. The tape appears under the catalog's "Nostalgia" section, with this description under the heading "Classic Pop": "This series features Pop's all-time greats from the top of the charts of the 1960s, '70s and '80s. This series differs from 'Classic Rock' by including only the smoother side of Pop music. These programs provide a timeless feel for an audience looking for the best popular music" (*MUZAK* 1992a: 29).

Tape 4016: Mixed Tempo/Classic Pop (track 4 of 4) (MUZAK *1992a*):
Creedence Clearwater Revival, "Have You Ever Seen the Rain?" (1972).
Jim Croce, "I Got A Name" (1973).
O'Jays, "Back Stabbers" (1972).
Hall & Oats, "She's Gone" (1973).
The Steve Miller Band, "Fly Like an Eagle" (1978).
Queen, "Bohemian Rhapsody" (1975).

Dave Loggins, "Please Come To Boston" (1974).
Olivia Newton-John, "Please, Mr. Please" (1977).
George Harrison, "My Sweet Lord" (1976).
Bob Seger, "Night Moves" (1976).
Leo Sayer, "You Make Me Feel Like Dancing" (1976).

This sequence of songs is notable because of the diversity of artists represented. The songs do, however, have several characteristics that bind them together. They all come from a six-year span: 1972–78. Although the songs were produced in differing styles, they all operate within a narrow range of production values and techniques endemic to 1970s pop, and a limited range of timbres. The ordering of the songs represents gradual changes in tempo from song to song, but no radical shifts in speed from one song to the next. To the devoted fans of any particular artist or group on the tape, the program may appear to juxtapose different music inappropriately; to a more casual listener, the differences will likely be overshadowed by the consistency among songs in overall production values (dynamic range, mix, timbre, use of reverb, and so on). The next track list, from a tape of "Mixed Light Symphony Favorites," is again organized by sound, tempo, and timbre. The selection appears in the "Classical" section of the catalog; "Light Symphony" is described as follows: "This series contains popular movements from the great symphonies. Favorite melodies recorded by some of the world's best orchestras provide a full sound without being imposing. Suitable for a sophisticated environment" (*MUZAK* 1992: 59).

Tape 6023: Mixed/Light Symphony Favorites (track 1 of 4) (MUZAK 1992a):
Johann Strauss, "Tales from the Vienna Woods, Op. 325" performed by The Vienna Johann Strauss Orchestra (1982).
Giuseppi Verdi, "La Donna E Mobile" performed by Andre Kostelanetz and His Orchestra (1989).
Boccerini, "Minuet from String Quartet, Op. 13, No. 5" performed by Academy of St. Martin-In-The-Fields (1980).
Edvard Grieg, "Morning from Peer Gynt Suite" performed by the Orchestre Philharmonique De Monte Carlo (1989).
Vivaldi, "Spring-Allegro 1" performed by the Vienna Philharmonic (1984).
Beethoven, "Symphony No. 7–3, Presto" performed by Staatskappelle Dresden (1987).
Mozart, "Overture from The Marriage of Figaro" performed by the City of London Sinfonia (1986).

The program limits itself to orchestral recordings since 1980, so that the timbres, mixes, dynamic ranges, and overall production values are consistent on the tape, as in the first example. One could chart similar tendencies throughout the various music programs. Although programming may be totally inconsistent along lines of genre, it is carefully regulated in terms of other musical characteristics that might be lumped together under the headings of "affect" or "mood." However, musical periods and histories can re-enter this framework: in the "Classic Pop" example, the selection's period determines the parameter of a production aesthetic; and as I discuss below, periods such as "the 1950s" can themselves become tropes in the rhetoric of programmed music. Programmed music is organized according to an aesthetics of production, where the recording itself is analyzed and programmed as much as the content or style of the musical selection the recording ostensibly represents. While many theories of recording treat the medium as an instrument to reproduce existing music, programmed music is more concerned with the substance and texture of the medium itself.

Foreground music utilizes these programming techniques to create consistent musical programs with which stores can then associate themselves. Retailers are encouraged to choose music styles to cultivate a business image considered most appealing to whatever demographic group of customers they hope to attract (*MUZAK* 1990). Although *MUZAK* and its main competitors produce mostly generalized programming, the choices of programs, the placement of speakers, the volume and the texture of the music are all determined at each individual site.

That this music resembles radio—but isn't radio—is further advantageous to stores. Playing radio stations in commercial establishments without paying royalties to musicians' unions is illegal. Thus many stores find it cheaper to subscribe to a programmed music service than to play the radio, because all such services pay their royalties in a lump sum. But a resemblance to radio is advantageous to businesses because radio stations are often instrumental in constructing local communities (Berland 1990). Not only is FM radio a part of regional community construction, but it also hails its listeners according to age, race, and class. In short, this resemblance mimics the use of FM radio in other spaces and thereby produces at least the possibility that a store will be able to associate itself with the other spaces in which the music resounds. This enables retailers to construct a "business image" through the music and possibly connect with other places listeners have heard the same music. Although the musical program does not necessarily conform to any rules of genre, it can appeal very strongly to consumer identity. Pierre Bourdieu has demonstrated the continuities between taste and social position (1984).[6] By knowing the tastes of a desired clientele, a store can position itself as within—or as a "logical" extension of—an already existing taste culture.

Generally speaking, stores within a particular chain all use the same or similar music programs to achieve a uniformity of corporate image, just as they use similar design and lighting techniques. However, this corporate image is itself an amalgamation of what the company wants potential consumers to think about it, and what the company wants potential consumers to think about themselves and the products they are browsing. In other words, programmed music can become a key that frames the experience of shopping in a store. Not only can it suggest a particular affective stance for listeners toward the store and their experience of it, foreground music can frame the context to suggest a whole range of possible responses to the commodities and experiences within, and a whole disposition toward those possible responses. This is no different than the use of music to frame other kinds of activities (see Goffman 1974 on framing; see Booth 1990 and Turino 1993: 218–232 on musical framing). Or as Regula Burkhardt Qureshi puts it, "extra-musical meanings in musical sound give music the power to affect its context in turn" (1987: 58). I will illustrate the variations on this process with four examples from the Mall of America.

(1) Victoria's Secret, a store specializing in lingerie, not only plays classical music in their stores but sells tapes of their music programs. "Romantic" selections like Mozart's Piano Concerto in E Flat, the allegro from Schubert's Symphony No. 5 in B Flat, or Beethoven's Romance No. 1 in G cascade over supple decor. While the store's merchandise and visual displays differentiate it in terms of gender, the music program, along with the decor, offers listeners an index of class and a coherent frame within which to experience the store and themselves.[7] The music plays to an American bourgeois identity by suggesting a refined, European, aristocratic taste. As a form of music that is generally associated with refined taste and prestige, it functions to legitimate the store as a respectable place to shop.[8] More generally, it helps to produce the atmosphere within the store. The store itself is decorated in plush style, and the lighting is particularly soft. While the music plays, sale displays encourage the visitor to take advantage of "a special opportunity to indulge yourself" in a lingerie purchase. The music suggests a continuity among the wide array of commodities available in the store (clothing, accessories, and perfume, as well as more general merchandise), some of which are only related to one another in that they are being sold in the same store.

The store is full of references to England and Europe (where displayed, the "Victoria's Secret" name is always pictured with a London street address). The programmed music, European classical music, supplements the rhetoric of the store's appearance. For many Americans, "Europeanness" can itself be an index of high-class status and refinement. To the other design features, however, classical music offers the possible pleasures of recognition: recognizing the music can be as important as enjoying the music, both for the cues it gives toward experiencing the store itself and for suggesting that proper customers of Victoria's Secret are people refined enough to recognize the music. In other words, knowledge of the music is a form of cultural capital: it suggests membership in a certain social stratum.

But Victoria's Secret goes one step further by selling the music it plays. Victoria's Secret can sell tapes of the music in their stores because their music is programmed by an independent contractor and was arranged especially for them, as opposed to their having simply purchased a "light classical" program from one of the major programmed music services. This has been a hugely successful venture for the store: according to a June 1995 *Forbes* article, they had sold over 10 million tapes and CDs since 1989 (Machan 1995: 133). The tapes offer their purchasers a chance to enjoy the music at home, but also to learn the music and thereby be able to recognize it upon return visits to the store (see Bourdieu 1984: 13–14, 272–273). In other words, should one find oneself outside the refined taste culture upon entering Victoria's Secret for the first time, one can undertake an education to culture the taste and refine the senses. This has a metonymic effect: insofar as the music works with other aspects of the decor, it suggests that one needs a cultivated, refined sensibility to enjoy all that Victoria's Secret has to offer. The liner notes to their tapes combine these purposes, offering rudimentary knowledge about composers' lives and trivia concerning the music, while extolling the sensibilities of composer and listener, patron and performer:

> This spectacular recording includes some of the world's best loved and most romantic music composed by Wolfgang Amadeus Mozart. You will thrill to the lyrical magic of this romantic collection recorded exclusively for Victoria's Secret by the London Symphony Orchestra. Victoria's Secret is proud to be among the distinguished individuals and corporations, who through their endowments, have been designated Diamond Patrons of The London Symphony. In his short life, Mozart lived amidst kings and courts, brilliance and despair, dying almost penniless at the age of 34. In this collection, you will celebrate Mozart at his most majestic and compelling. Inclusion of several intricate and intimate compositions, lesser known than most, gives this volume its singular quality. (Victoria's Secret 1993)

The irony, of course, is that the music in question is the most common and most easily recognized variety of European classical music. Thus the liner notes take time out to assure readers that *this* tape of Mozart is special. Victoria's Secret offers an experience of itself and the tools to sophisticate and heighten that experience. It offers potential consumers an image of themselves, as if that image could be actualized through the consumption of the experiences in the store and the commodities it offers.

(2) Compagníe Internationale Express, a more conventional fashion store, plays only French pop music, and it plays this music at a relatively high volume. Near the cash register, the store proclaims itself to be "a

world of French style." But the design within the store is quite common to fashion retailers in the Mall. The entire store is brightly lit, and track lighting allows individual displays to be highlighted. Most of the color comes from either the clothing or the displays in the store. Aisles are wide, but not too wide, so that one moves through the selection slowly. Most of the clothing for sale comes from places other than France. The window display features (at this writing) "grunge" fashions (faded flannel shirts and faded jeans) that are decidedly American. So "Frenchness" refers more to an affect the store would like to convey than to any trait that the store or its patrons actually possess. "Here again, the sign [in this case French pop programmed as foreground music] is ambiguous: it remains on the surface, yet does not for all that give up the attempt to pass itself off as depth" (Barthes 1972: 28). Or to make our French theorist speak to the matter at hand directly: the sound is like another layer of packaging laid over commodities. This packaging contains the real instructions for use—how to *feel* when using the products in the store. The French rock music envelops the commodities in an effort to stand in as their essence. As at Victoria's Secret, programmed music works with other environmental factors to confer certain meanings onto the merchandise and the people in the store—these meanings having no intrinsic connection to the people or products. However, there are crucial differences here: the volume of Express's music makes it more insistent—making conversation more difficult, and making the music much more present. But then, it has to be: while the programmed music in Victoria's Secret works in concert with other environmental factors to simulate an aura of sophistication and indulgence, the music here works all by itself. The music is authentically French, or at least *in* French; little else in the store is. The music suggests a particular experience of shopping or a way to experience a wide range of products. In suggesting these experiences, the music also offers a mode of experiencing the self. The actual nature of that experience depends on what one takes "Frenchness" to mean. Clearly, Express intends its "Frenchness" to connote sophistication of taste, affluence, luxury, and a touch of exoticism—after all, Paris is the capital of international fashion. But this is a particularly American representation of what it means to be French. Compagníe Internationale Express exemplifies this: they are headquartered in Columbus, Ohio.

(3) In addition to framing space and offering affective cues, foreground music can be used to help construct time and movement by its presence rather than its content. A Levi's store uses programmed music and video to these ends. The back of the store contains a giant nine-screen "video wall," and the store actually plays music videos. The store acquires these on laser

disc from a parent company, and true to form, the discs only appear coherent in terms of programming concerns (for instance, offering a program that is a combination of popular rap, alternative, and "top 40" musics). The store is itself otherwise unremarkable. The floors are a wood grain, and the store uses the usual bright lighting and track-lit displays. They sell a complete line of Levi's clothing. An eclectic range of musical genres would disallow any simple identification of music with a particular mood as in the above examples. Neither does the store sell itself as an experience; it relies primarily on its products. By essentially saying nothing, the store says everything it has to: as if the choice of basic decor was the result of the products speaking for themselves, and not a deliberate marketing strategy. Relieved of its ostensible content function, the programmed music and video wall can provide a generalized hip atmosphere—one that echoes or displays current fashion trends—while helping to structure the movement of people inside and outside the store through a process of distraction. Since the video wall is visible from the hallway, it is to the store's advantage that passers-by stop and watch the video. In so doing, they will also look at the available merchandise and sale displays. Moreover, for a visitor to stop shopping momentarily and watch the video is not a liability to the store: it is conventional wisdom among retailers that the longer people spend in a store, the more likely they are to make a purchase. In this way, Levi's is able to use the video screen's power to distract to its own advantage: the distraction may or may not gel with the rest of the shopping experience, but it affects that experience by prolonging it. Unlike music programmed according to stimulus progression, this music and video usage does not attempt to speed up or slow down customer movement by tempo, melody, or rhythm; rather, it functions as an interruption that becomes integrated into the shopping experience. By hailing people into the store, and by distracting them in the store (thereby increasing the duration of their stay), the programmed music and video juxtapose consumption and entertainment without having to fuse them in any meaningful way Or, to play on writers who treat the mall as a postmodern phenomenon: the music video serves to decenter shoppers (its "subjects"), but this decentering is not necessarily endowed with a resistant or subversive political potential. Rather, the discontinuity works in service of the one unifying "signifier of value" for the store: the point of sale. Any other effects of this decentering (as there may be) are incidental.

(4) Johnny Rockets, "The Original Hamburger," is located on the edge of the Mall of America's south food court. There it is sandwiched between an Asian fast-food restaurant and a larger sit-down bistro. It is far less enclosed and far less set off from neighboring businesses than are the clothing

stores discussed above. Through its volume, programmed music frames the experience of Johnny Rockets by physically differentiating the space from those around it and enveloping other noise within its frame. It spills out of the restaurant and touches every listener in the food court. The restaurant is a representation of a 1950s-style art-deco diner and includes a long counter and a row of seats in front of it. It is the only restaurant in the food court to provide its own seating, in addition to the tables and chairs in a large common area. The employees are all wearing 1950s-style soda jerk outfits (white shirts and hats) as the programmed music, 1950s pop (mostly love songs), comes blaring out of the restaurant. It is by far the loudest music in the food court. The menu is conventional—hamburgers, fries, sodas, and so on; one can get this food from a variety of sources at the Mall. Thus, rather than differentiating itself by its product, Johnny Rockets is about consuming an experience. On the west wall, a poster pictures a teenage man seated on a couch with two women seated very close to him and clearly enraptured with him. They are dressed in fifties clothes, and the slogan on the poster reads: "Johnny Rockets—Hospitality." The restaurant invites its customers to experience a nostalgic 1990s representation of a 1950s diner, complete with regressive gender roles and sexual mores. Here programmed music performs a double function: it is at once part of the experience to be consumed (as atmosphere) and simultaneously calls attention to the possibility of consuming that experience. Here the experience becomes crucial, because the product for sale is unremarkable in any other way. Yet it doesn't seem to offer the same cues for self-understanding as the clothing stores. It seems less about constructing an identity (being "1950s-ish") than suggesting that *anyone* can now enjoy the experience of a 1950s diner. Certainly, anyone in the food court can hear the music from the diner. The volume of the music here may be a more blunt and demanding message: it demands a reduction of all context to the nostalgia booming through the diner. Perhaps the music's blistering volume exists in quiet deference to the knowledge that the party never happened this way the first time.

In each of the above examples, programmed music plays a role particular to its context, but it also serves a more generalized function that could best be termed "articulation." Stuart Hall defines articulation as "the form of the connection that *can* make a unity of two different elements, under certain conditions. It is a linkage which is not necessary, determined, absolute and essential for all time" (1986: 53). Articulation is the process through which otherwise independent meanings, ideologies, or people are unified in some sense. In all four of my above descriptions, a musical program articulates specific meanings to the purchase of a commodity or service. Thus, underwear at Victoria's Secret becomes refined; "Frenchness" is conferred on flannel at Express; music and video distract visitors at the Levi's store, making them better and more valuable shoppers; and eating a hamburger becomes part of a "1950s experience" at Johnny Rockets. None of these connections are natural—rather, they have to be produced or performed.

Although the above are all examples of planned and deliberate uses of musical programs, intention on the part of the store is not necessary for this process to take place. Upon informally questioning the employees of an art gallery and a shoe store in the Mall, I found that in both cases, the music playing had been brought in from home by the stores' employees. Rather than specifying a particular kind of music to be played, the management of each outfit had simply defined a range of possible music and allowed employees to select the music they wanted to hear. On that particular day, Bob Seger resounded in the shoe store, and the Beach Boys could be heard in the art gallery.[9] Both of these musics could easily perform the same kind of articulatory functions I outlined in my previous examples, but any coherence with other environmental factors (such as decor) would be more coincidental. However this arrangement may prove to have other benefits, since it is the employees, not the customers, who must spend hours in a store; and the employees I spoke with tended to find preselected music programs repetitive. This could benefit the store as well: in giving up control over the sound space to the employees, it allows workers to take control of one aspect of their environment. In so doing, it offers underpaid service workers the comforting illusion of some ownership in the retail process. Even if the workers entertain no such illusions, it still offers a degree of comfort. Who wouldn't want to make a bad job a little bit better? Regardless of intention, the music can still function at the level of articulation, and it still builds acoustical space.

Building a Better Consumerism

Clearly, music's various functions within stores are always socially determined and constantly changing. My intent here has been to provide an illustration of how these processes work, rather than an exhaustive survey of programmed music's uses (or listeners' responses)—possible or realized—in the Mall. The Mall's handling of spatial difference is intimately tied to its handling of social difference. In considering the Mall of America's construction of space, elements are excluded and are juxtaposed together and enclosed. The structure and movement of the Mall's acoustical space—and more generally its social space—congeals around a clear set of priorities:

"interior space protects the germinal forces of a task to fulfill or a deed to do" (Deleuze and Guattari 1987: 311). Consumer culture is given a priority in the Mall by decree, not by consensus. Yet the Mall is far from a closed or seamless system. An awareness of cultural struggle and cultural difference plays a part in its very constitution. In particular, three issues concerning difference warrant some further exploration.

(1) If the Mall's play on a national consumerist identity requires a narrow mainstream to be represented, the programmed music refines that mainstream ideology in both its spatial and its interpretive functions. The bell curve and "the charts" determine the presence of any song or kind of music on a musical program. The ubiquitous *Billboard* charts depend on a laser scanning system that records sales in 11,000 music stores nationwide. *Billboard,* by their own admission, has to rely on larger retail chains, because smaller independent stores cannot afford the necessary survey technology (Ellis 1993). So the charts are determined by chain sales, but sales figures say little about taste. They don't reflect, for instance, the practice of sharing or "passing along" a recording common in many communities (see Rose 1994: 7–8). Moreover, buying power is not equal among all sectors of the population. Thus, a chain of effects occurs: social stratification is reproduced in the statistical—that is, commercial—distribution of taste; the statistical distribution of taste is foundational in the construction of programmed music; programmed music then becomes part of the Mall's architectonics; and the Mall thereby mobilizes social difference and makes it useful, all the while denying that it's anything but good clean American fun.

(2) The forms taken by Mall design and programmed music are connected with larger political concerns. 3M is very careful to assure potential clients that nothing "controversial or offensive" will ever appear on the soundtrack that might upset a customer (Pelisero 1993). In other words, they wish to keep social tensions and the differences that embody them out of listeners' minds. In reducing all traces of real social difference to taste preference, the programmers and the Mall must necessarily draw their lines of exclusion precisely along traditional axes of social difference, lest drawing different lines call attention to the process itself. Although music programmers and the Mall do differentiate by gender, age, and taste (which often stands in for race or class), all of these differentiations exist within a very limited field. In this particular case, difference is circumscribed by a mainstreamed construct of "Americanness" and, to a large extent, middle class whiteness.

The "nation" as it is used by Mall planners and music programmers is a narrow construct, corresponding exactly to empirical data but not to any living human being. (Of course, the data themselves are influenced by collection procedures that are biased toward a consumer public; again, see Meehan 1990). By deploying a range of social scientific norms in the service of building a better consumerism, the Mall of America and the programming of music function in the reproduction of a stratified society.[10] Decisions about what will be in the Mall or the soundtrack contain, implicit within them, three kinds of normative prescriptions: those pertaining to normative behaviors (signs in the hallways indicating appropriate behavior, and the repeated encouragement to identify oneself through consumption—sensing, looking, listening, and ultimately buying); those pertaining to normative tastes (the stocking of the stores, the songs that are used for programmed music); and those pertaining to normative differences (what kind of variation is allowed within and among stores and what kind of differences are and aren't allowed to be represented in the Mall—evangelical Christians can set up a kiosk to sell tapes and books, but Greenpeace cannot because mall management considers the latter to be political. Further, Devin Nordberg has noted that although the Mall does have an environmental booth, it is run by Browning Ferris Industries, a notorious corporate polluter [1993: 18]). The result is a perpetuation of taste through programming. One might ask whom the Mall is hailing through their programmed music or, more precisely, what kinds of identifications the Mall wants to encourage, and who is most likely to make those identifications.

Music programs correspond to the demography of the Mall's *desired*, rather than *actual*, visitors. While the Mall desires an affluent (and usually white) adult middle-class population, there is strong evidence to suggest that the real enthusiasts of the Mall are teenagers from a diversity of racial backgrounds (Karasov and Martin 1993: 27). But these teens must make use of an environment that is not immediately welcoming to them; or rather, that welcomes them as consumers first, and people second. African American teens, for instance, have reported being trailed by uniformed guards (ibid.)[11] Indeed, the Mall has displayed a great deal of ambivalence toward the population that seems to have taken to it most strongly: there are now signs outside of Camp Snoopy detailing expected appropriate behavior, and more security staff has been added. Such signs become a clear marker of difference: they prohibit "loud, boisterous behavior" precisely at the entrance to a sometimes deafening amusement park. Considering the loud, drunken revelry on Friday nights in the "bar district" and the joyful screams emanating from the roller coaster, it becomes clear that these prescriptions can be pretexts for enforcement of social boundaries rather than clearly delineated rules. Thus, the management of sound becomes one political strategy in the management (and collapsing) of difference.

(3) Listeners have to negotiate programmed music. This negotiation is rarely a conscious or intentional thing—how many people reflect on a store's wallpaper?—yet the contradictions embedded in mall and musical design leave open the possibility for alternative readings (I use this phrase with some caution). As statistical categories and behavioral ideals, the scientific norms animating music programming are neither internally consistent nor do they correspond to living individuals.[12] Such a system is necessarily clumsy, because no person or group of people is fully inside it at any given time. Since potential listeners' affective investments vary widely and are themselves overdetermined, there is always a range of possible engagements with programmed music, whatever the statistical and scientific rhetoric of marketing research might suggest. Thus, mall design can manifest itself in several ways for visitors: the intended result is a process of identification, where the (ideal) consumer identifies with the environment, the music, the spectacle, the mall, or simply (and most importantly) the commodities being offered. But any differentiating process—and here I return to programmed music specifically—can also alienate people. Because the acoustical landscape reflects the Mall's desired rather than its actual visitors, the environment could cause some cognitive dissonance. If this alienation does not chase people away, it can, at the very least, foster some kind of ironic distance. (But again, ironic distance alone is not necessarily a resistant or subversive stance, unless it is coupled with some kind of collective and active opposition. Any store will readily accept money from any customer, whatever level of irony the latter may read into the transaction.) By trying to paper over differences that may be entirely visible and audible in the hallways of the Mall (such as race), the environment itself may reveal its own biases more clearly. Ronald Radano views programmed music as an attempt to "domesticate" public spaces by placing familiar music in an unfamiliar space (1989: 452), but this familiarity does not guarantee a positive identification with the music.[13]

Several authors (such as Schaefer 1997; Lanza 1991; Jones and Schumacher 1992) have expanded these possible contradictions and tensions into a political program—they have felt the need to consider the ways a person might "resist" programmed music. Each winds up suggesting that a resistant response to programmed music would simply be to listen to it more closely. Joseph Lanza claims that "we can subvert the corporate canon by actually LISTENING with a fresh, ironic ear" (1991: 48). In a call for devoting more energy to discerning programmed music, these authors reproduce it as some kind of autonomous practice, removed from its surroundings. This is the very dichotomy that the *MUZAK* Limited Partnership sets up in their advertising: in the slogan "Music is art, Muzak is science," art and science are both constructed as socially autonomous truths. A call to ironic listening ignores any sense of context. Nobody is escorted from the Mall of America for listening to the music too closely. Jacques Attali claims that programmed music works to silence the listener, or more precisely, to hide the listeners' own silence from themselves (1985: 111). A finely tuned, ironic ear only reinforces this relationship. Or rather, those listeners who choose this political route will always wind up "resisting" all by themselves.

What disrupts the Mall environment is noise, the voicing of differences. The signs prohibiting "loud, boisterous behavior" located at the entrances and exits to the roaring amusement park are reminders that the Mall is attempting to construct a very specific kind of consumerism, and interference with that goal is grounds for ejection. This is a concern not only for visitors to the Mall who might not fit so neatly into its imagination of "America," but for the residents of the Mall itself—the stores.

Mall space is now an eminently familiar environment to many Americans, even to those who find it alienating. In addition to policing their images, stores have to worry about conflicts from outside communities arising in the Mall. In fact, there is a great deal of anxiety among stores in this respect. This is one reason that programmed music services all assure their potential customers that nothing "offensive" will appear in the soundtrack. But the soundtrack itself may be offensive. To illustrate: a popular urban legend has two adjacent stores in a shopping mall quarreling.[14] One plays light classical music and sells upscale clothing. The other sells the latest fashions and plays Top 40 music, which includes some rap. The former store fears that the latter's music will chase away its customers and petitions the mall management to have the latter keep their music at a lower level. The latter, of course, pleads that turning down their music would make their store design less effective in luring in potential customers. The moral of this story is simple: programmed musics in malls do not form a seamless and totally coherent system, nor do they always work together or as they're supposed to. Yet this divergence from designer intention should not be taken as cause for celebration. This is not *subversion* but *contradiction*—part of the everyday functioning of capitalist societies. Chance and coincidence play their part in composing the rhythms of acoustical space.

Conclusion: Ethnomusicology and the Problem of Reification

If all music is ethnic music (McAllester 1979: 183), then the ethnicity of programmed music is capitalism. Programmed music presupposes and builds upon an already-constituted commodity status for music and the experience of that music. In order for there to be programmed music, music must already have become a thing—it must be lived through its commodity status. The logic of programmed music follows Georg Lukács's description of reification all too perfectly: "The essence of the commodity-structure has often been pointed out. Its basis is that a relation between people takes on the character of a thing and thus acquires a 'phantom objectivity,' an autonomy that seems so strictly rational and all-embracing as to conceal every trace of its fundamental nature: the relation between people" (Lukács 1971: 83). In the case of programmed music, a relation of listening (itself highly structured by commodity circulation) is reified into a thing that can be bought and sold. That is why I have focused here more on rationalities underlying programming and design than on listener response. The latter is presupposed by the former.

If we follow Steven Feld (1988) and others in understanding sound in fundamentally social terms, then the problem of reification in mass mediated music should become a fundamental question for ethnomusicological inquiry. Ethnomusicologists tend to take mass-mediation as a problem or a point of departure—media are thought of as external to communities, impacting them or existing alongside them. In these accounts, the media begins as Other to the community. How the advent of a music industry affects local musicians, how it impacts on musical pedagogy, how music industries function on a global scale—these are the questions currently preoccupying ethnomusicological thought (see Chopyak 1987; Wallis and Malm 1984; Erlmann 1993; Guilbault 1993; Manuel 1993; Meintjes 1990). While there is certainly much research remaining to be done in these areas, in order to fully understand mass mediation in music, ethnomusicologists will have to move beyond a paradigm primarily concerned with distinguishing between tradition and change. This paradigm tends to understand mass-mediated music in terms of its difference from other music; its main concern with the mass media is in distinguishing between music that is mass mediated and music that is not (as is evident in the theoretical models proposed by Seeger 1987, Malm 1993, and Manuel 1993, for example). This approach tends to bracket mass mediation as a problem at the point of its advent, thereby overlooking the kinds of social conditions that emerge as a result of its proliferation. Ethnomusicology will have to consider formations like programmed music, which arise as a result of *the results of* mass mediation. To paraphrase an argument advanced by Ulrich Beck: in places like the Mall of America, mass mediation becomes reflexive; questions of the development and employment of media technologies are eclipsed by questions of the social, cultural, political, and economic "management" of the results of actually and potentially utilized media technologies (1992: 19). If ethnomusicology wishes to recover and critique modes of experience in a society fully saturated with the mass media, it will have to consider the phantom objectivity—the reification—of experience itself as a pervasive social phenomenon. In mass-mediated societies, this process is part of an endless chain in which the outside social world of recorded songs, mass-mediated images, and programmed spaces and schedules is folded into that which is most inside and private: the substance of affect and experience.

Notes

1. This essay is part of a larger work in progress on the culture and history of programmed music in work and leisure. I am deeply indebted to the following people, all of whom provided essential contributions to this project: John Archer, Greg Dimitriandis, Ariel Ducey, Lawrence Grossberg, Richard Leppert, Alex Lubet, Lauren Marsh, Roger Miller, Radhika Mongia, Negar Mottanedeh, Carrie Rentschler, Carol Stabile, Gary Thomas, Tom Turino, and Mike Willard. I'd also like to thank Leslie Ritter of *MUZAK*, Tom Pelisero of 3M, and Michael Ellis of *Billboard* for taking the time to answer my questions. Much of my analysis stems from personal observations during repeated visits to the Mall of America over the fall of 1992, the winter and spring of 1993, and follow-up visits in late fall of 1994 and late spring 1995 for the purposes of this study.

2. Although the history of programmed music is important for understanding its significance, a thorough historical account is beyond the scope of this article. There are several varying accounts of Muzak's industrial and cultural history. Jerri Husch (1984) and Jane Hulting (1988) both provide detailed accounts of the corporation and product's evolution. Both of these accounts focus on the history of the *MUZAK* corporation in their accounts of its development as a cultural form. Joseph Lanza (1994) focuses on Muzak (programmed music) as a kind of easy listening music. For shorter discussions of Muzak history, see Jones and Schumacher (1992) or McDermott (1990).

3. "Muzak" (which refers to a programmed music product) is a registered trademark and "stimulus progression" and "quantum modulation" are registered service marks belonging to the *MUZAK* Limited Partnership. The degree to which "Muzak" stands as a synecdoche for the industry can be illustrated by my conversations with both 3M and Audio Environments, Inc.: each time I raised more detailed technical questions, I was referred back to the *MUZAK* Limited Partnership. Much of my information on programming logistics thus comes from the *MUZAK* corporation, although the corporation does not provide most of the programmed music considered in this particular case study.

4. There is an important analogy to radio here, where broadcasters sell audiences to advertisers through programming and ratings (see Meehan 1990). However, as I will demonstrate below, programmed music presupposes the circulation of music through radio. It is thus a secondary mode of distribution, whereas radio is a primary mode of distribution. For instance, another economic function of radio is to advertise the music itself for purchase, thereby serving the needs of the music industry directly. Programmed music, on the other hand, pays royalties to musicians' unions because it then sells the music (and its popularity) as a service to clients. While programmed music could occasionally result in increased sales of recordings, this would be a purely accidental result.

5. I am engaging two important theoretical issues here. The first is territorialization, a term I borrow from Gilles Deleuze to connote the literal "embodiment" of

space. Territorialization is enclosure, but it is also the filling up of space, its endowment with a certain kind of meaning (or affect) and the exclusion of others. (For a more extended discussion of territorialization, see Deleuze and Guattari 1987: 310–350). The second issue is music's spatiality—its spatial character and behavior, as well as the ontology of space in which music resounds. While it is one of the questions that got me interested in programmed music to begin with, there is very little written on it (in addition to the above chapter from Deleuze and Guattari, see Attali 1985 and Grossberg 1991).

6. Bourdieu's analysis focuses mainly on class distinctions within France, although his model appears demonstratable along the many lines of affiliation and difference in American culture, such as class, gender, race, sexual preference, and age. In the American case, it seems less likely that one can correlate taste and class in any meaningful way, in part because Americans' experience of social class is itself so strongly mediated by other axes of difference (see, for example, Perlman 1993 for a summary to this position). Yet, given the importance of demographics in marketing (and the success of demographic models of marketing), dismissing connections between social difference and taste difference seems premature.

7. While a great deal of feminist scholarship has shown that when gender is not named, it is generally assumed to be male, programmed music is a messy case in this regard. Although *MUZAK* and 3M both offer a few programs pitched specifically to men or to women, the majority of their programs appear not to be intentionally geared toward a specifid gender. *MUZAK*'s own research literature verifies that the majority of shoppers are women, which may be one reason for the relative inattention to gender at the level of programming. Another possible reason for *MUZAK*'s approach may have to do with the nature of its service: stores are often already gendered by their merchandise, so music performing that function would be redundant (although redundancy can also be an effective design and marketing tool). Finally, there is the question of what it means to call programmed music gendered. If music is gendered by virtue of its intended audience, then we are left with rather ambiguous clues from the music programs. If we understand the "gender" of a music to be defined by formal characteristics in the music (melody, harmony, and chordal structure, as in McClary 1991; or timbre, rhythm, and tempo, as in Shepherd 1987a), then we have to add to this list the formal characteristics of the medium itself (production values, dynamic range, and so on), given that the latter are especially important in programmed music. But here we have to ask if listeners necessarily understand a characteristic of music as masculine or feminine just because it's described that way in the musicological or technical literature. Even if we turn it into a purely sociological question by instrumentalizing the concent of the music ("the text") and consider the genders of performers and audiences, this also becomes a muddled problem as music programs cut across audiences and genres.

8. My argument here is similar to one advanced by Jane Juffer (1996). She argues that Victoria's Secret uses indicies of class—both in decor and soundtrack—as part of an attempt by the store to distinguish itself from pornography and from other lingerie stores.

9. It is worth noting that this practice is strictly illegal, because it qualifies as a "public performance" of the music, and therefore would require the stores to pay royalties on each tape they played. The exception to this rule is the performance of tapes for sale (as in music stores), where the music being played is actually for sale.

10. I am loosely borrowing Michel Foucault's idea of *normalization* from his discussion of the deployment of norms. In some ways, it can be understood as the shadowy inverse of Weberian ideal-type analysis, where a norm is deployed in the service of comparison, differentiation, hierarchization, homogenization, and exclusion: "In a sense, the power of normalization imposes homogeneity, but it individ-

ualizes by making it possible to measure gaps, to determine levels, to fix specialties and to render the difference useful by fitting them one to another. It is easy to understand how the power of the norm functions within a system of formal equality, since within a homogeneity that is the rule, the norm introduces, as a useful imperative and as a result of measurement, all the shading of individual differences" (Foucault 1977: 183–84).

11. If one were looking for a population that "resists" this environment, the teens would be as close to that as possible, but we have to keep in mind the context of that resistance. The fact that it is cheaper to ride a bus out to the Mall of America in a distant suburb than to your friend's neighborhood inside the city limits demonstrates the systemic tendencies these teens are up against. A little decentering and a critical attitude may help urban teens survive (and in this sense they are a good thing), but ultimately these efforts alone will not lead to meaningful social change.

12. A passage by Jean Baudrillard is suggestive of a critique of *MUZAK*'s own appeals to "science" in the name of consumer engineering: "Besides, it will be noted retrospectively that the concepts 'class,' 'social relations,' 'power,' 'status,' 'institution'—and 'social' itself—all those too explicit concepts which are the glory of the legitimate sciences, have also only ever been muddled notions themselves, but notions upon which agreement has nevertheless been reached for mysterious ends: those of preserving a certain code of analysis" (1931: 4–5). Demographic science, for Baudrillard, becomes a trope of legitimation.

13. Radano only discusses background music. Although foreground music uses the same rhetoric of familiarity, I have shown here that it does not work in the same way.

14. I use the term "urban legend" (or perhaps suburban in this case) deliberately. I've heard the same story with a few mutations from several people. It is quite likely this happened somewhere at some time, but as an allegory, it works equally well to illustrate my point. "Anecdotes need not be true stories, but they must be functional in a given exchange" (Morris 1990: 15).

References

Attali, Jacques. 1985. *Noise: The Political Economy of Music,* translated by Brian Massumi. Minneapolis: University of Minnesota Press.
Barthes, Roland. 1972. "The Romans in Films." In *Mythologies,* translated by Annette Lavers. New York: The Noonday Press.
Baudrillard, Jean. 1983. *In the Shadow of the Silent Majorities,* translated by Paul Foss, John Johnston, and Paul Patton. New York: Semiotex(e).
Beck, Ulrich. 1992. *Risk Society: Towards a New Modernity,* translated by Mark Ritter. Newbury Park: Sage.
Berland, Jody. 1990. "Radio Space and Industrial Time: Music Formats, Local Narratives and Technological Mediation." *Popular Music* 9 (2): 179–92.
Booth, Gregory D. 1990. "Brass Bands: Tradition, Change and Mass Media in Indian Wedding Music." *Ethnomusicology* 34 (2): 245–62.
Bourdieu, Pierre. 1984. *Distinction: A Social Critique of the Judgement of Taste,* translated by Richard Nice. Cambridge: Harvard University Press.
Bruner II, Gordon C. 1990. "Music, Mood and Marketing." *Journal of Marketing* 54 (4): 94–104.
Chopyak, James. 1987. "The Role of Music in Mass Media, Public Education and the Formation of a Malaysian National Culture." *Ethnomusicology* 31 (3): 431–54.
Crawford, Margaret. 1992. "The World in a Shopping Mall." In *Variations on a Theme Park: The New American City and the End of Public Space,* edited by Michael Sorkin. New York: The Noonday Press.

Deleuze, Gilles, and Felix Guattari. 1987. *A Thousand Plateaus: Capitalism and Schizophrenia*, translated by Brian Massumi. Minneapolis: University of Minnesota Press.
Ellis, Michael. 1993. Telephone interview by author, 2 September.
Erlmann, Veit. 1993. "The Politics and Aesthetics of Transnational Musics." *World of Music* 35 (2): 3–15.
Feld, Steven. 1988. "Aesthetics as Iconicity of Style, or 'Lift-up-over-Sounding': Getting into the Kaluli Groove." *Yearbook for Traditional Music* 20: 74–113.
Foucault, Michel. 1977. *Discipline and Punish: The Birth of the Prison*, translated by Alan Sheridan. New York: Vintage Books.
Frow, John, and Meaghan Morris. 1993. *Australian Cultural Studies: A Reader.* Urbana: University of Illinois Press.
Goffman, Erving. 1974. *Frame Analysis: An Essay on the Organization of Experience.* Cambridge: Harvard University Press.
Greene, Alex. 1986. "The Tyranny of Melody." *Etc.* 43 (3): 285–90.
Grossberg, Lawrence. 1991. "Rock, Territorialization and Power." *Cultural Studies* 5 (3): 358–67.
Guilbault, Jocelyn. 1993. "On Redefining the 'Local' through World Music." *World of Music* 35 (2): 45–60.
Hall, Stuart. 1986. "On Postmodernism and Articulation." *Journal of Communication Inquiry* 10 (2): 45–60.
Hulting, Jane. 1988. "Muzak: A Study in Sonic Ideology." M.A. thesis, University of Pennsylvania.
Husch, Jerri A. 1984. "Music of the Workplace: A Study of Muzak Culture." Ph.D. diss., University of Massachusetts, Amherst.
Jones, Simon C., and Thomas G. Schumacher. 1992. "Muzak: On Functional Music and Power." *Critical Studies in Mass Communication* 9: 156–69.
Juffer, Jane. 1996. "A Pornographic Femininity?: Telling and Selling Victoria's (Dirty) Secrets." *Social Text* #48. n.p.
Karasov, Deborah, and Judith A. Martin. 1993. "The Mall of Them All." *Design Quarterly* (spring): 18–27.
Keil, Charles. 1984. "Music Mediated and Live in Japan." *Ethnomusicology* 28 (1): 91–96.
Lanza, Joseph. 1991. "The Sound of Cottage Cheese: Why Background Music is the Real World Beat!: *Performing Arts Journal* 13 (3): 42–53.
———. 1994. *Elevator Music: A Surreal History of Muzak, Easy-Listening, and Other Moodsong.* New York: St. Martin's Press.
Lefebvre, Henri. 1991. *The Production of Space*, translated by Donald Nicholson-Smith. Cambridge: Basil Blackwell.
Lukács, Georg. 1971. *History and Class Consciousness: Studies in Marxist Dialectics*, translated by Rodney Livingstone. Cambridge: MIT Press.
Machan, Dyer. 1995. "Sharing Victoria's Secret." *Forbes* (June 5): 132–33.
Malm, Krister. 1993. "Music on the Move: Traditions and the Mass Media." *Ethnomusicology* 37 (3): 339–54.
Manuel, Peter. 1993. *Cassette Culture: Popular Music and Technology in North India.* Chicago: University of Chicago Press.
McAllester, David. 1979. "The Astonished Ethno-Muse." *Ethnomusicology* 23 (2): 179–89.
McClary, Susan. 1991. *Feminine Endings: Music, Gender and Sexuality.* Minneapolis: University of Minnesota Press.
McDermott, Judy. 1990. "If It's to Be Heard but Not Listened to, Then It Must Be Muzak." *Smithsonian* 20 (10): 70–80.
Meehan, Eileen. 1990. "Why We Don't Count: The Commodity Audience." In

Logics of Television, edited by Patricia Mellencamp. Bloomington: Indiana University Press, 117–37.

Meintjes, Louise. 1990. "Paul Simon's Graceland, South Africa, and the Mediation of Musical Meaning." *Ethnomusicology* 34 (1): 37–74.

Morris, Meaghan. 1990. "Banality in Cultural Studies." In *Logics of Television,* edited by Patricia Mellencamp. Bloomington: Indiana University Press, 14–43.

Morse, Margaret. 1990. "An Ontology of Everyday Distraction: The Freeway, the Mall, and Television." In *Logics of Television,* edited by Patricia Mellencamp. Bloomington: Indiana University Press, 193–221.

MUZAK Limited Partnership. 1990. "The Right Music Style Can Successfully Promote the Image You Desire For Your Business." In *Muzak Research Review: Fashion Retail.* Seattle: Muzak Limited Partnership.

———. 1992a. *Tones Music Catalog 1992–93.* Seattle: Muzak Limited Partnership.

———. 1992b. "Quantum Modulation—The Story." *MUZAK Special Marketing Supplement* 4 (July/August).

Nettl, Bruno. 1978. "Introduction." In *Eight Urban Musical Cultures: Tradition and Change,* edited by Bruno Nettl. Urbana: University of Illinois Press, 3–18.

Nordberg, Devin. 1993. "The Mall of America: A Postmodern Factory of Service and Spectacle." Unpublished paper, University of Minnesota.

Pelisero, Tom. 1993. Telephone interview by author, 31 August.

Perlman, Marc. 1993. "Idioculture: De-Massifying the Popular Music Audience." *Postmodern Culture* 4 (1). Available electronically as REVIEW-7.993 from LISTERV@LISTERV.NCSU.EDU, or on diskette from Oxford University Press.

Petchler, Thea. 1993. "Errands into the Wilderness: The Mall of America and American Exceptionalism." Paper presented at the Ninth Annual Meeting of the Mid-America American Studies Association, 16–18 April, Minneapolis–St. Paul, Minnesota.

Qureshi, Regula Burkhardt. 1987. "Musical Sound and Contextual Input: A Performance Model for Musical Analysis." *Ethnomusicology* 31 (1): 56–86.

Radano, Ronald. 1989. "Interpreting Muzak: Speculations on the Musical Experience in Everyday Life." *American Music* 7 (4): 448–60.

Ritter, Leslie. 1993. Telephone interview by author, 2 February.

Rose, Tricia. 1994. *Black Noise: Rap Music and Black Culture in Contemporary America.* Hanover: Wesleyan University Press / University Press of New England.

Schaefer, R. Murray. 1977. *The Turning of the World.* New York: A. A. Knopf.

Seeger, Anthony. 1987. "Powering Up the Models: Internal and External Style Change in Music." Paper delivered at the Annual American Anthropological Association Meeting in Chicago, 21 November.

Shepherd, John. 1987. "Music and Male Hegemony." In *Music and Society: The Politics of Composition, Performance and Reception,* edited by Richard Leppert and Susan McClary. New York: Cambridge University Press.

Shields, Rob. 1992. *Lifestyle Shopping: The Subject of Consumption.* New York: Routledge

Turino, Thomas. 1993. *Moving Away from Silence: Music of the Peruvian Antiplano and the Experience of Urban Migration.* Chicago: University of Chicago Press.

Victoria's Secret. 1993. Liner notes to "Two Centuries of Romance: Mozart." *Victoria's Secret Timeless Tributes of Love* #6. Audiocassette.

Wallis, Roger, and Krister Malm. 1984. *Big Sounds from Small Peoples: The Music Industry in Small Countries.* New York: Pendragon.

CHAPTER FOURTEEN

Consuming Audio
An Introduction to Tweak Theory

―

Marc Perlman

Audio technology is a medium for music, and when we pay attention to it we tend to speculate about its effects on the music it transmits.[1] By now there are well-established traditions of commentary (many of them critical) about the impact of musical reproduction on musical production. Recording has commodified music, we are told, and driven a wedge between performer and audience. In popular music, multitrack recording dissolved the sense of group interaction; in classical music, audiences accustomed to the perfect accuracy of recordings demanded the same from live concerts, where interpretation consequently became cautious and stultified. Rock musicians, attempting to duplicate elaborate studio arrangements in concert performance, found themselves locked into synchrony with preprogrammed sequencers. And so on.

It is my purpose here neither to refine these arguments nor to refute them. Rather, I wish to change the subject slightly. I have nothing to say about technology as a *medium* for the *producers* of music; I will instead consider technology as part of the *culture* of the music *audience*. Audio technology, like other forms of technology, is not simply a tool used for a practical purpose; it bears cultural meanings and personal emotional investments. Furthermore, though it represents a realm of creative involvement and practical mastery for audio engineers, for most of its users it is something purchased, a *commodity*.

Scholars have already begun to study some of these aspects of music technology. Gay (1998), for example, investigates the frameworks of meaning New York rock musicians use to make sense of their equipment. Théberge (1997) studies the commercial nexus within which musicians act as consumers of technology. There has also been a desultory tradition of reflection on the influence of music technology on the listener (notably Adorno [1945] on radio, Eisenberg [1988] on the record, Hosokawa [1984] on the Walkman), but little of this work has focused specifically on the listener as consumer of technology, and in any event most of it is cultural criticism rather than being ethnographically oriented.

The most convenient point of entry into the universe of meaning surrounding consumer audio equipment is surely the world of the audiophile, the person who takes audio very seriously indeed, investing large amounts of time and money in acquiring, using, and thinking about audio technology. Audiophilia is ideally suited for this investigation because of its complexity and visibility. It is a hobby, a form of "serious leisure," and a form of consumption; it provides its devotees with a source of self-images and an arena of social relationships, ranging from informal friendship networks, clubs, annual trade fairs, and Internet discussion groups to the "community" of consumers of specialist audiophile magazines.

My topic, then, can be defined as the culture of audiophile consumption. In what follows I bring together my observations of audiophile life and thought with the findings of anthropologists, sociologists, and other scholars who have investigated the social meanings of consumption. I argue that in audio consumption, like consumption in general, the purchaser engages in appropriative work to turn the commodity into a possession. In the past, when audio technology was more mechanically based, more modular, and more accessible to the user, such work could engage directly with the technical details of the product. As the technology became increasingly sealed off from the user, he—audiophiles are overwhelmingly male—responded with non-technical interventions. These "tweaks," with no scientifically accepted relationship to the technological principles of the audio device, continued to allow the user to appropriate, domesticate, and personalize store-bought equipment. However, they brought him into conflict with the technology's chief source of epistemic authority: audio engineering.

The Audiophile

The audiophile community cannot be sharply delineated; it shades off on one side into the great mass of casual consumers and on the other into the professional coterie of engineers. Furthermore, it displays relatively little consensus of opinion; indeed, in some ways it is easier to define it by the topics it debates rather than the positions it unanimously affirms. For the purposes of this paper I will provisionally define the audiophile as someone interested in what the industry calls "specialty" audio, what is usually marketed as "the high end."

In 1994, Richard, a salesman in a specialty audio store, defined "high end" for me as any system that costs $5,000 or more (others would no doubt draw the line elsewhere). In monetary terms, there is essentially no upper limit to the high end; it is possible to spend $175,000 on a pair of speakers. Such extravagant products are probably more ogled-at than purchased even within the innermost circles of audiophilia, but in general it seems true that audiophiles, despite their small numbers, hold a relatively elite position in society and wield disproportionate economic power.

Audiophiles are mostly professional men, affluent and well-educated. In a survey conducted in 1988 by *Stereophile* magazine, of 9,000 respondents 47.4 percent had salaries above $50,000; the average income reported was $58,900. In a smaller 1991 survey of 702 subscribers the average income was $80,700. In the latter survey 55.8 percent of the responses reported household incomes above $50,000, and 25 percent reported income above $100,000. Of the 1988 respondents, 80.6 percent had at least an undergraduate degree, and 57.2 percent worked in professional or technical fields. In both surveys, 99 percent of the respondents were male (Atkinson 1988, 1992).

Audiophiles typically have extensive collections of recordings; in its 1988 survey, *Stereophile* found its readers owned an average of 620 LPs and 100 CDs.[2] Audiophiles follow one or more specialist periodicals, including *Audio, Stereophile,* and *The Absolute Sound.*

Although there are distinct schools of audiophilic thought, audiophiles distinguish themselves as a whole from the buyers of mass-marketed audio, the patrons of Circuit City and readers of *Stereo Review* whose homes are filled with "rack systems" made by Japanese companies. All of these are subject to audiophilic scorn. The manufacturers are exposed to generic Japan-bashing (O'Connell 1992: 22); magazines like *High Fidelity* are accused of selling out to the companies that advertise in them (Pearson 1973: 4); their customers and readers are characterized in more or less unflattering terms (e.g., "comatose"; Parsons 1994: 27).

Thus from one point of view the audiophile world is defined by equipment, and audiophiles set themselves apart from mass-market audio consumers through the esoteric commodities they buy, with brand names unknown to the average consumer (Krell, Thiel, B & W, VAC, etc.). But many audiophiles are wary of defining themselves entirely in terms of mechanical and electronic devices. They are sensitive to the charge that they care only about expensive gadgets. They publicly present themselves as people who truly care about music, who take it much more seriously than the average person. Indeed, they commonly describe themselves as addicted to music. Listening becomes a passion: as one audiophile (I'll call him Jack) told me, he sometimes stays up all night listening to his records. Richard, the high-end audio salesman, said, "Once you get bitten by the bug, that's the end. It's almost a disease—music is such an emotional involvement, it pushes so many buttons for people."

Audio and Consumption

Regardless of the importance of audiophilia's technological and emotional aspects, audio is very much a realm of consumption, inhabited by a staggering variety of competing products. In 1995, *Audio* magazine's annual directory listed 6,222 models of equipment, including 3,019 kinds of speakers, 818 kinds of amplifiers, 503 kinds of preamps, 358 kinds of headphones, 168 kinds of phono cartridges, 110 kinds of turntables, 102 kinds of analog cassette decks, and 93 kinds of ambience/surround sound processors. To understand the nature of audiophilia, therefore, we need to understand some basic facts about the way commodities are consumed.

Commodities awaiting sale are objects designed for consumption, depersonalized goods that belong to no one in particular. But pure commodities—permanently anonymous and perfectly interchangeable objects to which no personal identity or social meaning ever sticks—probably do not exist. Even money, supposedly an impersonal, undifferentiated medium defined entirely by its exchange value, acquires personal and social meanings in people's lives (Zelizer 1997). Commodities, too, absorb sentimental investments and social significance. People who buy commodities do not merely gain ownership over them in exchange for cash; they appropriate them through rituals of acquisition and display. Appadurai (1986: 13) therefore prefers to speak of commodity *situations* rather than commodities, situations into and out of which objects can move throughout their life cycles.

The social and cultural import of the commodity was famously argued by Mary Douglas and Baron Isherwood in their 1979 study, *The World of Goods.* "Commodities are good for thinking," they insisted, borrowing Lévi-Strauss's famous phrase; consumers use goods to make "visible and stable the categories of culture" (Douglas and Isherwood 1996: 41, 38). Consumption also provides a medium for social interaction (1996: 51):

Enjoyment of physical consumption is only a part of the service yielded by goods; the other part is the enjoyment of sharing names . . . Take football, or cricket: the fan internalizes a reel of his names inside his head. He knows the famous victories, infamous losses, and draws; he loves to talk about historic games, good referees, vast crowds, inspiring captains, good years and bad, the present and the old days. Inside him are grades of passionate judgment. Another enthusiast need only utter two words to betray the vast amount of sharing that is possible for them both. These joys of sharing names are the rewards of a long investment of time and attention and also of cash.

What sorts of meanings do commodities acquire? We can broadly categorize them as social and personal, corresponding to two approaches in the study of consumption. The first—indebted to the work of Veblen and Bourdieu—traces the meanings of objects to the social strata with which they are associated (Carrier 1990: 579). The other approach—associated with the names of Mauss, Kopytoff, and Carrier—looks at the private uses of objects, the ways they lend meaning to, and draw meaning from, personal relationships (Carrier 1990: 581).

I will deal here only briefly with the social meanings of audio equipment. As is clear from the survey results reported above, high-end audio is closely associated with demographic variables such as socioeconomic status and gender. The readers of *Stereophile* belong disproportionately to the higher income brackets, and they are relatively well educated, thus ranking high on the scales of both economic and cultural capital. (Of course, not all readers of audiophile magazines can afford to spend tens of thousands of dollars on audio equipment. Many of them participate in the high-end world by buying second-hand equipment at a steep discount, some assemble their own electronics, and some are purely spectators, reading the audiophile press and hanging out at audio salons without ever making a purchase. The audiophile press functions for the latter group as the purveyor of a fantasy world, but it is still a high-status world of economic power.)

The gender connotations of audiophilia are equally evident: audiophiles are overwhelmingly male. This is unsurprising, given the historic identification of engineering in general with men, and audio technology's specific associations with masculinity over the past several decades (Keightley 1996).

For my present purposes, however, the audiophile's personal relationship with the equipment is of more interest. By expending energy in the process of acquisition, by modifying the object acquired, and by means of rituals of display, the audiophile can personalize the commodity he buys; he can, that is, appropriate it.

Appropriation is the work of imprinting the identity of the owner on the thing owned, converting an anonymous commodity into a personal possession (Carrier 1990: 583). Appropriation can involve physical or symbolic alteration of the object. The work of appropriation can be promoted by special consumption spaces that "provide a coherent identity for much of what is bought and sold in them" (Carrier 1990: 583). It is also furthered by a personal relationship between seller and buyer (Carrier 1990: 588).

Another way commodities are appropriated is through the buyer's exercise of choice. Some consumers invest a great deal of work in choosing what to buy. When this effort is directed at discovering the best price or value and avoiding the seductions of advertisers, impulse buying, and so on, we may call it rational consumption. This kind of consumption is a way of investing the buyer's identity into the act of shopping, and hence into the commodities purchased (Carrier 1990: 586). (Strictly speaking, however, it injects the commodity with a *certain aspect* of the shopper's identity: his or her identity as the careful, rational agent, in control of him or herself, keeping utility always in view, immune to the blandishments of fashion, peer pressure, or even convenience.)

These facets of appropriative work are all evident in audiophilia. High-end audio boutiques or "audio salons" provide special consumption spaces. Audiophiles often develop long-term relationships with a salesperson at such stores. Indeed, some audiophiles make personal contact with the manufacturers of their equipment. Jack was proud of the fact that his preamp was made by two men in a shop in West Virginia, and he could get on the phone to them whenever he wanted.

The type of appropriation I wish to focus on here, however, is the physical modification of the equipment, for it is through this sort of work that the audiophile most obviously invests himself in his equipment. There is a term for this in audiophilia; it is called "tweaking."

Tweaking

Jay was one of the system engineers at the campus computer center. I noticed a cartoon from an audiophile magazine on his door when I went to apply for an email account and struck up a conversation. He told me he liked *The Absolute Sound* better than *Stereophile*, which he thought was less critical of the components it reviews. He admired Harry Pearson, editor of *The Absolute Sound*, because he was one of the first people who refused to accept engineers' claims about audio equipment, and trusted his ears instead.

Jay then told me about a little experiment he tried with his equipment. He had been reading about a company that makes small ebony discs which, when attached to audio components, improve the sound. The reviewer tried them on some high-end speakers and was amazed at the difference. Jay wanted to try it on his own speakers, but since they weren't as expensive as the ones the reviewer used, he thought he might not need fancy ebony discs, so he taped pennies onto the sides of the speaker cones. It made a difference! Jay was listening to a saxophone concerto (he used to play sax), and the pennies let him hear the sound of the reed. They improved the definition, they made it sound like there was a real instrument there, not like a watercolor. He couldn't understand how it could work. He didn't know all the physics of it. All he could imagine was that the wood sides of the speakers vibrate sympathetically at some frequency, and the pennies damp it. Though how thick, braced wood can be vibrating, he had no idea. If Jay told some engineer about this, he'd say Jay was on drugs!

Tweaking is modifying purchased equipment; a tweak is "any small, fussy thing that improves the sound of an audio system" (Denby 1990: 44). The term is not unique to the audio world; it is applied to computers, cars, and bicycles as well, as any search of magazine indexes will reveal. It has, however, taken on some remarkable manifestations in the audiophile community, and it plays a special role in the conceptual universe of audiophilia. Tweaking audio equipment is an act of devotion to sound, a demonstration of one's willingness to "labor in its service" (Rothstein 1985). The personal effort involved makes tweaking a way of investing the equipment with one's self, that is, a process of appropriation.

The recent prominence of tweaking is due in part to the historical shift from phonograph-based to CD-based audio systems. These technologies differ greatly in the scope they allow for consumer appropriation. The phonograph lent itself especially well to user modifications, since it had several mechanically separable parts (stylus, cartridge, tonearm) and adjustable operating parameters (tracking force, anti-skating, cartridge alignment, and vertical tracking angle). Records and styli could be tended in various ways: dusted with special brushes, lubricated with special fluids, cleaned with special cleaners.

Electronic components, lacking such mechanical interfaces, do not demand, or even allow, such simple but conspicuous demonstrations of the user's solicitude. However, in earlier decades at least, such components could be assembled by the consumer. When building an amplifier from a kit, there is no clear line of distinction between the act of acquisition and the act of tweaking: one simply modifies the unit during or after assembly, as many do-it-yourselfers did with their Dynaco kits (Hodges 1985). Some store-bought units were susceptible to similar kinds of modification; for example, in the 1970s many audiophiles added ribbon tweeters to their speaker systems (Lofft 1982: 65).

However, a different way of tweaking electronic components such as amplifiers and preamps eventually emerged (apparently in the late 1970s), one that involved minimal or no technical skills or knowledge. Some of these used easily available materials, some required specially marketed devices, but most were inert and non-electronic (indeed, non-mechanical) — for example, placing bricks or bags of sand on top of components or insulating mats or rubber feet under them (Lofft 1982: 65–66).

The introduction of the CD eliminated the scope for user adjustment permitted by the LP: "the closed nature of digital systems seems to prohibit the tinkering and finicky adjustment which make analog such involving fun." The CD seemed to deprive the buyer of "the thrill of custom-tailoring" his system, since he could no longer mix and match turntables, tonearms, and cartridges (Berger 1985). For some audiophiles, digital's untweakability was a reason to avoid the format (Tellig 1990: 62). Others embraced the CD but found new, non-technical tweaks to apply to it. Consequently, the formerly obscure category of the non-technical tweak increased sharply in visibility (Denby 1990).

A great variety of such tweaks proliferated: applying magic marker to the edge of CDs, or rubber rings to their upper surface; spraying them with Armor All, or putting them in the freezer. New tweaks for other components continually emerged: placing small bits of foil on the electronic parts of a CD player or amplifier (to counteract the adverse effects of "gravitational energy"), resting speakers on tennis balls, and so on (Kessler 1991).

Tweaks like these, involving cheap, readily available materials, require no sophisticated industrial laboratories or million-dollar research budgets, and their use preserves the consumer's autonomy. However, manufacturers soon recognized in tweaking a market they could exploit, and they started producing tweaking devices—some of them quite costly—such as eight-pound metal blocks encased in oak, to be placed over power transformers (*Brick* 1982); electrical units to demagnetize a CD before playing it (Scull 1996); vibration-damping and isolation devices ("platforms, stands, weights, bases, and feet"; Colloms 1993: 131); and devices to "tune" a room or a component.

As an example of the latter, consider the Shun Mook "resonance-control" devices. (These "Mpingo Discs" are the type of product that inspired Jay's experiment with pennies, described above.) Made of ebony, about one and one-half inches in diameter and one-half inch thick, they cost $50 each. They can be applied to turntables, CD players, or preamps. (Other "tuning" devices are meant to be applied to speaker cabinets.) They were devised by Dr. Yu Wah Tan, who gave the following rationale for his decision to make them from African ebony: "most fine instruments, like the cello or double bass, use ebony in the finger boards. The *right* resonance! Resonance is energy, so you can't eliminate it, you can only *transform* it." As Scull explains, "The basic idea is to preserve and tune these resonances rather than damage the sound by trying to eliminate them . . . Good resonances are like the ones in a violin . . . Bad resonances are the mechanical ones found in preamps, for instance, that are outside the musical spectrum . . . [T]he Shun Mook products attempt to preserve the 'good' resonances and evacuate harsh, mechanical, non-musical resonances . . ." (Scull 1994: 119).

Tweaks and the Contest for Epistemic Authority

The effectiveness of such tweaks is controversial, and many audio engineers (as well as the self-described "sensible" or "rational" audiophiles) hold them up to derision. Some argue that, devoid of either "solid theory" or "empirical evidence," tweaks like these merely "cater to the vanity of the audiophile" (Weaver 1994).[3] Indeed, when subjected to engineering tests they routinely fail. For example, in March 1990 the International CD Exchange newsletter recommended that CD owners paint the edges of their disks with green magic marker to improve the sound. An engineer, David Ranada, decided to test this tweak. He digitally measured the output of treated versus untreated disks and found no change. Hans Fantel, a *New York Times* audio reviewer, then spread the word that this tweak was ineffective and even dangerous (Fantel 1990).

These pronouncements in the name of science do not silence the tweakers, however. They respond chiefly by appealing to the ear. The defenders of these tweaks privilege the evidence of their own senses over the findings of the officially accredited authorities on technology, with their elaborate equipment (Denby 1990: 44): "The proponents of the various tweaks are not claiming that the improvements will show up in measurements. Yet every music lover I know has heard the differences. Obviously, there are improvements as well as degradations in digital sound that aren't measurable. Digital has its unsolved mysteries."

As this quotation shows, awarding ultimate authority to the ear means limiting the authority of scientific knowledge; it means accepting the presence of *mystery* in human hearing or in audio equipment. Indeed, the manufacturers of tweaks sometimes use the inexplicability of their devices as a promotional resource. The following advertising copy[4] shows how Shun Mook sells the Mpingo discs by emphasizing their resistance to rational explanation:

We invite you to place a Shun Mook spatial control kit consisting of 9 Mpingo Discs on any system in our store and experience a musical miracle. You will hear cleaner highs, richer harmonics, a more holographic sound stage and tighter bass. While your mind will not deny the experience you will hear, it will have trouble understanding just how 9 small discs that measure only 1.75" by 0.5" created this dramatic improvement. It's enough to drive anyone bananas! We know, because we're still trying to figure it out.

The critics of tweaking are not convinced by these arguments. They neither find the appeal to the ear conclusive nor admit that current scientific knowledge is powerless to account for these phenomena. They typically dismiss the "audible improvements" as due to placebo effects—they believe audiophiles think they hear improvements where none exist because they feel they should be able to. To distinguish truly audible differences from imaginary ones, standard engineering methodology employs double-blind comparison tests. These tests typically show that most people who claim to hear differences cannot do so under controlled conditions (O'Connell 1992: 14).

Defenders of tweaks respond in various ways. It is possible, for example, to admit that the audible improvements in sound that tweaks produce are due to placebo effects but to argue that, from the point of view of the user, placebo effects are as real as any other sort of effect (Willis 1994: 49): "Belief and perception are two mutually dependent interactive variables—they feed and influence each other . . . If a baseball player believes that not washing his "lucky socks" will improve his batting average, then it will . . . The power of an audiophile's degaussed and cryogenically treated CD is just as real as the ballplayer's lucky socks, and just as dependent on his belief to work its wonders." It is also possible to dismiss the incompatibility between these devices and scientific theory by noting that science— *Western* science—is only one belief system among many. Using science to judge the products of other belief systems is therefore ethnocentric (Willis 1994: 51): "Shun Mook's products are being summarily rejected because the belief system they're based in is different from traditional Western thought. This isn't fair. Until we become familiar enough with the belief system in which they were developed to have a "feel" for what they can or cannot do, we cannot really evaluate them."

Conclusion

The evolution of audio technology has thus turned the appropriative gesture of the tweak into an occasion for epistemic controversy. The tweak debate is only one instance of a general pattern of contestation of authority in the world of audio. There are similar arguments over the merits of tube versus transistor amplifiers, single-ended versus push-pull circuit designs, LP versus CD, 96kHz versus 44.1kHz sampling rates, and over esoteric cables. All of these debates pit the findings of audio engineering against the evidence of audiophile ears; all reveal a tension between two varieties of epistemic legitimation, which elsewhere I call "golden-earism" and "meter-readism" (Perlman 2002).

Golden-earism's rhetoric of resistance to scientific authority is not unique; many of its argumentative strategies can be found in other arenas of epistemic contest—for example, in alternative medicine (where appeals to personal experience underwrite challenges to the biomedical establishment). Furthermore, these strategies serve a variety of purposes within audiophila. They can preserve the *jouissance* of consumption in the face of an ascetic technological rationalism. Also, by valorizing supposedly superannuated formats like the LP, they allow the audiophile to protest the accelerated lifecycle of audio products driven by capitalism's destabilizing need for endless growth. And as we have seen, these strategies can provide an epistemic haven for the consumer's appropriate activity. They reassure the audiophile that he can still invest himself in his equipment, still participate in the adventure of musical reproduction, even in the face of technological developments that would seem to deny him the opportunity and authority to do so.

Notes

1. This essay develops some of the themes of a larger research project on audiophilia. I wish to thank Robert Lancefield for originally alerting me to the existence of the audiophilia debates, and to René Lysloff for encouraging me to present my interim findings at the 1994 meetings of the Society for Ethnomusicology. I am grateful to "Richard," "Jack," and "Jay" for welcoming me into their homes, offices, or stores; to Susan Smulyan for some enlightening discussions on the history of technology; to Peter Kulchyski, Dominic LaCapra, and Susan Buck-Morss for their perceptive criticisms; and to Trevor Pinch, Bruno Latour, and Luc Boltanski for their reassuring enthusiasm. I alone, however, am responsible for the interpretations offered herein.

2. The reported predominance of analog recordings is in part attributable to the age of these collections and the novelty of CDs (introduced in 1983), and in part to the explicit refusal of 20 percent of the respondents to buy digital recordings (Atkinson 1988). In the 1991 survey, only 39.2 percent of the respondents reported buying LPs in the past twelve months, compared with 93 percent who purchased CDs (Atkinson 1992).

3. The critic goes on to offer his own interpretation of the significance of tweaking: "That is, after all, the nature of tweaking. It is saying, 'I can improve this multi-kilobuck equipment by using this cheap little fix'" (Weaver 1994: 18). Here we see one way the proponents of rational consumption make sense of the activity of appropriation: they can attribute it to moral weakness ("vanity") on the part of the consumer.

4. This advertisement appeared in *The Absolute Sound* 19 (93): 167.

References

Adorno, T. 1945. "A Social Critique of Radio Music." *Kenyon Review* 8 (2): 208–217.
Appadurai, Arjun, ed. 1986. *The Social Life of Things.* Cambridge: Cambridge University Press.
Atkinson, John. 1988. "Stereophile and You." *Stereophile* 11 (10): 69–77.
———. 1992. "Who Are You?" *Stereophile* 15 (6): 7–10.
Berger, Ivan. 1985. "Philosophies to Count On." *Audio* 69 (11): 8–10 (November).
Brick. 1982. "The VPI Brick." *Stereophile* 5 (4): 13–14, 30.

Carrier, James. 1990. "Reconciling Commodities and Personal Relations in Industrial Society." *Theory and Society* 19: 579–598.

Colloms, Martin. 1993. "Harmonix Tuning Devices by Combak." *Stereophile* 16 (7): 131–140.

Denby, David. 1990. "Twin Tweaks." *New York* 23 (27): 44–45 (16 July).

Doris, Frank. 1994. "The Single-Ended Triode Amplifier: A Renewed Interest." *The Absolute Sound* 19 (96): 93–101 (June/July).

Douglas, Mary, and Baron Isherwood. 1996 [1979]. *The World of Goods*. With a new introduction. London: Routledge.

Eisenberg, Evan. 1988. *The Recording Angel*. London: Picador.

Fantel, Hans. 1990. "Brush Aside the Idea of Painting CD's." *New York Times,* section 2, p. 26 (3 June).

Gay, Leslie. 1998. "Acting Up, Talking Tech." *Ethnomusicology* 42 (1): 81–98.

Hodges, Ralph. 1985. "David Hafler and Audio Civilization." *Stereo Review* 50 (3): 94 (March).

Hosokawa, S. 1984. "The Walkman Effect." *Popular Music* 4: 165–180.

Keightley, Keir. 1996. "'Turn It Down!' She Shrieked: Gender, Domestic Space, and High Fidelity, 1948–59." *Popular Music* 15 (2): 149–177.

Kessler, Ken. 1991. "Industry Update." *Stereophile* 14 (1): 53–57.

Lofft, Alan. 1982. "Sense and Nonsense in High-End Hi-Fi." *Stereo Review* 47 (10): 62–69 (October).

Marvin, Carolyn. 1988. *When Old Technologies Were New*. New York: Oxford University Press.

O'Connell, Joseph. 1992. "The Fine-Tuning of a Golden Ear: High-End Audio and the Evolutionary Model of Technology." *Technology and Culture* 33: 1–37.

Parsons, William. 1994. "In the Digital Quicksand." *Absolute Sound* 19 (93): 26–27.

Pearson, Harry. 1973. "Viewpoints." *Absolute Sound* 1 (1): 4–6.

Perlman, Marc. 2002. "Golden Ears and Meter Readers: The Contest for Epistemic Authority in Audiophila." Presented to the *Sound Matters* conference at the University of Maastricht, the Netherlands, 15–17 November 2002.

Rothstein, Edward. 1985. "The Quest for Perfect Sound." *New Republic* 193: 29–37 (30 December).

Scull, Jonathan. 1994. "Shun Mook Resonance-Control Devices." *Stereophile* 17 (2): 119–125.

———. 1996. "Four Tweaks and a Freebie." *Stereophile* 19 (2): 177–183.

Tellig, Sam. 1990. "The Audio Anarchist." *Stereophile* 13 (5): 62–67.

Théberge, Paul. 1997. *Any Sound You Can Imagine: Making Music / Consuming Technology*. Hanover: Wesleyan University Press/University Press of New England.

Willis, Barry. 1994. "The Art of Scientific Illusion." *Stereophile* 17 (5): 49–51.

Zelizer, Viviana. 1997. *The Social Meaning of Money*. Princeton: Princeton University Press.

CHAPTER FIFTEEN

Fairly Used
Negativland's *U2* and the Precarious Practice of Acoustic Appropriation

David Sanjek

Pay Attention to the Man behind the Curtain

"He's paranoid, he's not, I'm obsessional, David's delusional. We're a band of modern noisemakers."[1] With these words, Don Joyce characterizes the members of the group Negativland and their collective musical enterprise. Lack of personal acquaintance with these individuals precludes any judgment of the accuracy of Joyce's psychological profile, yet the frequency with which Negativland has faced vigorous litigation since their formation in 1979 would drive anyone a bit batty. In addition, Joyce's characterization of their material as noisemaking seems an apt but incomplete designation for their body of work. Virtually everyone who comments upon Negativland finds themselves at loose ends coming up with appropriate and all-encompassing terminology. Stephen Ronan dubs them "an industrial/media/humor band" (1989: 89); Neal Strauss characterizes the group as "less a rock band than a watcher of rock bands" (Weisbard and Marks 1995: 266); Robert Christgau asserts that "they would have been called comedians or just wise guys in prepostmodern times" (1990: 288). Christgau may be closest to the mark. If nothing else, one can argue that the members of Negativland have been inveterate "cut-ups" in both senses of that phrase. They routinely lay waste to the absurdity of our collective cultural enterprise and the legal structures that define and deform its activities as well as disassemble and recombine the omnipresent sonic barrage into commentaries upon contemporary culture that combine hilarity and high theory.

Negativland's 1990 release *U2* is the most well known example of their satiric deconstruction: a 13-minute single that fuses recitation of the lyrics to the Irish band's "I Still Haven't Found What I'm Looking For" with obscene outtakes of the radio broadcaster Casey Kasem's "Top 40" program, thirty seconds of the original recording, and a host of other musical and non-musical materials.[2] The results were packaged with a cover that visually referred to the spy plane flown by Francis Gary Powers that was shot down over the Soviet Union in 1960. However, for most people, the recording *U2* remains as elusive a phenomenon as any memories of Powers's surreptitious activities. The single is, in point of fact, non-existent, commercially unavailable, a matter of rumor and clandestine distribution. Its manufacture and distribution was unceremoniously halted by a body of forces including the band U2's label (Island Records), Negativland's former label (SST Records), and Kasem himself. Negativland was accused of abrogating the copyright statutes by sampling U2 as well as Kasem without their permission, in addition to possibly interfering with sales of the group's recordings through the misleading design of the album jacket. Negativland refused to capitulate to these claims and managed over time to convince all but Kasem that *U2* was a work of parody that by no means meant to compete with the band U2 and, therefore, its samples were protected by the fair use clause of the copyright statutes as well as the 1994 Supreme Court decision in the case of 2 Live Crew vs. Acuff-Rose Music Inc.[3]

During the period 1990–95, Negativland responded to the litigation against them by a series of actions that ranged from concerted use of the press as a platform for public agitation to the publication of *Fair Use: The Story of the Letter U and the Numeral 2*. This volume includes a CD that features a collage piece entitled "Dead Dog Records" and comments on the fair use principle by one of the characters in Negativland's mythology, Crosley Bendix. In the text, Negativland states, "There is a very important cultural battle afoot to decide who will have the ultimate control over *what art will consist of*" (1995: 190). Response to both the recording *U2* and the subsequent actions by Negativland to defend and eventually reacquire the rights to their material have ranged from animosity to approbation. A counter-commentary to a statement published by Negativland in the music industry journal *Billboard* accused them of posturing in their own defense and promulgating "nonsense" (Negativland 1995: 158). Supporters of the group have been so adamant in their promotion of the right to acoustic appropriation that the FBI investigated death threats issued against Casey Kasem (Negativland 1995: 132). Whichever side of the debate one assumes, the release and recall of *U2* gives credence to Ray Pratt's argument that "In the United States sound itself is oppositional. It creates a new space, reinforcing within each person the possibility of an aggressively oppositional posture before existing social reality, but also . . . imposing a new order in sound on a context once oppressively closed" (1994: 210).

While the narrative that Negativland presents of the *U2* controversy stresses the machinations of corporate and legal entities behind the scenes, it must be added that the group was no less manipulative in creating, promoting, and defending their recording. Even if Don Joyce argues that it is necessary for creators to remove themselves from the effects of their work—he states to Stephen Ronan, "I think the artist has to distance himself from that in order to produce anything that's of any critical import or value. 'Cause it always involves somebody else's feelings"— *U2* amounts to a deliberate though purposeful contravention of the letter of the law and a potential abrogation of the rights to privacy of those individuals sampled in the recording, most obviously Casey Kasem (Ronan: 94). Their release of a media "virus" into the culture at large may be ideologically persuasive, but they cannot overlook the concurrent potential for their own infection.[4] Therefore, rather than laud or lambaste *U2* and the practice of acoustic appropriation, I wish instead to critique and problematize Negativland's work by raising a series of inquiries about the form and function of sampling: its heroic dimension, documentation in Craig Baldwin's collage film *Sonic Outlaws,* treatment of virtuosity, creation of narrative, relationship to marginality, incorporation of history, cultivation of boredom, incorporation of politics, and deliberate interjection of noise into the public consciousness. In no way will I exhaust either *U2* as a recording or the practice of sampling as a compositional tool and critical apparatus. Instead, I hope this inquiry will bring about a new appreciation of the complexity of acoustic appropriation as well as the potential contractions it embodies.

When the Going Gets Weird, the Weird Turn Pro

It has been argued that our age is devoid of heroism, that efforts at grandiosity amount to nothing more than illusory self-aggrandizement. The era of grand narratives that recounted such efforts is said to be defunct. In an often-quoted formulation, Jean-Francois Lyotard observes, "The narrative function is losing its functions, its great heroes, its great dangers, its great voyages, its great goals" (1984: xxiv). On the other hand, Negativland's efforts to recapture *U2* would seem to be the exception to this purported rule. Their dogged refusal to capitulate to either a corporate giant (Island

Records), an independent entrepreneur (Greg Ginn, owner of SST Records) or a self-possessed broadcaster (Casey Kasem) constitute a heroic enterprise that virtually achieved its goal: recuperation of *U2*. In a fax to Greg Ginn, Negativland states, "We believe that public opinion dramatically shifts on the basis of experience, not on theory" (1995: 157). Their experience can be said to prove that conviction carried to its ultimate conclusion can move moguls.

At the same time, the heroic dimension of Negativland's activities possesses more than a bit of irony, for one of the ultimate goals of *U2* is to deflate if not demolish the grandiosity of the band U2's material in general and Bono's lyrics in particular. The recitation of the words to "I Still Haven't Found What I'm Looking For" by band member David Wills (known in Negativland's mythology as "The Weatherman") in a flat and adenoidal voice that Neil Strauss accurately describes as "needle-nosed" erases any trace of poetic effusion from Bono's writing (Weisbard and Marks: 267). In one memorable passage, Wills elaborates upon the song's metaphor of "honey lips" by comparing the figure of speech to the melting of cheap plastic. Furthermore, the juxtaposition of Wills's recitation with the excerpts of Casey Kasem in which he castigates the Irish band for their, to his mind, absurd name demolishes whatever cultural capital U2 or its fans believe the quartet to possess. At the same time, which body of performers, one wonders, comes off as more critically defensible or ideologically acute: those with pretensions to verbal expressiveness and a conviction that power chords can change the universe, or technologically savvy smart-alecks with a microchip on their shoulders? In the end, *U2* ironically illustrates that Negativland has become that which they critique: heroes in spite of themselves.

It Will All Come Out in the Editing

In 1995, the San Francisco–based collage filmmaker Craig Baldwin released *Sonic Outlaws,* a documentary on acoustic appropriation that features Negativland in addition to other "culture jammers" such as the Tape Beatles, Emergency Broadcast Network, Barbie Liberation Organization, and critic/composer Douglas Kahn. Negativland coined the term "culture jamming" on their 1984 release *Jamcon 84*. Mark Dery defines the process as follows:

"Jamming" is CB slang for the illegal practice of interrupting radio broadcasts or conversations between fellow hams with lip farts, obscenities, and other equally jejune hijinx. Culture jamming, by contrast, is directed against an ever more intrusive, instrumental technoculture whose operant mode is the manufacture of consent through the manipulation of symbols . . . Part artistic terrorists, part vernacular critics, culture jammers . . . introduce noise into the signal as it passes from transmitter to receiver, encouraging idiosyncratic, unintended interpretations. (1993: 6–7)

The antecedents for this activity are numerous and cross a number of formal boundaries, from deliberate agit-prop to musical composition, the visual arts, and motion pictures. The type of cinema most germane in this regard is collage film: the deliberate recycling of found footage—some cast off or disregarded by the purveyors of the cinematic apparatus, some possessing widespread recognizability—into a novel form of montage that calls attention to and thereby critiques the footage's original context. At its best, collage film acts in more than a merely self-referential manner by being "media-referential" as well; these works, William Wees argues, "cannot avoid calling attention to the 'mediascape' from which they come, especially when they also share the media's formal and rhetorical strategies of montage" (1993: 25).

Craig Baldwin has produced collage film for the last two decades. His two works immediately preceding *Sonic Outlaws*, *Tribulation 99: Alien Anomalies under America* (1991) and *O No Coronado!* (1992), are media- as well as historically referential. Each film pointedly critiques a historical paradigm: the former, conspiratorial theorizing in the post–World War II period, and the latter, the colonization and subjugation of the Americas. Baldwin interweaves imagery from the mass media, both factual and fictional, to support and illustrate his theoretical points. He masterfully combines a wicked, sarcastic streak with a devotion to ideological critique. By so doing, he challenges the predetermined meaning of the images he combines and thereby "subjects the fragments of media-reality to some kind of deconstruction, or at very least to a recontextualizing that prevents an unreflective reception of representations as reality" (Wees 1995: 47). Individual juxtapositions of ideas and images retain global meanings and thereby avoid coming across as singleminded smart-aleck posturing. As the satirist Kurt Tucholsky argues, it is all too easy to produce nothing more than "critical cross-sections" of mediated materials that "cut the cheese without hitting the maggots" (Kahn 1988: 113).

The formal construction of *Sonic Outlaws* mirrors the compositional practices and agitational activities of Baldwin's protagonists, most particularly Negativland. He supports his storyline by "saturating it with the very strategy practiced by his subjects: unrelated sound and image are reborn and implicated in a barrage of poetic imagery that substitutes 'satellite dish' for 'stream of consciousness'" (Chang 1995: 85). That narrative conforms to a familiar ideological pattern in the critical analysis of popular music: the competition between corporate gigantism and entrepreneurial individualism. Many of the visual quotations Baldwin draws from Grade B and Z horror or sci fi films illustrate this paradigm, as does Negativland's own rhetoric in their interpolated interviews. However, the danger to this tactic is that by depicting the corporate forces pitted against audio appropriators as engaging in sledgehammer tactics, Baldwin and Negativland potentially personalize the issue and imply that systematic transformation of the copyright system may result from individual enterprise and chutzpah. Little is argued as to how coalitions might be organized or the means by which the law itself becomes transformed. Negativland and Baldwin indisputably root out some of the maggots, but reducing media corporations as large as Island or as small as SST to a demonic Other amounts to a rite of exorcism that may not protect or promote the right to artistic self-determination.

One Producer Is As Good As Another

In her discussion of the film musical, Jane Feuer distinguishes an instance of alienation between performer and audience central to all forms of mass entertainment. "Instead of a community where all, at least potentially, may perform," she writes, "relations of production are alienated from those of consumption. The performers do not consume the product and the consumers do not produce it" (1993: 2). In an effort to overcome that divide, the film musical posits "community" as an ideal concept and endeavors to elide if not erase the separation between performer and audience, to cancel the sense of professionalism, so to speak, by disguising the effort and expertise required to create a work of art. "Wouldn't it be nice," she asks, "to combine the charisma of the professional with the folksiness of the amateur?" (1993: 14). This utopian proposition resembles the Do-It-Yourself (DIY) agenda of punk, post-punk, hip-hop and deejay culture whereby the access to technology—whether guitar, turntable, or sampler—permits individuals to create their own commodities and not merely be passive consumers of the products released into the mass market. By erasing the customary and time-honored distinction between audiences and performers, the DIY model confirms Walter Benjamin's assumption that an apparatus is "better the more consumers it is able to turn into producers—that is, readers or spectators into collaborators" (1986: 234). The manner with which reproductive technologies in general and the digital sampler in particular encourage this process reinforces the critic's belief that "mechanical reproduction emancipates the work of art from its parasitical dependence on ritual"—in this case, the ritual valorization of technical virtuosity (Benjamin 1969: 224).

Ironically, in the original suit issued by Island Records against Negativland, it is asserted that *U2* is of such "poor quality"—e.g., lacking in either technical virtuosity or engineering expertise—that it might tarnish the group's reputation by causing unsuspecting consumers to assume that U2 permitted an unfinished or inadequate recording to be marketed to the general public (Negativland 1995: 15). One only need listen to *U2* in order to dismiss this assertion. Even if one did not know the amount of time and effort that Negativland puts into each composition—the process can take as long as a year—the density of assembled sound and the organizational sophistication with which those sounds are assembled belie any accusation of crudity or incompetence. At the same time, that very sophistication may not call into question the DIY agenda, but it does indicate that the erasure of the distinction between audience and performer is no simple matter. Bill Martin argues that musical virtuosity occurs only when individuals can surmount whatever difficulties confront them by the possession of consummate skills and a large musical vocabulary (1998: 101–103]. Those who dismiss the practice of acoustic appropriation as either nonmusical or a denial of traditional musical skills refuse to acknowledge that, as Chris Cutler believes, the digital sampler confronts us with "a musical instrument which is a recording device and a performing device—*whose voice is simply the control and modulation of recording*" (1995: 77). The ability to manipulate skillfully such an instrument requires its own form of virtuosity. Therefore, digital technology may allow the gulf between producer and consumer to be narrowed, but not every consumer can successfully sample. In the end, Negativland stands as the unique creators of *U2*, something that might not easily by accomplished by "you, too."

Desperately Needing Narrative

While critics of acoustic appropriation deplore what they perceive as the medium's lack of formal authority, the very practice of sampling demands an acute sense of form. The ability to decipher and detach a unique riff, rhythm, or verbal expression from an original context and then determine how it might facilitate the creation of a separate composition attracts individuals with a sensitive set of ears as well as a connoisseur's proclivity for search and retrieval. One of the most notable contemporary acoustic appropriators, DJ Shadow, whose *Entroducing* (1997) is wholly sampled, disdains those who pass up the hunt and allow others to do their ransacking for them.

Beat shopping is a culture. I never use reissues or bootlegs or compilations. I didn't use any CDs on the album. And I never . . . I don't just run out and buy Blue Note breaks, if you know what I mean. I spend a lot of time, almost all my time, trying to find the obscure thing. I figure the bigger my library is, the better I'm going to be able to tap into the exact sound that I want. (Rule 1997: 55)

While these practices may be dismissed as a restless search for novelty, what I observe instead in Shadow's quest is the effort to initiate a process whereby the acoustic appropriator can isolate what Susan Sontag designates as the "microstructures" in a piece of recorded information (Bann 1995: 120). This term refers to those seemingly trivial phenomena secreted within larger entities or practices that elude any effort at momentous themes or grand designs. By constructing material out of what otherwise might be considered inconsequential detritus, Sontag and others believe we can circumvent the powers-that-be associated with macrostructures as well as grant the force of agency to subordinate individuals and forms of expression.

Isolating narratives within narratives and thereafter using those elements to construct a coherent and original composition requires a talent for both deformation and reformation. *U2* demonstrates that Negativland possesses both sets of skills. It contains a critique of the group, the media through which they are promoted, and the process by which the owners of copyrighted material protect their property and endeavor to preserve its meaning. At the same time, some of Negativland's other works fail to cohere or present a argument whose ideological weight comes across as equal to its target. Their most recent recording, *Ideppiss,* a critique of mass marketing as practiced by the Pepsi Company, repeats its thesis almost as incessantly and without variation as does the corporation's advertising. Isolating "microstructures" requires that one determine how an individual piece of acoustic information possesses a surplus of possible meaning, even a meaning that the original owners or creators fail to acknowledge. The piece of acoustic appropriation into which that surplus of meaning is subsumed cannot become subservient to its sources. If one of acoustic appropriation's aims is social critique through the deconstruction of public sound and speech, then the techniques available by means of the digital sampler are a tactic to that end and not a technical aide to be indulged in for its own sake. Otherwise, as Tim Hodgkinson states, "the performer risks being revealed as the agent of the technology, with the performance degenerating into a public ritual of submission to alienated social power" (1996: 8). "Let's imagine," he requests, "that the point is not to use *all* sound, but only *all sound that matters*" (ibid.). Should we abandon such a process of deliberation, we might think we are constructing narratives, but we are merely placing one detail after another.

Is Anybody Listening?

The fact that an accelerated use of digital technology has resulted in the potential fragmentation of time-honored narrative models bears a social analogy. Human collectivity increasingly amounts to a contradiction in terms when faultlines of class, race, sexual orientation, even access to technology and information cause once stable communities to become isolated and inhibited. Among the many social bodies affected by this erosion of bonds of association is the innercity demimonde. C. Carr laments, "I've found it harder to crack the art margins lately . . . The climate for things experimental, for things adversarial, has only worsened; the damage to these 'autonomous zones' has been so intensively exploited that it's hard to become 'invisible'" (1993: 314).[5] The diaspora of the avant-garde from no longer supportable urban landmarks—the Left Bank, Greenwich Village, Haight Ashbury, and others—has resulted in a nomadic, transitory culture: "the boho as hobo," Carr observes (1993: 320). In response to a pervasive sense of social dispossession, a widespread fascination with the utopian possibilities of the Internet and digital technology has arisen. The advocates of cyberculture promise transcendence from all forms of inhibition, physical and metaphysical. For those in Bohemia whose allegiance to a civic order remains voluntary at best and tenuous at least, the possibility of what Mark Dery characterizes as "a technoeschatology of their own—a theology of the ejector seat" possesses obvious attractions (1996: 8). Of course, replacing social bonds with digital associations not only abandons those without access to the requisite technology but also requires an indissoluble commitment to those who manufacture and profit from that equipment. It is impossible to reach out and touch someone where there is no one at the other end of the line or they are simply unable to accept the charges. If Bohemia has always been inescapably marginalized, why should we assume that its members will be automatically affiliated, let alone capable of countercultural subversion, once online?

Furthermore, as Andrew Ross reminds us, the very notion of a "subversive" was invented by the FBI, not the avant-garde. It is a matter of individual identity and social function, Ross states,

> familiar to people who have a very secure sense of identity in the world. And the option of obliterating that identity and then romanticizing what results is something of a luxury, a privilege, That's not to discount it, but it is to say that there is an uneven effect of that kind of resource across social sectors. (Reynolds and Zummer 1997: 58)

One of the most crucial forms of privilege that contributes to those opportunities for subversion through the use of digital technology is the possession of cultural capital. Without an extensive body of knowledge brought about through ready access to the present day deluge of forms of expression, online and offline, it is virtually impossible to appreciate, let alone emulate, the transformation of a sampled stream of expression into critical commentary. As Dick Hebdige asserts, "in order to *e*-voke, you have to be able to *in*-voke. That's the beauty of quotation. The original version takes on a new life and a new meaning in a fresh context" (1987: 14). The transformation brought about by acoustic appropriation remains sterile and devoid of any resonance or potential social impact if a sampled quotation fails to evoke any comprehension of its original context. When neither memory nor the possession of cultural capital stimulates critical consciousness, digital technology cannot effectively reconnect those bonds of association our nomadic society fragments.

    U2 succeeds so well as a piece of acoustic appropriation because it builds upon and calls into question the widespread public awareness of U2, Casey Kasem, and the marketing of popular music through the mass media. As a result, it epitomizes the process Douglas Kahn (1988: 120) dubs "*mimikry*," whereby the presence of a total entity permits attention to the variety of meanings that entity possesses without assuming the acquisition of specialized knowledge on the part of the public. This tactic "reflects back not only upon its own act but brings the social existence of the original, with which it is inseparable, into scrutiny." Other compositions by Negativland, particularly *Helter Stupid/The Perfect Cut* (1989) and *Guns* (1991), persuasively illustrate "*mimikry*" and, like *U2*, build upon the ubiquitous presence of the mass media, the relentless processing of commercial entertainment, and the persistent equation of popular culture with organized aggression. On other occasions, such as *Free* and *Ideppiss,* the sampled material exists in a vacuum and fails to appeal to a preexistent body of public knowledge. The results are, therefore, self involved and alienating, the chatter of another set of "hobos" trekking along the "boho" highway.

What Time Is It?

    The practice of acoustic appropriation invariably fuses and confuses any number of domains, principal amongst them that of time. Sampling neither recognizes nor respects the boundaries of temporality. As a result, the fine distinctions we construct to delineate rhythmic structures, musical time periods, or definable genres fall by the wayside. In the process, fragments of spaghetti Western soundtracks might collide with elements of Jamaican dub over which the mellifluous sounds of an R&B vocalist waft while synthesized percussion clatters and crashes. History itself lands in the cuisinart, as time is seized and wrestled to the floor. The ability of the digital sampler to plunder sound and divorce it from its original context invites if not requires such an aggressive stance. Chris Cutler remarks, "as a pirated cultural artifact, a found object, as debris from that sonic environment, a plundered sound also holds out an invitation to be used *because* of its cause, and because of all the associations and cultural apparatus that surround it" (1995: 74). The manner in which some forms of acoustic appropriation—dub, house, techno, or that loose baggy monster of a musical niche marketed as electronica—appeal to if not invite a hallucinogenic state of consciousness illustrates the degree to which this form of music steps out of time altogether, denies its boundaries, runs our master narratives about temporality straight into the wall.

It is also apparent that such an approach can unleash all of the dubious, aggressive, and potentially imperialistic qualities inherent in the process of freewheeling appropriation. If one considers a sound to be there for the taking, then one frees any individual fragment from all manner of once inalienable forms of protection and boundaries of definition. Chris Cutler accurately assesses the environment that results: "Leakage, seepage, adoption, osmosis, abstraction, contagion: these describe the life of sound work today" (1995: 79). That "life" can lead to the reduction of all forms of sound to a kind of freewheeling sonic smorgasbord, what Andrew Ross derides as the "retro-quotationism of the style market" (1991: 153). Issues of power and property can subsequently be elided or ignored, all cast aside in the quest for grooves. The flipside of the joyful ecumenicism that accommodates all forms of expression without reducing the field to a canonical hierarchy can be a kind of unintended free market mentality. The deejay, sampler, or mixologist without realizing it takes on the values of "the creative maverick universally prized by entrepreneurial or libertarian individualism" (Ross 1991: 162). When that appropriated material comes from a range of time periods, "history itself," René Lysloff argues, "is the object of economic exploitation and expansion, offering a virtually limitless supply of natural and cultural resources while also providing an abundance of cheap industrial goods. In a nutshell, the past becomes the future's third world" (1997: 206).

Greg Tate has lauded the conflation of time periods as specifically practiced in hip-hop culture by dubbing the process "ancestor worship" (1992: 130). The frequent incorporation of, say, the drumbeat from James Brown's "Funky Drummer" can be applauded, he infers, as giving the Godfather his props. On the other hand, in a discussion of the contemporary R&B vocalist R. Kelly, Paul Gilroy questions to what degree the intertextual tradition "makes the past audible in the here and now but subserviently. History is conscripted into the sacrifice of the present"; the resulting "narrative shrinkage" comes across to him as "simultaneously insubordinate and reverent" (1997: 87). Gilroy's observations lead one to ask how often a sampled musical quotation brings about a reinvestigation or reconsideration of the material's source. Furthermore, how often does that quotation lead to progressive impulses that might serve libertory ends and not just the liberation of the groove? Gilroy laments that sampled musical citations "do not play with the gap between then and now but rather use it to assert a spurious continuity that adds legitimacy and gravity to the contemporary" (1997: 111). A less than fine line might be argued to exist between "ancestor worship" and the insatiable inquiry after recyclable material on the part of acoustic appropriators or the culture industries that support them.

U2 conflates temporality in a critical manner by juxtaposing the pomposity inherent in Bono's lyrics and the band's stance toward their audience with Casey Kasem's derogatory comments upon their very name as well as his assumption that the group has little to no professional future. At the same time, the association of the group and Negativland's recording with the actions of Francis Gary Powers exists in a kind of temporal freefall. A parallelism can be said to exist between the group's clandestine activities and those of the pilot, but to equate cold war anti-Communism with the appropriation of a pop group's lyrics and rhythm tracks lacks a certain substance. An equivalent kind of overloaded historical parallel occurs in the composition "Dead Dog Records," included in *The Letter U and the Numeral 2*, wherein a passage of blues is juxtaposed with the comments of a British musician—perhaps one of the Rolling Stones?—about the influence of African American music upon their recordings. While the passage accurately illustrates Negativland's assumption that all music is based upon some form of theft, it fails to inform the analogy with much sense of the manner in which class, race, and region influenced the appropriation in the worst sense of African American forms by Western Europeans. The point Negativland wishes to make is inarguably significant, but its collapsing of temporality empties the argument of some of its vitality and certainly a great deal of its ideological substance. Conflate the deep South of the 1920s and 1930s and the postmodern present all they wish, Negativland still haven't found what they're looking for.

Daring to Be Dull

From the beginning of its career, Negativland has been drawn to the seductive phenomenon of boredom. What functions for most people as little more than "an all-purpose index of dissatisfaction" elicits from the group an insatiable appetite for the trivial and the commonplace (Spacks 1995: 249). The allure of the banal positively possesses Negativland. They are able to transform things that many might feel to be mind-numbing or devoid of any critical intelligence into the dadaism of daily life. Whereas contemporary society endeavors to inculcate a virtual entitlement to satisfaction, an obstacle-free opportunity to fulfill every desire, Negativland opts instead for an entitlement to be intrigued by the insignificant. Unsurprisingly, of all the dreck and detritus that comprises our quotidian existence, the group is most drawn to the babble of the mass media. Their compositions routinely incorporate the white noise that emanates from television and radio: public service announcements, advertisements, talk radio and shock TV commentators, the high-energy pronouncements of disk jockeys. Group member Don Joyce has amassed a virtual warehouse of such material, which he routinely interjects into either their recorded compositions or his weekly radio program, "Over the Edge," broadcast over Berkeley's community-sponsored radio station KPFA. In addition, David Wills ("the Weatherman") obsessively tracks and records random conversations through the use of a scanner.

The interpolation of such material serves two purposes for Negativland. First, as an act of culture jamming, it permits the group, and its fans, to make use of the media that engulfs them and thereby vitiate if not eradicate its influence. In the comments included in *The Letter U and the Numeral 2*, Crosley Bendix states that the "emotional relevance" of acoustic appropriation amounts to an "opportunity for self-defense against media coercion." Culture jamming constitutes, he asserts, a "sometimes nasty but wholly appropriate response to a society in decline and denial." The perpetuators of this process take the promise of digital technology to heart and staunchly demand, Mark Dery observes, to be "interactive rather than passive, nomadic and atomized rather than resident and centralized, egalitarian rather than elitist" (1993: 14). By refusing to become overwhelmed and thereby paralyzed by the boring blather of the mass media, the acoustic appropriator insists upon "a true plurality in which the univocal world view promulgated by corporate media yields to a multivocal, polyvalent one" (ibid.: 16). Refuse must be reused. The lamebrained becomes a form of liberation. *U2* itself has its origin in a desire to recuperate the banal obscenities of Casey Kasem and find a functional use for his absurd anger over having to make a dedicatory address to a dead dog, Snuggles. His cussing becomes a critique of the vaunted position we allow media figures to occupy as well as the sanctimonious manner with which the public media attempts to mediate private passions.

Second, and equally important, Negativland is attracted to tedium for its own sake. They find substance and a source of amusement in what others feel to be mind-numbing. This does, however, mean that the acoustic material and the individuals who are thereby recorded are to a significant degree taken for granted, their very words abstracted from either context or commentator. When David Wills is shown using a scanner in *Sonic Outlaws,* he responds to Craig Baldwin's question about the technology's potential violation of privacy by stating, "It probably does, but I just use it to provide stuff for the group. Find sound and then turn it over to them and let them deal with it." Another member of Negativland characterizes the ham radio jammers who inspired the practice of culture jamming as "guys going on and just causing lots of trouble" and their activities as "all pretty juvenile but we liked it as a metaphor." Furthermore, some of Negativland's recordings, particularly those in the *Over the Edge* series drawn from their radio program, seem positively suffused with the juvenile. Volume 6, *The Willsaphone Stupid Show,* consists largely of tape recordings made by David Wills with microphones set about his parents' home interspersed with passages of the prepubescent Wills announcing weather patterns. What some might consider the surrealism of suburbia comes across more often as an endless litany of monotonous details about mom, baked goods, and the family pets. In his commentary on fair use, Crosley Bendix states, "Appropriation by its very nature results in something funny, and 'funny' can be just as important in life and culture and art as all that serious stuff that will get you ideological followers or a grant." *The Willsaphone Stupid Show* forces one to ask what the joke might be and who its target is.

Jane Feuer argues, "To dare not to be entertaining is the ultimate transgression, the ultimate form of reflexivity as critique. For to be unentertaining means to think about the base upon which mass entertainment is constructed" (1993: 92). Most often, Negativland's courting of boredom accomplishes this task. In and of themselves, Casey Kasem's scatological comments are of limited significance and provide a limited form of entertainment. Encased in the collage that constitutes *U2,* they take on a whole other meaning and accumulate a value unavailable when they stand alone. The recontextualization of his words brought about by Negativland forces us to pay attention, and in this case the cost is well worth the effort. The art critic Dave Hickey assesses the value of the investment in a work as being directly proportional to "the amount of risk the investor takes on behalf of the work in question" (1997: 110). Predicating a composition upon tedium amounts to a considerable gamble, one Negativland has seen fit to take time and time again. Sometimes they come up boxcars, sometimes craps.

The Luxury of a Loud Megaphone

In *Sonic Outlaws,* Negativland discusses their KPFA radio program "Over the Edge," broadcast since 1981. The station, a member of the listener-sponsored Pacifica Foundation, is powered by 59,000 watts and reaches virtually a third of the state of California. No wonder Negativland jokes that they possess the luxury of a loud megaphone, but the question remains, what has the group done with that opportunity and is there a deliberate or even implicit political framework to their activity? Andrew Ross reminds us that technologies such as radio or sound recordings are "much more than hardware objects or technical extensions of the human body. Technologies are also intentional linguistic processes" (1991: 3). Those technologies are not imposed upon consumers but form a deliberately chosen portion of their lives. Attributing either libertory or onerous characteristics to them is beside the point, for the ability of mediated forms of communication to enhance or imprison any one of us remains a matter of choice. Ross adds, "No frame of technological inevitability has not already interacted with popular needs and desires; no introduction of new machineries of control has not already been negotiated to some degree in the arena of popular consent" (1991: 98).

The longstanding broadcast of "Over the Edge" underscores the fact that a critical mass of listeners have tuned in to enjoy the deformation of the mass media that Negativland perpetuates. Furthermore, despite the fact that court records attached to the prosecution of *U2* indicate that the sale of their recordings is far from significant, the vigorousness of those who came to the group's defense illustrates that strength of conviction outweighs a weakness of numbers. At the same time, the rationale that Negativland mounted for their appropriative procedures could be accused of being excessively libertarian, more concerned with the group's ability to function than with a change in the system of production and consumption that reigns in the mass media. Negativland argues in their essay on fair use, "Has it occurred to anyone that the private ownership of mass culture is a bit of a contradiction in terms?" But, the weight of their efforts comes down to saying show me the master tapes! (Negativland 1995: 195). The group's efforts to tweak the nose of those in power—as when they published Greg Ginn's financial records or asked a member of U2, guitarist The Edge, to underwrite their campaign against Island Records—illustrate an ingenious ability to take the upper hand. Nonetheless, one feels compelled to ask, in the words of Mark Dery, whether "This is D. H. Lawrence's 'revolution for fun,' staged by reality hackers who want to upset the apple cart not for lofty principles, but to see which way the apples will fall" (1993: 34).

Dery's observations remind one that in his writings on Charles Baudelaire, Walter Benjamin distinguishes between "occasional" and "professional" conspirators. The former dedicate themselves to the piecemeal skirmishes of the provocateur while the latter commit themselves to long-term social transformation (1976: 11, 12). A number of individuals who argue for the libertory potential of digital media believe that "professional" agitation is no longer tenable, because "If the mechanisms of control are challenged in one spatial location, they simply move to another location" (Critical Art Ensemble 1996: 9). Collective social activism confronts not only the inertia of the present-day power structure but also the increasing fragmentation of oppositional groups into compartmentalized and incompatible niches. "Occasional" agitation may be the only workable alternative, yet each chink in the wall of consensus reality enables another challenge to the status quo. That should not blind us to the fact that there is "something inescapably American about this philosophy of Boundless Expansion, Self-Transformation, Intelligent Technology, Spontaneous Order, and Dynamic Optimism, reconciling as it does the mechanist reductivism of artificial intelligence with the evangelical zeal and relentlessly peppy can-do of the human potential movement" (Dery 1996: 305).

Negativland's advocacy of acoustic appropriation does, at times, embody this relentlessly upbeat sensibility, particularly when the social frame of reference of the material being collaged comes across as lacking in depth or focus. Furthermore, it can be argued that thumbing one's nose is not the same as raising one's fist. The recording of *U2* and the efforts to retrieve its ownership provide us with a vigorous and laudable effort at critiquing the copyright regime. Yet, as the process of sampling requires that one have access to products released by the mass media, it inescapably binds one to the system one wishes to disempower. Therefore, the most significant political dimension to Negativland's activities may well be not so much the ideological claims they espouse but the model they offer for others to follow. The group argues, "These now all-encompassing private locks on mass media have led to a mass culture that is almost completely 'professional,' formularized, and practically immune to any form of bottom-up, direct-reference criticism it doesn't approve of" (Negativland 1995: 251). Others have followed and will follow in their wake, despite the considerable obstacles that block any works of acoustic appropriation.[6] Simon Reynolds states that "Sampling should not be the cue for pop to eat itself so much as to breach itself. Sampling can mean disorientation, expansion, the disruption and death of the song, sonic architecture, futurism. Who is brave, willing or capable enough to embrace these prospects?"

(1990: 171). Quite possibly, it is Negativland and those other "unprofessionals" who wish to man the sonic barricades.

Turning the Barrage back upon Itself

At the beginning of this essay, I quoted Ray Pratt's assessment that "In the United States sound itself is oppositional" and asserted that *U2* was proof of his argument. I would like to conclude by developing further how *U2* and the process of acoustic appropriation create a "new space" and impose a "new order in sound on a context once oppressively closed." While Pratt does not directly refer to the seminal volume by Jacques Attali, *Noise: The Political Economy of Music*, his analysis mirrors Attali's in many ways. For Attali, music is the domestication of noise, and the organization of disparate sounds into a coherent composition acts as a means of consolidating the community or totality that supports that music and hears in it something that compliments the community's notion of social order. Noise in its literal form is the acoustic equivalent of chaos or anarchy and, therefore, "it is necessary to ban subversive noise because it betokens demands for cultural autonomy, support for differences or marginality" (Attali 1985: 6). But what kind of "noise" is it that *U2* constitutes and how might the manner in which Negativland created the piece offer a means of articulating a form of cultural autonomy?

The powers-that-be felt the recording to constitute Attali's notion of "noise" in two ways. First, the very collage of elements was perceived as dissonant in the extreme, a reflection of inadequate musicianship and a paucity of technique. In other words, *U2* did not sound the way a recording was supposed to sound. As mentioned earlier, the original lawsuit by Island Records argued that purchasers of *U2* could be falsely led to assume that the band had lost their very ability to produce a competent recording. Guitarist, former member of Black Flag, and SST Records president Greg Ginn echoed this line of argument when he accused Negativland of being amateurs in the worst sense of that term: "The fact that Negativland is but an occasional hobby for it's [*sic*] members has allowed them the freedom to take well deserved pot shots at the music industry" (Negativland 1995: 51). They are not, unlike Ginn, *real* musicians. The manner in which Negativland composes and performs music deforms the normative definitions of those processes that are recognized by professional musicians and recording companies.

And yet, it is the second manner in which the recording constitutes "noise" that illustrates Attali's notion of the word as well as Pratt's argument. *U2* constitutes a potent form of "noise" in that it forces the listener to absorb and critically assess a barrage of sound. To ironically call upon the title of one of Negativland's earlier releases, the power of *U2* results from the fact that that there is no escape from noise. At the same time, the "noise" embodied by *U2* amounts to a particular form of agitation that demands an appropriate process of cognition, one that must be distinguished from the form of cognition customarily associated with sampled or appropriated music. The British music critic Simon Reynolds argues that the forms of "noise" and "bliss" that can be distinguished in certain sampled music are one and the same thing: "A rupture/disruption in the signifying system that holds [a] culture together" (1990: 13). Ultimately, any form of music that embodies "noise," he believes, should bring about "self-subversion, overthrowing the power structure in your own head," in order to annihilate "the mind's tendency to systematize, sew up experience, place a distance between itself and immediacy" (1990: 59). This kind of cognitive dissonance can be illustrated by the use of Rickie Lee Jones's comments about her Western childhood in Orb's "Fluffy Little Clouds," found on the album *Live 93*. Specifically what the singer says matters very little, and our apprehension of her words results in no significant or intended transformation of our notions of geography or memory. Her language amounts to one of many elements in the piece, no more or less important than the samples from Ennio Morricone's score for Sergio Leone's *Once upon a Time in the West* (1968). Orb renders the content of Jones's speech into ephemeral discourse as tenuous as those very clouds she describes.

In the case of *U2*, each piece of language, even the most trivial, is important, and our apprehension of how each fragment fits into the final work demands of us a *cognition of dissonance*. We are made aware that even the most patently absurd comments, like those addressed to a dead dog, Snuggles, can possess significance by virtue of the refusal on the part of the composer/musician to treat that acoustic information as ephemeral. In an ironic sense, despite Negativland's desire to deform and deconstruct language and media discourse, it is a positively logocentric ensemble. (One might even accuse them of being what Reynolds damns as "logocrats,": artists whose aim is "providing sociological rational [*sic*] in advance" for their work [1990: 110].) They remind us that, much as we must consider and critique the political economy of popular culture, we must pay equal attention to "the political economy of the everyday, of how we choose to watch or listen or make culture" (Davies 1995: 128). Their fight over fair use and the right to release *U2* amounts to not merely a legal debate but the legal arm of a larger process whose aim is to direct our attention to the acoustic barrage all about us. Maybe Don Joyce was right, and Negativland are just a merry bunch of "modern noisemakers." For, in the end, if *U2* teaches us anything, it forces us to remember that we cannot escape, but must instead embrace, "noise."

Notes

1. Don Joyce makes these comments when Negativland is first introduced in Craig Baldwin's *Sonic Outlaws*.
2. See Negativland, *The Letter U And The Numeral 2* (261–65) for a verbal transcription of the "1991 A Capella Mix" and "Special Radio Edit Mix" of U2. As far as I'm aware, this is only the second transcription in print of a sampled composition, the other being the Cold Cut remix of Eric B. & Rakim's "Paid In Full" included in *Signifying Rappers* (Costello and Wallace 1990). The recording can be found on the soundtrack to the 1988 Dennis Hopper–directed film *Colors* (Warner Brothers 9 25713–1).
3. I have addressed this case and the decision's effect upon the practice of sampling in "Ridiculing the 'White Bread Original': The Politics of Parody and the Preservation of Greatness in 2 Live Crew vs. Acuff-Rose Music Inc.," a talk delivered at the 1995 American Studies Association meeting and in expanded form in 1998 at a New York University Music Department colloquium.
4. The notion of a "media virus" has been promulgated by the analyst Douglas Rushkoff. He describes the term as the infusion of new ideas, a series of "information 'bombs,'" into the public sphere; once attached to that sphere, the virus "injects its more hidden agendas into the datastream in the form of ideological code" (1994: 8, 10). These viruses prevent our culture from becoming monolithic or vapid and can be most readily detected in the predilections and pastimes of adolescents. These children of the media, "screenagers" to Rushkoff, are profoundly altering our society (1996: 3). In response to those changes, we should acquire a radical form of open-mindedness, "*pro*noia" rather than paranoia in Rushkoff's language, that regards all forms of technology as inherently socially productive and ideologically acceptable (1996: 154). I have critiqued Rushkoff's work in "'I Ain't Afraid Of No Kids': Douglas Rushkoff and the Ascendance of the Digital Sublime."
5. The "temporary autonomous zone" or T.A.Z. is the construction of Hakim Bey, who defines the phenomenon as "like an uprising which does not engage directly with the State, a guerrilla operation which liberates an area (of land, of time, of imagination) and then dissolves itself to reform elsewhere/elsewhen, *before* the state can crush it" (1991: 101). The T.A.Z. has been widely influential amongst many counter-cultural communities, both online and offline.
6. The most recent advocate of acoustic appropriation is the work of one "Philo T. Farnworth," whose CD *Deconstructing Beck* takes apart and critiques the work of this successful contemporary artist, who himself employs, albeit with the required permissions, an ample number of samples (Bunn 1998: 31).

References

BOOKS AND ARTICLES

Attali, Jacques. 1985. *Noise: The Political Economy of Music*. Translated by Brian Massumi. Minneapolis: University of Minnesota Press.

Bann, Sally. 1993. *Greenwich Village 1963: Avant-Garde Performance and the Effervescent Body*. Durham, N.C.: Duke University Press.

Benjamin, Walter. 1976. *Charles Baudelaire: A Lyric Poet in the Era of High Capitalism*. Translated by Harry Zohn. London: Verso.

———. 1969. *Illuminations*. Translated by Harry Zohn. New York: Schocken Books.

———. 1986. *Reflections*. Translated by Edmund Jephcott. New York: Schocken Books.
Bey, Hakim. 1991. *T.A.Z. The Temporary Autonomous Zone: Ontological Anarchy, Poetic Terrorism*. New York: Autonomedia.
Bunn, Austin. 1998. "Free Samples." *Village Voice* (March 3): 31.
Carr, C. 1993. *On Edge: Performance Art at the End of the 20th Century*. Hanover, N.H.: Wesleyan University Press/University Press of New England.
Chang, Chris. 1995. "Property Is Theft." *Film Comment* 31/5 (Sept.–Oct.): 85.
Christgau, Robert. 1990. *Christgau's Record Guide: The 1980s*. New York: Pantheon.
Costello, Mark, and David Foster Wallace. 1990. *Signifying Rappers: Rap and Race in the Urban Present*. New York: Ecco Press.
Critical Art Ensemble. 1996. *Electronic Civil Disobedience and Other Unpopular Ideas*. Brooklyn, N.Y.: Autonomedia.
Cutler, Chris. 1995. "Plunderphonics." In *Sounding Off! Music As Subversion/Resistance/Revolution*. Edited by Ron Sakolsky and Fred Wei-Han Ho. Brooklyn, N.Y.: Autonomedia: 67–89.
Davies, Ioan. 1995. *Cultural Studies and Beyond: Fragments of Empire*. New York: Routledge.
Dery, Mark. 1993. *Culture Jamming: Hacking, Slashing and Sniping in the Empire of Signs*. Westfield, N.J.: Open Magazine Pamphlet Series.
———. 1996. *Escape Velocity: Cyberculture at the End of the Century*. New York: Grove.
Feuer, Jane. 1993. *The Hollywood Musical*. 2nd ed. Bloomington: Indiana University Press.
Gilroy, Paul. 1997. "'After the Love Has Gone': Bio-Politics and Ethno-Politics in the Black Public Sphere." In *Back To Reality? Social Experience and Cultural Studies*. Edited by Angela McRobbe. Manchester: Manchester University Press: 83–115.
Hebdige, Dick. 1987. *Cut 'n' Mix: Culture, Identity and Caribbean Music*. London: Methuen.
Hickey, Dave. 1997. *Air Guitar: Essays On Art and Democracy*. Los Angeles: Art Issues Press.
Hodgkinson, Tim. 1996. "Sampling, Power and Real Collisions." *Resonance* 5/1: 6–10.
Kahn, Douglas. 1988. *John Heartfield: Art and Mass Media*. New York: Tanam Press.
Lyotard, Jean-Francois. 1984. *The Postmodern Condition: A Report on Knowledge*. Translated by Geoff Bennington and Brian Massumi. Minneapolis: University of Minnesota Press.
Lysloff, René. 1997. "Mozart in Mirrorshades: Ethnomusicology, Technology, and the Politics of Representation." *Ethnomusicology* 41/2 (spring/summer): 206–219.
Martin, Bill. 1998. *Listening to the Future: The Time of Progressive Rock, 1968–78*. Chicago: Open Court.
Negativland. 1995. *Fair Use: The Story of the Letter U and the Numeral 2*. Concord, Calif.: Seeland.
Pratt, Ray. 1994. *Rhythm and Resistance: The Political Uses of American Popular Music*. Washington, D.C.: Smithsonian Institution Press.
Reynolds, Robert, and Thomas Zummer. 1997. "Cybernetic Capitalism and Surplus Intelligence: An Interview with Andrew Ross." In *Crash: Nostalgia for the Absence of Cyberspace*. Edited by Reynolds and Zummer. New York: Thread Waxing Space.

Reynolds, Simon. 1990. *Blissed Out: The Raptures of Rock.* London: Serpent's Tail.
Ronan, Stephen. 1989. "Is There Any Escape from Stupid? An Interview with Don Joyce and Mark Hosler of Negativland." *Mondo 2000:* 89–100.
Ross, Andrew. 1991. *Strange Weather: Culture, Science and Technology in the Age of Limits.* New York: Verso.
Rule, Greg. 1997. "DJ Shadow and Akai MPC = History." *Keyboard* (October): 51, 53, 55–56, 59–60.
Rushkoff, Douglas. 1994. *Media Virus! Hidden Agendas in Popular Culture.* New York: Ballantine.
———. 1996. *Playing the Future: How Kids' Culture Can Teach Us to Live in a World of Chaos.* New York: Harper Collins.
Sanjek, David. 1998. "'I Ain't Afraid of No Kids!' Douglas Rushkoff and the Ascendance of the Digital Sublime." *The Review of Education/Pedagogy/Cultural Studies* 20/1: 173–81.
———. 1995, 1998. "Ridiculing the 'White Bread Original': The Politics of Parody and the Preservation of Greatness in 2 Live Crew vs. Acuff-Rose Music Inc." Presentation before the American Studies Association in 1995 and the New York University Music Department in 1998.
Spacks, Patricia Meyer. 1995. *Boredom: The Literary History of a State of Mind.* Chicago: University of Chicago Press.
Tate, Greg. 1992. *Flyboy in the Buttermilk: Essays on Contemporary America.* New York: Simon and Schuster.
Wees, William. 1993. *Recycled Images: The Art and Politics of Found Footage Films.* New York: Anthology Film Archives.
Weisbard, Eric, with Craig Marks, editors. 1995. *The SPIN Alternative Record Guide.* New York: Vintage.

RECORDINGS

Colors (soundtrack). 1988. Warner Brothers 925713–1.
D.J. Shadow. 1997. *Entroducing.* Mo Wax/FFRR 697–1245–123–2.
Negativland. 1987. *Escape from Noise.* SST 133/RecRec 17.
———. 1993. *Free.* Seeland 009.
———. 1991. *Guns.* SST 272.
———. 1989. *Helter Stupid.* SST 252/ RecRec 29.
———. 1997. *Ideppiss.* Seeland 017.
———. 1984. *JamCon '84.* Seeland 004.
———. 1995. *Sex Dirt.* Over the Edge, vol. 8. Seeland 015.
———. 1991. *U2.* [Originally SST 272.]
———. 1993. *The Willsaphone Stupid Show.* Over the Edge, vol. 6. Seeland 011.
Orb. 1993. *Live 93.* Island 162–535–004–2.

FILMS

Baldwin, Craig. 1992. *O No Coronado.*
———. 1995. *Sonic Outlaws.*
———. 1991. *Tribulation 99: Alien Tribulations under America.*

CHAPTER SIXTEEN

Afterword
Back to Basics with the Roland 303

Andrew Ross

One of the things this essay collection demonstrates is that advanced technologies are often sold on the premise that they can deliver elemental experiences which are no longer available through the technologies they are seeking to supplant. According to this pitch, our current toys have formed an obstructive, mediating layer that the new ones will leap over and restore access to the authentic stuff. This is a key principle of techno-primitivism, and the best contemporary example I know of is the dance music genre that is labeled "trance."

Trance is now so popular in so many First World nations that it can't help but be reviled by discerning music critics in every one of these nations. Not since disco's critics saw Moroderania spreading like the bubonic plague has an industry formula swept the boards with less resistance. U.S. mass taste has been fiercely isolationist in its indifference to dance music, but large sectors are now turning into treacle before the lush, sweet soundscapes of Paul Oakenfold, Paul Van Dyk, Sasha and Digweed, and the other anthem champs. As passionately as its devotees flip their lids, the discerning critic decries the trance formula as an engine of easy sentiment, its practitioners as manipulators of middlebrow escapism, and their product as the last word in cheese. Fair enough as a verdict of taste, but quite irrelevant to any understanding of the sheer popularity of the music's quasi-symphonic euphoria.

Nor can it help forty-somethings, like myself, whose crush on the genre has all the makings of a top-notch guilty pleasure. Physically outdone, these days, by the challenge of the nightclubber's graveyard shift, my use of the music as an emotional regulator, at home, in the office, or in transit, has become both routine and devotional. I'm convinced that the attachment has little to do with envy of a scene in which I cannot participate, or nostalgia for the blissed-out yearnings that are mandatory in late adolescence. My own memories of that period carry way too much guilt and too little pleasure for that. Besides, the wonderment of the time had a firmly medievalist cast that is quite removed from the galactic odysseys of trance. Indeed, the first album I really coveted was "In the Court of the Crimson King." Its jugglers, fire witches, jesters, prophets, and knightly tournaments were of a piece with those Tolkienesque (and Lovecraftian) currents of the hippie mind that fed into progressive rock and managed to survive in an abridged edition through the Celtic twilights and Teutonic death-fests of heavy metal. The pixie-dust and sword & sorcery were straightforward, late Romantic fantasies about a simpler, more flamboyant past, but there were no illusions about the advanced technology that delivered them. According to the recipe for this techno-peasant cocktail, what made the debut of the Moog synthesizer so "progressive" (for King Crimson, Yes, Emerson, Lake, and Palmer, and their like) was not its futurist feel but its ability to conjure up associations with ancient tongues and Nostradamian auguries.

Thirty years later, the trancey descendants of the Moog are still in possession of the conjurer's bag of tricks. The transcendental journey is still the coin of the realm. But the medievalism has dissolved. Perhaps the gaming boom of Dungeons and Dragons and all its descendants has drained the market for the time being. Youth's wistful historicism has given way to a geographical imagination that carries the clear imprint of popular ecology. Where fey chevaliers and eloquent griffins used to hold court, Gaia is now the presiding deity. No doubt this has a lot to do with the centrality of environmentalism in youth consciousness—the sublime goal of planetary care is much more appealing and less threatening than the impure politics of local statecraft. Since trance is almost wholly instrumental, this one-worldism is commonly evoked in the ethereal soundwash (in the "oceanic feeling" that Freud described as the source of religious musings about eternity), or in the iconography, trappings, and accessories of its rolling raver fans.

Vocal samples in trance are few and far between, and they tend to verbalize a mood rather than a message. Every so often, however, a dreamy female voice utters something that requires thought, as in the breakout vocal on the Lost Tribe (aka Matt Darey) classic, "Gamemaster," where the Gaian profile is quite explicit. "Embracing the goddess energy within yourselves will bring all of you to a new understanding and value of life . . . Like a priceless jewel buried in dark layers of soil and stone, Earth radiates with brilliant beauty into the caverns of space and time. Perhaps you are aware of those who watch over your home and experience it as a place to visit and play with reality. You are becoming aware of yourself . . . as a Gamemaster." It's hard for me not to think of this as a software upgrade of Richard Brautigan's 1968 famous poem, much admired by the early Bay Area cyberculture, about the dream of "a cybernetic meadow / where mammals and computers / live together in mutually / programming harmony / like pure water / touching clear sky," and all of it "watched over by machines of loving grace." In the Lost Tribe sample, the kinship between technology and ecotopia gets looped into a benign mind-warp—melodramatic for sure, but in no doubt, finally, about who is in control. The machine is in the garden, and the garden might also be a product of the machine.

Trance is commonly described as a global music, even a global movement, unifying many of electronica's subgenres: hard house, psy, techno, hybrid, progressive house. Naturally, the term "global" refers to mostly white people in the overdeveloped world, and to the process by which trance draws inspirational sources from the underdeveloped world. Deejays do their R&D in wild places, like the beaches of Goa and Thailand, the deserts of Australia or Israel, or wherever "native spiritual energies" can best be tapped, and then the polished product is released in metropolitan mega-clubs. While the communitarian rituals of dance music are essentially urban, the expansive ambitions of a genre like trance often require larger, open spaces, and nomadic journeying to prove their point. In the techno-primitivist setting of freedom events like Burning Man, the push toward things primordial has given rise to a whole new generation of McKenna-drenched talk about deejay shamanism and higher being. There has always been loose, usually urbane, talk about tribalism among dance music communities, but the children of trance want to get much closer to what traditionally tribal peoples are supposed to possess. Again, this might be seen as an expression of the geographic imagination, since it is timelessness, and not history, that is being sought out.

In proto-trance hits, indigenous chants were often directly sampled. Future House of London's "Papua New Guinea," Enigma's "Return to Innocence" (discussed at length in these pages by Timothy Taylor), and Deep Forest's shopping spree of UNESCO recordings of rainforest peoples were all massive commercial wins. But the genre fully matured on the world traveler circuit, way beyond Ibiza and Nevada, so it is less about raiding and poaching native sources (like the corporate pharmacologist) than about simulating a journey for Western youth to faraway places. The synth-driven peaks, valleys, and plateaus of the trance formula are really a kind of narrative that reflects the yearnings of modern Euro-individualism, a *Bildungsroman* for the age of the Roland 303. It is a solo quest narrative, in other words, and not a story about migration, or syncretism, of the sort that non-Western peoples tend to have experienced. No doubt, trance will encourage some reverse-ethnography, or fusion music *a la* Yothu Yindi, if only because the commercial opportunities are too ripe to pass up. Even more likely are scenarios of the sort described here in Janet Sturman's diagnosis of *tecno-macondismo,* whereby local music scenes will use the digital technologies to recombine their own regional styles, preserving folk expression through the media of modern relevance.

For myself, I can't help spinning the CDs. Trance puts my head on a platter but keeps my brain working. That's not a lot to ask for, but little else in my life will do the trick. It's a plain expression of functional music, doing its job well.

Contributors

KAI FIKENTSCHER is Assistant Professor of Ethnomusicology at Ramapo College of New Jersey and holds degrees in jazz studies from Berklee College of Music and Manhattan School of Music, as well as in ethnomusicology from Columbia University. He has taught on African American expressive culture and arts at Columbia University, New York University, Tufts University, Amherst College, and Rhode Island School of Design. He is the author of *"You Better Work!" Underground Dance Music in New York* (Wesleyan University Press, 2000). Living in New York, he also freelances as a music producer, deejay, guitarist, lecturer, audio consultant, and sound technician.

TONG SOON LEE is Assistant Professor of Ethnomusicology at Emory University. His research interests include Southeast Asian music, diasporic musical cultures, nationalism, and cultural studies. He has done research on the Chinese and Islamic communities in Singapore and has published in *Ethnomusicology, Asian Music*, the revised edition of the *New Grove Dictionary of Music and Musicians,* and *Tongyang Úmak,* and he has forthcoming contributions in *Identities,* the *Encyclopedia of Popular Music of the World,* and *Historical Encyclopedia of Southeast Asia*. Prior to joining Emory, Tong Soon lectured at the University of Durham, UK, from 1998–2001.

RENÉ T. A. LYSLOFF teaches ethnomusicology at the University of California, Riverside and has conducted extensive research in rural Java. He has published numerous articles in *Ethnomusicology, Asian Music Journal, Asian Theatre,* and other journals and collections (including the *Garland Encyclopedia of Music*). His recent article in *Ethnomusicology,* entitled "Mozart in Mirrorshades: Ethnomusicology, Technology, and the Politics of Representation," attempts to develop a crosscultural perspective on how new media and communications technologies permeate musical experience, knowledge, and practice.

MATTHEW MALSKY is a composer on the faculty of Clark University, where he is the director of the Computer Music Studio and the co-director of the Group for Electronic Music. His writings have appeared in the *Leonardo Music Journal* and the *Journal of the International Computer Music Association*. His compositions for acoustic instruments and live computer processing have been widely performed.

CHARITY MARSH is a Ph.D. candidate in Popular Music Studies and Ethnomusicology at York University in Toronto. She is currently completing her dissertation, "Disrupting Toronto's Raving Bodies: Politicizing Myth and Freedom within Rave Culture." Recent publications include essays in *Canadian Music: Issues of Media and Technology*, *The Journal of Scandinavian-Canadian Studies*, and *Women and Music in America since 1900: An Encyclopedia*.

MARC PERLMAN, an ethnomusicologist with interests in the musical traditions of Indonesia and Burma (Myanmar), teaches at Brown University. He has also done research on pitch perception and on conceptualizations of performance in Western art music. His book, *Unplayed Melodies: Javanese Gamelan and the Cognitive Anthropology of Music Theory*, is forthcoming from University of California Press.

THOMAS PORCELLO is Assistant Professor of Anthropology at Vassar College, where he teaches courses in linguistic anthropology as well as the anthropology of music and sound. In addition to conducting intensive fieldwork in recording studios throughout Texas and the southwestern United States, he has worked professionally as a sound engineer since the early 1990s.

ANDREW ROSS is Professor and Director of the Graduate Program in American Studies at New York University. He is the author of several books, including the forthcoming *No-Collar: The Humane Workplace and Its Hidden Costs* (Basic Books) and, most recently, *The Celebration Chronicles: Life, Liberty and the Pursuit of Property Value in Disney's New Town* (Ballantine). He has also edited several books, including, most recently, *No Sweat: Fashion, Free Trade, and the Rights of Garment Workers*.

DAVID SANJEK is Director of the BMI Archives and former Chair of the U.S. Branch of the International Association for the Study of Popular Music. Recent publications include essays in *Key Terms in Popular Music and Culture* and *CINESONIC: World of Sound in Film*. His work *Always On My Mind: Music, Memory and Money* will be published by Wesleyan University Press.

JONATHAN STERNE is Assistant Professor of Communication at the University of Pittsburgh, with interests in the theory and history of mass communication, and music and communication. He is the author of *Audible Past: Cultural Origins of Sound Reproduction* (Duke University Press, 2003).

JANET L. STURMAN is Associate Professor of Music at the University of Arizona, where she teaches courses in ethnomusicology, music literature, and general studies. Much of her work concerns the impact of Iberian musical traditions upon music in the Americas and the subsequent transformations as the result of contact with indigenous and African practices. She is the author of *Zarzuela: Spanish Operetta, American Stage* (University of Illinois Press, 2000). In the summer of 1997 she was granted a USIA faculty exchange grant to teach and conduct research at the Universidad de los Andes in Bogotá, Colombia.

TIMOTHY D. TAYLOR is a musicologist who teaches in the Department of Music at Columbia University in New York City. His publications include *Global Pop: World Music, World Markets* (Routledge, 1997) and *Strange Sounds: Music, Technology, and Culture* (Routledge, 2001), as well as numerous articles on various popular and classical musics.

PAUL THÉBERGE is Canada Research Chair in Music at the Institute for Comparative Studies in Literature, Art and Culture, Carleton University. He has published widely on issues related to music, technology, and culture and is author of *Any Sound You Can Imagine: Making Music / Consuming Technology* (Wesleyan University Press).

MELISSA WEST is a Ph.D. candidate in the Communication and Culture program at York and Ryerson Universities in Toronto, Ontario. Her major scholarly areas of interest are popular music, feminist/gender theory, and interactive multimedia. Her dissertation, "Marketing Madonna: The Commodification of Ethnicity, Gender, and Sexuality" discusses the mechanisms of production behind Madonna's career.

DEBORAH WONG teaches ethnomusicology at the University of California, Riverside, specializing in the musics of Thailand and Asian America. She is author of *Sounding the Sacred: History, Aesthetics, and Epistemology in Thai Performers' Rituals* (Chicago University Press, 2001) and has published numerous articles and reviews in *Ethnomusicology, Asian Music, The Drama Review, Asian Folklore Studies, College Music Symposium, Notes,* and *American Ethnologist*. She contributed "The Asian American Body in Performance" to *Music and the Racial Imagination*, ed. Philip V. Bohlman and Ronald Radano (University of Chicago Press, 2000) and "Just Being There: Making Asian American Space in the Recording Industry" to *New Perspectives in American Music,* ed. Kip Lornell and Anne Rasmussen (Schirmer, 1997). She has also written on Thai popular musics for the Southeast Asia volume of *The Garland Encyclopedia of World Music.*

Index

ABC (American Broadcasting Company), 246
ACA (American Composer's Alliance), 235, 249, 250
Accordion, 156, 161, 165, 220
Acoustics of industrial societies, 317
Adhan, 109–120
AEG (Allgemeine Electriziäts Gesellschaft), 241
Aesthetics of authenticity, 5. *See also* Authenticity
Africa, 102–104
African Americans and minstrelsy, 212–216, 224
Alchemy, 137
Allies, World War II, 241
Alternative (music), 183
Amateur performance, 209
Ambient (music), 185
American Marconi Company, 239
American Telegraphone Company, 238
Ami, 69, 71, 76, 83
Ampex Electric, 245, 246
Anderson, Benedict, 40, 56, 142
Anderson, Laurie, 198
Annenberg School of Communication, 127
Appadurai, Arjun, 11, 64, 115, 117, 349
Appropriation, 47, 48, 102–103, 350–352
Architectronics, 316–341
Architecture, music as. *See* Music, as architecture

Armstrong, Craig, 188
Armstrong, Louis, 245, 292
Aronowitz, Stanley, 6
Arrow Music Press, 249
ar-Rumi, Jalal ad-Din Muhammad Din, 73
Art world, 208
Articulation, 334
Aruacho, 154, 172
ASCAP, 250
Association of Recording Copyright Owners (Taiwan), 70
Astaire, Fred, 245
Aterciopelados, 174
Attali, Jacques, 57, 374
Audio engineering, 267–268, 270–272, 347, 354
Audio Environments, Inc., 318
Audiophilia, 17, 347–356
Audio tape, magnetic (analog), 264
Audio technology, 346–356
Augé, Marc, 146–147
Authenticity, 4, 47, 98–102

Babbitt, Milton, 293
Background music, 323–326
Baldwin, Craig, 361, 362, 363, 371
Barros, José, 171
Barthes, Roland, 265, 266, 267
Bartok, Bela, 106
BASF (Badische Anilin und Soda Fabrik), 241, 246

Batak, 128
Baudelaire, Charles, 373
Baxter, Les, 107
BBC, 230, 240
Beatles, the, 293
Beaudrillard, Jean, 30, 45, 57
Beck, Ulrich, 340
Becker, Howard, 208
Beckett, Samuel, 236, 237, 238
Bell, Mark, 186, 189, 194
Benjamin, Walter, 115, 363, 373
Bennington College, 235
Bentley, Jerry H., 65
Bernstein, Howie, 188, 189
Big Fish Audio, 99
Billboard, 306, 336
Björk, 13, 182–198
Black Flag, 374
Blacking, John, 58
Blattnerphone: Blattner, Louis, 239, 240
Block, Martin, 292, 293
BMI, 250, 253
BMP, 102
Bogotá, 158, 164, 165, 166, 171
Bohemia, 366
Bois, Pierre, 67, 70
Bonnot, Miss Louise, 204, 210, 212, 224, 225
Boulez, Pierre, 251
Bourdieu, Pierre, 280–281, 329
Bourgeois identity, 330
Bower, Thomas M., 208, 209
Bozak, 298
Brautigan, Richard, 381
Brennan, Timothy, 66
Brown, James, 368
Brunner, José Joaquim, 155
Bui Minh Cuong, 139–140, 144–145
Burgan, Mary, 210

Cage, John, 251, 252
Cahill, Thadeus, 238
Cai luong, 135
Campesino, 170
Camras, Marvin, 240
Canby, Edward Tatnall, 247
Canclini, García, 176, 177
Caruso, Enrico, 291
Cassette recording, 127, 128
Cassette tape, 241
Castells, Manuel, 64
Catlin, Amy, 128
CBS Orchestra, 249

CD (compact disc), 13, 102, 103, 128, 132, 137, 142, 154, 224, 244, 276, 307, 308, 353–356 passim
CD-ROM, 13, 96, 97, 106, 135, 138–150 passim
Chanan, Michael, 249
Chang, Emil, 70
Chen, Bobby, 71
Chernoff, John Miller, 101
Chickering pianos, 207
Chinese Folk Arts Foundation, 67–69
Chion, Michel, 239
Chiptune, 37–38
Chow, Rey, 118
Christensen, Dieter, 169
Chung, Eddie, 96
Clearmountain, Bob, 96
Cleveland, 304
Clifford, James, 148–149, 277
Cockrell, Dale, 214
Coding music, 33, 46–47, 53
Coloa, Inc., 140
Colombia, class and social structure in, 158, 169
Colombian popular music, 153–177
Columbia-Princeton Electronic Music Center, 253
Columbia University Composers Forum, 234, 235, 249, 253
Commodity, 349–351
Community, 27–28, 31, 33–44, 54–59, 347
Compagníe International Express, 331, 332, 335
Consumption, 346–356
Coon songs, 208, 212
Cooper, Jack, 292, 293
Copyright and fair use, and Negativland, 359
Cowboy songs, 220
Cowell, Henry, 235
Cracking software, 37
Creative Labs, 97
Cretu, Michael, 67, 68, 72–73, 76
Crooner, 245
Crosby, Bing, 245, 246, 247
Cross-fader control, 299
Culture jammers/culture jamming, 361, 362, 370
Culture of simulation, 28, 29
Cumbia, 154, 158, 162, 171, 172, 176
Cutler, Chris, 368
Cyborg, 195–197
Cylinder phonograph, 238
Czukay, Holger, 293

Dance (and dancing), 294, 295, 296, 297, 300, 309
Dance music, 190, 290, 302
DAT, 100, 131
Davis, Mike, 30
Deejay, 16, 102, 290–310
Deep Forest, 81, 83
De León, Arnoldo, 217
Demo scene, 37
Denny, Martin, 107
DePino, David, 300, 301
Derrida, Jacques, 273
Dery, Mark, 303, 370, 372, 373
DeVries, Marius, 188
Díaz, Diomedes, 167, 169
Dictaphone, 238
Didgeridoo, and Big Fish Audio, 99
Difang, 81. *See also* Kuo Ying-nan
Digital music modules, 27, 33
Digital recombinancy, 45
Digital sampling. *See* Sampling
Disc jockey, 290, 305
Disco (disco music), 295, 296, 301
Discotheque, 295, 296
Disney, 75, 98
"Disney effect," 31
DIY (do-it-yourself), 34, 58, 363, 364
DJ Shadow, 364, 365
Douglas, Mary, 349
Dub, 305
Duc Huy Group, 135
Durán, Alejandro, 165
Duy Cuong, 135, 137–140, 142, 149
Dylan, Bob, 6
Dynaco, 352
Dynamic range, 244

East-West Communications, 97, 98
Easy Street Records, 308
Ecofeminism, 192
Edison, Thomas, 238
Education, music. *See* Music, education
Einstein, Albert, 280
Electronica, 182, 183, 185–186, 192, 381
Elektronische Musik, 251
Elliot, Missy, 198
Ellisen, E. Patrick, 70
Emerson, Lake, and Palmer, 293
EMI, 70, 168
E-mu Systems, 97, 99
eMuezzin, 120
Encarta, Microsoft, 75
Enigma, 11, 67–78, 81, 83

Entertainment industry, 156, 160, 176, 177
Escalona, Rafael, 163, 164, 168
Escobar, Arturo, 30
Esoniq Corporation, 97
Esso, 96
Estefan, Emilio, 176
Esthero, 198
Ethnomusicology, 4; of technoculture, 1–2
Ethnotechno, 76, 81

Fantel, Hans, 354
Fast cuts, 299
Feld, Steven, 84, 105, 107, 340
Fender Stratocaster, 282
Feuer, Jane, 371
Fikentscher, Kai, 16
Films, 156; Hollywood, 254
Fischer, Michael, 277
Fitzgerald, Ella, 241, 242
Flick, Larry, 306
Ford Foundation, 253
Foreground music, 326, 332
Fornell, Earl, 218
Foster, Hal, 29
Foucault, Michel, 273
Frequency range, 244
Frith, Simon, 6, 17, 74, 106

Gabriel, Peter, 107
Gaia, 380
Gaira Música Local, 174
Galveston, 204–225
Gaming boom, 380
Garafalo, Reebee, 54
Gaskin, Barbara, 101
Gay, Leslie, 14, 346
Geertz, Clifford, 12
Gender and technology, 182–198, 284–285, 330, 350. *See also* Women and technology
General Electric, 240
Geneva Accord, 138
Geneva Conference, 139
German immigration, 206, 217, 218
Gibbons, Walter, 297
Gibbs, Jason, 131
GigaSampler, 97
Gilard, Jacques, 164
Gilroy, Paul, 368, 369
Ginn, Greg, 361
Glennie, Evelyn, 189
Global economy, 156, 176, 177
Global Village, 104
Globalization, 11, 64–67, 83, 84, 107

Index / 389

Glocalization, 11, 64–67, 83, 84
Gnutella, 9
Goa, 381
Goankar, Dilip Parameshwar, 30
Goodman, Nelson, 264
Goodwin, Andrew, 186
Grasso, Francis, 296, 297, 301, 302
Green, Alex, 317
Greenpeace, 337
Grier, Katherine, 210
Grint, Keith, 187
Grossberg, Lawrence, 126
GUI (graphical user interface), 28, 42
Gunderson, Edna, 191

Haaren, L. F., 223
Habitus, 119
Hackers, 37
Hacking, 37, 38
Hagen, Andrew, 96
Hall, Stuart, 66, 148, 334
Hamm, Charles, 209, 212
Hammond B3 organ, 47
Ham radio, 238
Hanslang and Reinsfield, 189
Hanslick, Eduard, 264
Haraway, Donna, 196–197
Hart, Mickey, 68
Harvey, Steven, 296, 303
Hay, James, 126
Hebdige, Dick, 58, 367
Henry, Pierre, 251
Henwood, Doug, 65, 66
"High end" (sound system), 348
Hip-hop, 182, 198, 299, 300, 302, 303, 307
Hmong America, 128
Hogan, W. A., 222
Holiday, Billie, 245
Hollywood, 98. *See also* Films, Hollywood
Holst, Eduard, 222–223
Hooper, Nellee, 189
Horn, Trevor, 293
Hornbostel, Erich M. von, 292
Horton, Donald, 21
Hsu Ying-Chou, 67, 69
Humphries, Tony, 299
HyperEngine, 137
HyperPrism, 137

Iceland, 192–195
Iconography, 214
Identity, 102–106, 350–351

Ilio Entertainment, 97, 100
Imam, 111
Infinity Looping, 137
Inner time, musical, 273–275
International CD Exchange, 354
Internet, 24, 17–28, 31, 32, 40, 42, 44, 49, 54, 55, 57, 59, 74, 75, 104, 132, 145, 347
Internet fieldwork, 24–27, 34
Isherwood, Baron, 349
Islamic call to prayer, 12, 109–120. *See also Adhan*
Islamic community, 109–120
Islamic Council of Singapore, 120
Islamic sacred space, 110–112, 115, 117
Island Records, 360, 361, 363, 364, 372, 374

Jameson, Frederic, 28–29
Java, 24–25
Javanese shadow theater, 5
Jazz, 269
Jim Crow, 214, 217
John, Elton, 98
Johnny Rockets, 333, 334, 335
Johnson, James Weldon, 210
Joik, 74
Jones, Rickie Lee, 375
Jones, Steve, 127–128
Journey, 326
Jungmann, A., 204

Kampung, 111
Kant, Immanuel, 164
Kaplan, E. Ann, 29, 171
Karaoke, 128, 141
Karasov, Deborah, 321
Kasem, Casey, 17, 359, 360, 361, 367, 369, 370, 371
Keane, Webb, 56–57
Keil, Charles, 317
Kicon, 132, 143
Kirshenblatt-Gimblett, Barbara, 27, 40
Kisliuk, Michelle, 278, 279, 280
Kottak, Conrad, 127
KPFA (radio station), 372
Kroker, Arthur, 45
Kuan Yew, Lee, 111
Kuo Ying-nan, 68–73, 76, 82
Kurzweil, 96

Lacksman, Dan, 81–82
La Golondrina, 217, 218, 219
Lamer, 38

Landa, 26
Language use, 212, 216
Lanza, Joseph, 338
Laser disk, 128
Lebermann, H. A., 222
Lee, Robin, 70
Lee, Roger, 83–84
Leonard, Patrick, 188
Leone, Sergio, 375
Levan, Larry, 300, 302
Levi's clothing, 332, 333, 335
LFO, 186
Lhamon, W. T., 212
Light bulb, 12
Listening/listener, 319, 338
Literacy, 212, 216
Little Saigon, 128–132, 141
Lockard, Joseph, 55
Lomax, John, 220, 221
Lost Tribe (Matt Darey), 380, 381
Lott, Eric, 212
Loudspeaker, 109–117
LP (long-play or 12-inch) records, 224, 239, 297, 304, 305, 353, 355
Luening, Otto, 233, 235, 249, 253, 254
Lukás, Georg, 340
Lum, Casey Man Kong, 126
Lynch, David, 243
Lyons, Fred, 213, 214, 215, 224
Lyotard, Jean-Francois, 30
Lysloff, René, 10

Mackenzie, Donald, 15–16
Macondismo, 155, 171, 174
Madonna, 13, 182–198, 325
Magic Stone Music, 69, 81
Magnetic audio tape (magnetic tape recorders), 293–254. *See also* Pfleumer, Fritz; Project for Music for Magnetic Tape
Maison des Cultures du Monde, 67–69
Mall of America, 16, 316–340
Malm, Krister, 101, 130
Malsky, Matthew, 15
Mambo Music, 74
Mandarin Road, 149
Mangione, Chuck, 241, 242
Manual, Peter, 319
Marcus, George, 277
Márquez, Gabriel García, 155, 156, 165, 174, 175
Marre, Jeremy, 165, 166
Martin, George, 293

Martin, Judith A., 321
Martín-Barbero, José, 170
Marvin, Carolyn, 205
Masey, Graham, 188, 189
Mass media, 156, 159, 168, 169, 340
Masterbits, 97
Mauzey, Peter, 235
MAZ, 49
McDonalds, 96; and McDonalization, 177
Mead, G. H., 276
Medellín, 166, 171
Mediation, 116, 128, 284–285
Medicine, 137
Memorex Corporation, 241, 242, 246
Menser, Michael, 6
Mestizaje, 161, 169
Metallica, 9
Miami, 175, 176
Michelsen, López, 163
Microphone, condenser, 244
Middleton, Richard, 3
MIDI, 131, 140, 141
Mimesis, 281
Minaret, 110, 117
Miniaturization, 117
Ministère de la Culture et de la Francophonie/Alliance Française, 67
Minstrel songs, blackface minstrelsy, 208, 212, 213, 214, 215
Mitchell, William J., 24
Mixer, audio, 298, 303
Mod (Digital Music Module), 10, 33
Modernity, 65
MODPlug Central, 39, 48, 50–52
Mod scene, 27, 34–40, 40, 43, 44, 58
Mod tracker, 33
Molina, Freddy, 164
MOMA (Museum of Modern Art), 235, 249, 250, 251, 253
Momposina, Totó de, 173, 174
Moog synthesizer, 47, 107, 380
Morales, David, 302
Morricone, Ennio, 375
Mosque, 111, 113, 116
Motherhood, 191–192
Moulton, Tom, 304
Mowitt, John, 143–144
Mpingo disc, 354
MP3, 9, 17, 50, 53–54, 224
Muezzin, 120
Mullin, John, 246
Multi-track tape recorder, 292, 293

Murch, Walter, 233, 234, 235, 236, 242, 243, 251, 252, 254
Music: as architecture, 317–341; and commercial space, 317–341; in or as culture, 4; education, 204, 211
Musical America, 247
Musicking, 58
Music publishing, 204–225
Music Teachers National Association, 211
Music video, 153, 154, 155, 171, 172, 173
Music Workstation, 137
Musique concrete, 251, 252
MUZAK (Limited Partnership), 17, 317, 318, 320, 323, 324, 327, 328, 329, 338
Myers, Helen, 3, 5

Napster, 9, 17
National identity, construction of, 160, 171–174
Native America, 74–75, 99
Nature and technology dichotomy, 182–185, 191, 193–198
Nazi Party, 240, 246
NBC (National Broadcasting Company), 246
Negativland, 17, 358–375
Negus, Keith, 191
Nemesys Music Technology, 97
Nestle, 96
Nettl, Bruno, 321
New Age pop, 81
New Formosa Band, 71, 78–81
Newport Folk Festival, 5–6
New wave, 183
New York City, 296, 298, 299, 304, 308
Nineteenth century, 161, 204–225
Nintendo, 98
Nirvana, 325
Noel, Terry, 301, 302
Non-places, 146
Northstar Productions, 95
Northstar Recording, 96

O'Connor, Sinead, 191
Oliver, Daniel, 235, 251
Oliver Ditson Publisher, 205, 208
Olson, Harry, 247
Olson-Belar Sound Synthesizer, 253
Orb, 375
Orbit, William, 186, 187–188
Orchestra for RCA Victor, 249
Orr, John, 246
Overlays, 299

P.A. system, 298, 299
Pacifica Foundation, 372
Pan-Asian Hip, 96
Panasonic Technics SL-1200, 298, 299
Paradise Garage, 300
Paris, 296
Paseo, 162
PDC Productions, 135
Penley, Constance, 13, 126
Peoples Action Party, 110–111
Perlman, Marc, 17
Persing, Eric, 96
Pfleumer, Fritz, 240, 241, 246
Pham Duy Enterprises, 135
Pham Duy, 125–150
Phenomenology, 264–268, 272–274
Piano, 7–8, 206, 207, 210, 211
Pink Floyd, 293
Plagiarism, 49–50, 83
Plug-in, 36
Polan, Dana, 30
pop ballads, 154
Porcello, Thomas, 15
Post, David, 10
Postmodernism, 28–31, 83
Postmodernity, 146, 276, 279
Poulsen, Vlademar, 238, 239, 240
Powers, Francis Gary, 359, 369
Practice, theory of, 119, 280–281
Pratt, Ray, 374
Pre-echo, 266. *See also* Print-through
Price, Sally, 73
Primitive, 99–101
Print-through, 264–267, 282
producers, 293, 303
Programmed music, 16, 318–339
Progressive rock, 380
Project for Music for Magnetic Tape, 251
Propeller Island, 100
Proteus, 3, 99, 137
Proteus World, 137
Puya, 162

Quantum modulation, 326, 327
Qur'an, 118
Qureshi, Regula Burkhardt, 330
Qwerty keyboard, 16
QY20, 189

R&B, 198
Race records, 17
Rack Mount, 137
Radano, Ronald, 338

Radio, 109–120
Radio broadcasting, 211; and Adolf Hitler, 241
Radio Indochine, 135
Ranada, David, 354
Rapée, Erno, 98
Ravel, Maurice, 194
Raver fans, 380
RCA Corporation, 253
Real, concept of: Lacan, 236, 237, 238, 239, 242, 243
Recording, 239, 245; history, 290, 291, 292; industry in Colombia, local and international, 158, 161, 174–177; realism in recording, 244; studios, 293; technology, 266–268, 276, 279, 346, 348. *See also* Audio technology
Reed-Maxwell, Katherine, 212
Reification, 339, 340, 341
Reitmeyer, William F., 204, 207, 210, 212, 224, 225
Remixing, 48, 81, 305, 306
Representation, 277–282
Reyes, Adelaida, 129–131, 139
Reynolds, Simon, 373, 375
Rheingold, Howard, 54
Riesman, David, 234
Ripper, 49
Ripping, 47–53
Ripple effect, 13–14
Rives, Leonora, 223
Rockefeller Foundation, 253
Rockwell International, 97
Rodgers, Susan, 128
Roland 303, 380, 382
Roland Corporation, 95
Rosenberg Library, 204
Ross, Andrew, 13, 18, 126, 366, 368, 372

Sachs-Hornbostel, 143
Saigon, 135
Salinger, Pierre, 75
Salsa, 167
Salvation, 296
Samaranch, Juan Antonio, 69
Samper, Daniel, 163
Samper, Ernesto, 165
Sample Cell II, 137
Sampling, 33, 47, 48–49, 67, 68, 74, 94–107, 137, 149, 189, 355, 367, 373
Sanctuary, 296
Sanjek, David, 17
Sanjek, Russell, 205, 210

Schafer, R. Murray, 99, 119
Schismogenesis, 107
Schizophonia, 84, 99, 105, 107, 119
Schtung Music, 96; Schtung Records, 96
Schutz, Alfred, 266–267, 272–277, 280
Scotch (brand tape), 245
Sears, Roebuck & Co., 15
78-rpm recordings, 291
Shakira, 174
Shannon, Doug, 304
Sheet music, 127, 204–225, 291
Shields, Rob, 322
Shun Mook, 353–355 passim
Sigurason, Njall, 193
Simonelli, Victor, 306, 306
Simulacrum, 45, 242, 276
Simulated environments. *See* "Disney effect"
Singapore, 109–120 (passim)
Singapore Airlines, 96
Singapore Muslim Action Front, 113
Singapore Muslim Assembly, 113
Slip-cueing, 299, 301
Slobin, Mark, 11, 127
Small, Christopher, 58
Softcity, 24
Son (Cuban), 162
Song sheets, 204–225
Sonic Outlaws, 361, 362
Sonolux, 167, 168, 174, 175
Sontag, Susan, 365
Sony, 70, 96, 168, 174
Sony Music Taiwan, 83
sound, original and reproduction, 242, 243
Sound Blaster, 97
Sound Designer II, 137
Sound Edit, 137
Sound effects, 96
Sound Factory, 301
Sounds Good, 102
Soundware, 97
Soviet Union, 359
Space, 147
Spector, Phil, 292
Spectrasonics, 96, 100, 101
Spinning, 298
Spirituals, 216
SST Records, 361, 363, 374
Steel band recording, 239
Steinway pianos, 207
Stereophonic, 244
Sterne, Jonathan, 16
Stewart, Dave, 101
Stewart, Kathleen, 277–278, 279

Stille, Kurt, 239
Stockhausen, Karl-Heinz, 105, 293
Stok, Gavin, 74
Stokowski, Leopold, 249, 250, 251, 253
Strictly Rhythm Records, 308
Sturman, Janet, 13
Subculture, 58–59
Suburban ethnomusicology, 320, 321
Sullivan, Andrew, 10
Supermodernity, 146–147
Surau, 111
Sweetwater Sound, 96
Swing, 297
Synthesizer, 97, 132, 185

Tape bow, 198
Tape recording, and hobbyist, 248, 252
Tapesichordist, 235, 251
Tate, Greg, 368
Taubmann, Howard, 250
Taylor, Timothy, 11, 94
Techno (music), 182
Technoculture, 2–3, 32, 125. *See also* Technology and culture
Technoculture, 204–225; women and, 210, 211
Technological change, 14
Technologies, regimes of, 13
Technology and culture, 6–18 passim, 125–126, 130–131, 346–347. *See also* Technoculture
Technology and identity, 153–177; definition of, 159. *See also* Gender and technology; Identity; Women and technology
techno-peasant, 380
Tecno-macondismo, 13, 153–177
Tecno-vallenato, 156, 159, 171
Telegraphone, 238
Teleharmonium, 238
Telenovelas, 168, 169, 170
Television, 156, 168, 169, 170
Texas *Sängerbund*, 219
Texas-Mexicans, 216, 217
Textaphone, 240
Théberge, Paul, 2, 11, 14, 16, 119, 208, 211, 347
Thomson, Mrs. Robert, 220, 221, 222
Thorp, Jack, 221
Thos. Goggan and Bro., 204–225
3M Corporation (Minnesota Mining and Manufacturing), 245, 318, 323, 336
Time + Space, 97

Time Bandit, 137
Tin Pan Alley, 206
Titon, Jeff, 278–279
TLC, 198
Toll, Robert, 214
Tracker/tracking, 33
Trance, 380, 381, 382
Transectorial interdependence, 208, 224
Tran Van Khe, 141
Trax in Space, 39
Tricky, 188, 189
Turkle, Sherry, 28, 31, 32, 42–43
Turntable, 298, 303, 309
Tutor, David, 251
Tweaking, 17, 351–356
12-inch singles, 290, 304, 307, 308
2 Live Crew vs. Acuff-Rose Music Inc., 359

United Trackers, 39
Upar, 166
Urban ethnomusicology, 321
Urei 1620 mixer, 298
Usenet, 39
Ussachevsky, Vladimir, 15, 233–235 passim, 242, 243, 251–254 passim
Utopia, multimedia, 155, 160
U.S. Army Signal Corp, 245
U.S. Naval Research Laboratory, 239
U2, 358–375

Valledupar, 164, 165
vallenato, 153–177; dance rhythms, 162; history, 161–165; *colitas*, 163; *conjunto*, 161; *nuevo*, 168; *parrandas*, 163; *piquerias*, 163
Vangelis, 48
Varèse, Edgar, 293
Vasquez, Junior, 301, 306
VH1, 185
Vically, Bell Telephone Labs, 240
Victoria's Secret, 330, 331, 332, 335
Videotape, 128
Viet Minh, 131, 135
Viet, 147–148
Vietnam, 125–150
Vietnamese Americans, 125–150
Vietspace, 132–133
Vintage Keys, 137
Virgin Records America, 71
Virtual community, 10, 54–59
Vitous, Miroslav, 96
Vives, Carlos, 13, 153–177

Wade, Peter, 158, 164, 165, 166, 168
Wajcman, Judy, 15–16
Wallerstein, Immanuel, 65, 66, 83
Wallis, Roger, 101, 130
Waltz songs, 208
Wartella, Ellen, 126
Watson, Nessim, 55
Wave Convert, 137
Weinert, Ellie, 74
West, Melissa, and Charity Marsh, 13
Whitaker, Mark, 278, 279
Whiteman, Paul, 245
William C. Peters Piano Company, 206, 208
Wills, David, 361
Wilson, Morton, 96
Winner, Langdon, 8, 16, 18
Wire recording, 238, 239, 240
Wittgenstein, Ludwig, 278
Women and technology, 210, 211. *See also* Gender and technology)
Wong, Deborah, 13
World Wide Web, 9, 24, 40, 41, 42, 57. *See also* Internet
Wu Ching-Kuo, 69

Yu Wah Tan, 353
Yúdice, George, 175

Zimmer, Hans, 96, 98
Zizwek, Slavoj, 236, 237, 238, 239, 242, 243

MUSIC / CULTURE

A series from Wesleyan University Press

Edited by George Lipsitz, Susan McClary, and Robert Walser

My Music
by Susan D. Crafts, Daniel Cavicchi, Charles Keil, and the Music in Daily Life Project

Running with the Devil:
Power, Gender, and Madness
in Heavy Metal Music
by Robert Walser

Subcultural Sounds:
Micromusics of the West
by Mark Slobin

Upside Your Head!
Rhythm and Blues on Central Avenue
by Johnny Otis

Dissonant Identities:
The Rock'n'Roll Scene in Austin, Texas
by Barry Shank

Black Noise:
Rap Music and Black
Culture in Contemporary America
by Tricia Rose

Club Cultures:
Music, Media and Subcultural Capital
by Sarah Thornton

Music, Society, Education
by Christopher Small

Listening to Salsa:
Gender, Latin Popular Music, and
Puerto Rican Cultures
by Frances Aparicio

Any Sound You Can Imagine:
Making Music/Consuming Technology
by Paul Théberge

Voices in Bali:
Energies and Perceptions in
Vocal Music and Dance Theater
by Edward Herbst

Popular Music in Theory
by Keith Negus

A Thousand Honey Creeks Later:
My Life in Music from Basie to
Motown—and Beyond
by Preston Love

Musicking:
The Meanings of Performing
and Listening
by Christopher Small

Music of the Common Tongue:
Survival and Celebration in
African American Music
by Christopher Small

Singing Archaeology:
Philip Glass's Akhnaten
by John Richardson

Metal, Rock, and Jazz:
Perception and the Phenomenology of
Musical Experience
by Harris M. Berger

Music and Cinema
edited by
James Buhler, Caryl Flinn,
and David Neumeyer

"You Better Work!"
Underground Dance Music
in New York City
by Kai Fikentscher

Singing Our Way to Victory:
French Cultural Politics and Music
during the Great War
by Regina M. Sweeney

The Book of Music and Nature:
An Anthology of Sounds, Words,
Thoughts
edited by David Rothenberg and
Marta Ulvaeus

Recollecting from the Past:
Musical Practice and Spirit Possession
on the East Coast of Madagascar
by Ron Emoff

Banda:
Mexican Musical Life across Borders
by Helena Simonett

Global Noise:
Rap and Hip-Hop outside the USA
edited by Tony Mitchell

The 'Hood Comes First:
Race, Space, and Place in Rap
and Hip-Hop
by Murray Forman

Manufacturing the Muse:
Estey Organs and Consumer Culture
in Victorian America
by Dennis Waring

The City of Musical Memory:
Salsa, Record Grooves, and Popular
Culture in Cali, Colombia
by Lise A. Waxer

Angora Matta: *Fatal Acts of*
North-South Translation
by Marta Elena Savigliano

Music and Technoculture
edited by René T. A. Lysloff and
Leslie C. Gay, Jr.

ABOUT THE EDITORS

René T. A. Lysloff is Assistant Professor of Music (Ethnomusicology) at University of California, Riverside, where he teaches both gamelan and Javanese rural musical tradition.

Leslie C. Gay, Jr., is Associate Professor of Ethnomusicology at the University of Tennessee in Knoxville. Both editors have published in journals such as *Ethnomusicology*, *American Music*, and *Worlds of Music*.